Theory and Application of Rare Earth Materials

Changzhen Wang

# Theory and Application of Rare Earth Materials

Changzhen Wang
Northeastern University
Shenyang, China

ISBN 978-981-19-4177-1  ISBN 978-981-19-4178-8 (eBook)
https://doi.org/10.1007/978-981-19-4178-8

Jointly published with Science Press
The print edition is not for sale in China mainland. Customers from China mainland please order the print book from: Science Press.
ISBN of the Co-Publisher's edition: 978-7-03-048452-9

Translation from the Chinese Simplified language edition: "Xi Tu Cai Liao Li Lun Ji Ying Yong" by Changzhen Wang, © Science Press 2016. Published by Science Press. All Rights Reserved.
© Science Press 2023
This work is subject to copyright. All rights are solely and exclusively licensed by the Publisher, whether the whole or part of the material is concerned, specifically the rights of reprinting, reuse of illustrations, recitation, broadcasting, reproduction on microfilms or in any other physical way, and transmission or information storage and retrieval, electronic adaptation, computer software, or by similar or dissimilar methodology now known or hereafter developed.
The use of general descriptive names, registered names, trademarks, service marks, etc. in this publication does not imply, even in the absence of a specific statement, that such names are exempt from the relevant protective laws and regulations and therefore free for general use.
The publishers, the authors, and the editors are safe to assume that the advice and information in this book are believed to be true and accurate at the date of publication. Neither the publishers nor the authors or the editors give a warranty, expressed or implied, with respect to the material contained herein or for any errors or omissions that may have been made. The publishers remain neutral with regard to jurisdictional claims in published maps and institutional affiliations.

This Springer imprint is published by the registered company Springer Nature Singapore Pte Ltd.
The registered company address is: 152 Beach Road, #21-01/04 Gateway East, Singapore 189721, Singapore

# Contents

**1 Quantum Chemical Basis of Chemical Bonds** .................... 1
  1.1    Introduction ............................................... 1
  1.2    Quantum Chemical Basis ................................. 2
  1.3    Schrödinger Equation ..................................... 3
  1.4    Electron Configuration of Atoms ......................... 5
  1.5    Size and Valence State of Rare Earth Atoms and Ions ......... 6
  1.6    Specific Properties of the Lanthanides ..................... 6
  1.7    Valence Bond Theory ..................................... 7
  1.8    Structures of Multi-atomic Molecules ..................... 9
  1.9    Hybrid Orbitals ......................................... 11
  1.10  Cluster Compounds ..................................... 11
  1.11  Molecular Spectroscopy ................................. 12
  References ...................................................... 15

**2 Occurrence State and Resources of Rare Earth Elements** ......... 17
  2.1    Situation of Rare Earth Elements in the Periodic Table ........ 17
  2.2    Discovery of the Rare Earth Elements ..................... 18
  2.3    Occurrence State of Rare Earth Minerals .................. 19
  2.4    Rare Earth Resources and Optimization Strategies ........... 19
  2.5    Geochemistry of Non-mineral Rare Earth
        Occurrence States ........................................ 23
  2.6    Distribution Characteristics in Beihai Peninsula .............. 24
  2.7    Distribution Characteristics in Xiamen West Sea ............. 24
  2.8    Tracing of Ancient Turquoise Origin Based on Rare Earth
        Elements ................................................ 25
  2.9    Rare Earth Content of Loess Plateau Surface Soil ............ 25
  2.10  Rare Earth Accumulation, Distribution, and Migration
        in Soil .................................................. 26
  References ...................................................... 26

v

# 3 Properties, Compounds, and Complexes of Rare Earth Elements .... 29

3.1 Physical Properties of Rare Earth Elements ............... 29
3.2 Chemical Properties of Rare Earth Metals ............... 30
3.3 Reactions of Rare Earth Metals with Non-metals ............ 31
3.4 Complexes of Rare Earth Elements ..................... 36
References ........................................... 45

# 4 Rare Earth Invar Alloys, Intermetallic Compounds, and Composite Oxides ..................................... 47

4.1 Invar Alloys ......................................... 47
4.2 Thermodynamics of Intermetallic Compounds ............. 52
4.3 Use of the Gibbs–Duhem Equation to Calculate Activity ...... 55
4.4 Thermodynamic Study of Rare Earth Compounds ........... 60
4.5 Thermodynamic Study of Single Oxides .................. 62
4.6 Thermodynamic Study of Composite Oxides .............. 63
4.7 Thermodynamic Study of Non-stoichiometric Compounds .... 66
4.8 Thermodynamic Study of Non-oxide Systems .............. 70
References ........................................... 72

# 5 Rare Earth Nanomaterials ................................. 75

5.1 On Nanomaterials .................................... 75
5.2 Basic Features and Characteristics of Nanoparticles ......... 76
5.3 Structure of Nanoparticles ............................. 76
5.4 Nano Block Materials ................................. 77
5.5 Nanocomposites ..................................... 78
5.6 Preparation of Rare Earth Nano-Oxide Materials ........... 79
5.7 Preparation of Multi-dimensional Rare Earth Materials ....... 82
5.8 Composites and Assembly of Rare Earth Nanomaterials ...... 84
5.9 $CeO_2$ Nanofibers .................................... 85
5.10 Study of Low-Dimensional Nanostructured $Gd_2O_3$ ........... 86
5.11 Composites of Dendrimers and Inorganic Nanoparticles ...... 87
5.12 Fluorescence-Based Immunoassay of Rare Earth Nanoprobes ......................................... 87
5.13 Rare Earth Nanoparticle Memory ....................... 88
5.14 Effects of Rare Earth Coating .......................... 88
References ........................................... 89

# 6 Solid Electrolytes Based on Rare Earth Oxides and Fluorides ..... 91

6.1 Brief Description .................................... 91
6.2 $ZrO_2(+Y_2O_3)$ Stable Solid Electrolyte (YSZ) ............... 92
6.3 $Bi_2O_3$-$Y_2O_3$ Solid Electrolyte ......................... 94
6.4 Rare Earth Fluoride-Ion Conductors .................... 95
6.5 Rare Earth Solid Electrolytes at High Temperature .......... 95
6.6 Interaction Between Element Atoms in Metal Melts .......... 102
References ........................................... 105

Contents      vii

**7    High-Temperature Proton Conductors Containing Rare Earths** ....... 109
     7.1    Brief Introduction ......... 109
     7.2    Preparation of High-Temperature Proton Conductors ......... 110
     7.3    Causes of Proton Conduction in Perovskite Materials ......... 111
     7.4    Discussion on Some Perovskite Proton Conductors ........... 112
     7.5    Infrared Spectroscopy ......... 117
     7.6    Application of Perovskite Proton Conductors ......... 118
     References ......... 122

**8    Role and Application of Rare Earth Elements in Steel** ........... 125
     8.1    Early Recognition ......... 125
     8.2    Rare Earth Oxides, Sulfide, Carbides and Nitrides ........... 126
     8.3    Deoxidation, Desulfurization Effect ......... 127
     8.4    Vahed and Kay's Work ......... 131
     8.5    Thermodynamic Experiments in Liquid Iron ......... 131
     8.6    Thermodynamic Studies on the Nucleation of Rare Earth Compounds ......... 135
     8.7    Effects of Rare Earths and Sn, Pb, As, Tb, Bi, etc. ........... 136
     8.8    Solid-Soluble Rare Earths in Iron and Steel and Alloying Action ......... 137
     8.9    Role of Electroslag Containing Rare Earth Slag ......... 140
     8.10    Permeability of Slag for Electroslag Remelting ......... 142
     8.11    Rare Earth Applications in Some Steel Types ......... 142
     8.12    Surface Infiltration of Rare Earths in Steel ......... 144
     8.13    Production of Rare Earth-Containing Steel ......... 145
     References ......... 145

**9    Application of Rare Earths in Nodular and Vermicular Cast Iron** ......... 147
     9.1    On Nodular and Vermicular Cast Iron ......... 147
     9.2    Double-Peak Effect of the Taylor Expansion Equation ......... 151
     9.3    Process Conditions for Stable Production ......... 152
     9.4    Metallurgical Factors Affecting Production ......... 152
     9.5    Thermodynamic Study of the Effects of Arsenic, Antimony, and Bismuth ......... 154
     References ......... 162

**10    Role of Rare Earths in Non-ferrous Metals and Alloys** ........... 165
     10.1    Role of Rare Earths in Ni, Cu, and Al Solutions ......... 165
     10.2    Role of Rare Earths in Mg Alloy Solutions ......... 172
     10.3    Properties of Rare Earth Alloys ......... 172
     10.4    Effect of Rare Earths on Non-ferrous Alloys ......... 176
     10.5    Application of Rare Earth-Magnesium Alloy Bone Grafting Materials ......... 177

viii                                                                           Contents

|  | 10.6 | Application of High Pressure to Form New Alloys | 179 |
|  | References | | 181 |

**11  Application of Rare Earths in Agriculture, Forestry, and Animal Husbandry** ......................................... 183
- 11.1  Brief Description ......................................... 183
- 11.2  Rare Earths in Agriculture, Forestry, and Animal Husbandry ............................................... 183
- 11.3  Study of Rare Earth Bio-inorganic Chemistry ............... 191
- 11.4  Use of Rare Earth Compounds as Drugs .................... 193
- 11.5  Glue-Powder-Modified Pavement Asphalt ................... 194
- 11.6  Biological Effect of Rare Earths on Algae ................. 194
- 11.7  Application of Rare Earths in Animal Breeding ............. 195
- 11.8  Other Rare Earth Research in Different Fields .............. 196
- References ..................................................... 196

**12  Rare Earth Catalysts and Catalytic Activity** .................... 199
- 12.1  Brief Description ......................................... 199
- 12.2  Characteristics and Definitions of Catalytic Activity .......... 199
- 12.3  Basic Features of Catalytic Activity ....................... 201
- 12.4  Composition and Functions of Catalysts ................... 201
- 12.5  Adsorption and Multiphase Catalytic Reactions ............. 203
- 12.6  Metal Oxide Catalysts and Their Catalytic Activity .......... 204
- 12.7  Molecular Sieve Catalysts and Their Catalytic Activity ...... 208
- 12.8  Preparation and Application of Industrial Catalysts .......... 209
- 12.9  Research and Application of Rare Earth Catalysts ........... 210
- References ..................................................... 216

**13  Rare Earth Hydrides and Hydrogen Storage Alloys** ............. 217
- 13.1  Brief Description ......................................... 217
- 13.2  Hydride Types and Hydrogen Storage Alloy Components ..... 218
- 13.3  Structures of Rare Earth Hydrides ........................ 219
- 13.4  Thermodynamic Properties of Rare Earth Hydrides .......... 219
- 13.5  Magnetic Properties, Conductivity, and Structure of Rare Earth Hydrides ........................................... 220
- 13.6  Principles of Metal Hydride Hydrogen Storage ............. 221
- 13.7  Characteristics of Hydrogen Storage Materials ............. 222
- 13.8  Rare Earth Hydrogen Storage Alloys and Their Optimization ............................................. 223
- 13.9  Preparation of Rare Earth Hydrogen Storage Materials ...... 225
- 13.10  Storage and Application of Hydrogen Storage Materials ...... 226
- 13.11  Experimental Studies on Hydrogen Storage Materials ........ 228
- 13.12  Yttrium Hydride as a Hydrogen Sensor Reference Electrode ... 230
- References ..................................................... 231

Contents ix

**14 Theory and Application of Superconducting Materials** .................................... 233
    14.1 Brief Description .......................................... 233
    14.2 The Main Properties of Superconductors ................... 234
    14.3 Bardeen-Cooper-Schrieffer (BCS) Theory of Superconductivity ....................................... 235
    14.4 Types of Superconductors ............................... 236
    14.5 Oxide Superconductors ................................. 237
    14.6 Preparation of RE123 Oxide Superconductors ............... 239
    14.7 Second Generation (2G) High Temperature Superconducting (HTS) Wires ........................... 241
    14.8 Application of Superconducting Materials .................. 246
    14.9 Research and Application of Rare Earth HTS ................ 247
    References ...................................................... 248

**15 Rare Earth Magnetic Materials** .............................. 249
    15.1 Brief Description .......................................... 249
    15.2 Magnetic Properties of Substances ......................... 249
    15.3 Strong Magnetism and Ferromagnetism .................... 252
    15.4 Magnetic Origin of Rare Earth Elements .................... 255
    15.5 Magnetic Properties of Rare Earth Metals and 3d Transition Metal Compounds ............................ 257
    15.6 Types of Rare Earth Permanent Magnetic Materials .......... 258
    15.7 Phase Diagram of Rare Earth Permanent Magnet Materials .... 260
    15.8 Application of Alloy Phase Transition ...................... 263
    15.9 Preparation of Rare Earth Permanent Magnet Materials ....... 265
    15.10 Application of Rare Earth Permanent Magnet Materials ....... 266
    15.11 Rare Earth Super Magnetostrictive Materials ................ 267
    15.12 Magnetic Refrigeration and Magnetic Storage Materials ...... 268
    References ...................................................... 269

**16 Rare Earth Luminescent Materials and Laser Materials** .......... 271
    16.1 Brief Description .......................................... 271
    16.2 On Luminescent Materials ................................. 271
    16.3 Spectral Properties of Rare Earth Ions ..................... 278
    16.4 f-f and f-d Transitions of Rare Earth Ions .................. 279
    16.5 $Eu^{2+}$ Spectrum ........................................ 281
    16.6 Charge-Transfer Band of Rare Earth Ions ................... 282
    16.7 Energy Transfer Between Rare Earth Ions in Crystals ........ 282
    16.8 Application of Rare Earth Luminescent Materials ............ 284
    16.9 Rare Earth Long-Afterglow Luminescent Materials .......... 286
    16.10 White Light-Emitting Diodes ............................ 287
    16.11 Rare Earth Laser Materials .............................. 288
    References ...................................................... 290

# 17 Rare Earth Functional Ceramics ... 293
17.1 Brief Description ... 293
17.2 Piezoelectric Ceramics ... 293
17.3 Rare Earth Transparent Electrooptical Ferroelectric Ceramics ... 302
17.4 Rare Earth Dielectric Ceramics ... 304
17.5 Rare Earth Semiconductor Ceramics ... 308
References ... 312

# 18 Gemstones, Jade, and Rare Earth Optical Glasses and Ceramics ... 313
18.1 Brief Description ... 313
18.2 Origin and Classification of Precious Gemstones/Jade ... 313
18.3 Rare Earth Optical Glasses ... 316
18.4 Rare Earth Polishing Materials ... 324
18.5 Rare Earth Ceramic Color Glazes ... 325
18.6 Rare Earth Heating Materials ... 327
References ... 330

# 19 Scandium and Its Materials Applications ... 333
19.1 Brief Description ... 333
19.2 Scandium Resources ... 334
19.3 Scandium Extraction and Purification ... 335
19.4 Methods of Oxidized Hydrate Extraction and Separation ... 337
19.5 Preparation of Scandium Metal ... 341
19.6 Scandium Properties ... 343
19.7 Application of Scandium and Its Materials ... 346
References ... 350

# 20 Rare Earth Research, Production, Policy, and Future Development ... 351
20.1 Current Situation ... 351
20.2 Research Areas ... 356
20.3 Production and Market ... 357
20.4 Policy and Management ... 359
20.5 Further Discussion ... 362
References ... 367

# Abbreviations

| | |
|---|---|
| DOTA | Dodecane tertraacetic acid |
| DTPA | Diethylenetriaminepentaacetic acid |
| EMF | Electromotive force |
| FCC | Fluid catalytic cracking |
| JAEA | Japan Atomic Energy Agency |
| JSPS | Japan Society for the Promotion of Science |
| NCEI | National Centers for Environmental Information |
| NdFeB | Neodymium-iron-boron permanent magnets |
| NIR | Net import reliance |
| NMG | n-methyl glucosamine |
| NMR | Nuclear magnetic resonance |
| RE/REE | Rare earth elements |
| USGS | U.S. Geological Survey |
| XRD | X-ray diffraction |
| YSZ | Yttrium-stabilized zirconium |

# List of Figures

| | | |
|---|---|---|
| Fig. 4.1 | Battery assembly diagram | 51 |
| Fig. 4.2 | Relationship between activity-temperature ($\mathbf{a}$), activity-mole fraction ($\mathbf{b}$) | 51 |
| Fig. 4.3 | Relationship between free energy-temperature ($\mathbf{a}$), free energy-molar fraction ($\mathbf{b}$) | 51 |
| Fig. 4.4 | Diagram of $\ln p_{O_2}$ changes with $x$ in the $SnO_{2-x}$ | 68 |
| Fig. 4.5 | The relationship between $\ln p_{O_2}$ and $\ln x$ in $SnO_{2-x}$ | 69 |
| Fig. 6.1 | $La_2O_3$–$Al_2O_3$ system phase diagram | 96 |
| Fig. 6.2 | Relationship between $E_{La}$-[La] and $E_0$-[La] | 99 |
| Fig. 6.3 | Relationships between $\alpha_{La}$-[La] and $p_{O_2}$-[La] | 101 |
| Fig. 8.1 | Relationship between interfacial energy and supersaturation of Fe–rare earth compounds | 137 |
| Fig. 8.2 | $La_2O_3$ isoreactivity lines expressed as mass fractions at 1600 °C | 140 |
| Fig. 8.3 | $La_2O_3$ isoreactivity line in mole fraction at 1600 °C | 141 |
| Fig. 8.4 | $Ce_2O_3$ isoreactivity line in mole fraction at 1600 °C | 141 |
| Fig. 8.5 | $CeO_2$ isoreactivity lines expressed as mass fractions at 1600 °C | 141 |
| Fig. 9.1 | $E_Y$–[Y] curve | 149 |
| Fig. 9.2 | $\ln \alpha_Y$–[Y] curve | 149 |
| Fig. 9.3 | $E_0$–[Y] curve | 149 |
| Fig. 9.4 | Relationship between oxygen and Al content of Fe-Al-O system (1600 °C) | 150 |
| Fig. 9.5 | Relationship between oxygen activity coefficient and P content in copper melt | 150 |
| Fig. 10.1 | Relationship between $-\lg K'_{YS}$ and Y, S concentration in Ni liquid | 167 |
| Fig. 10.2 | Relationship between $-\lg K'_{CeS}$ and Ce, S concentration in Ni liquid | 167 |

xiii

| Fig. 10.3 | Relationship between $-\lg K'_{LaS}$ and La, S concentration in Ni liquid | 167 |
| Fig. 10.4 | Relationship between $-\lg K'_{Ce_2O_3}$ and [% Ce] in Cu solution (1200 °C) | 169 |

# Chapter 1
# Quantum Chemical Basis of Chemical Bonds

## 1.1 Introduction

In this book, we will discuss the rare earth materials which include the theory and application (Wikipedia 2021). We will introduce some important concepts in this first chapter, including quantum chemical basis, Schrödinger equation, electron configuration of atoms, valence bond, lanthanide family, multi-atomic molecules cluster, and molecular spectrum, and so on.

The quantum chemistry and chemical bonds is the basis of the relevant research (Jin 1955). It requires an understanding of the knowledge of quantum and bond and how their properties will provide a framework for the study (Gombas 1963; Tang 1964; Huang 1994).

The definition of covalent bonds proposed by Pauling (1939) has been widely used by chemists, but many compounds synthesized after Pauling's time, such as organic metals and cluster metals, are not well explained by Pauling's covalent definitions. Xu and Wang (2010) proposed new definitions and quantum chemical expressions for covalent bonds using density functional theory to quantitatively calculate covalent methods, essentially developing concepts of structural chemistry. They also collected and calculated the energy levels of various atomic and molecular orbitals in an attempt to generate a quantitative energy map.

Material sciences have gradually been deepened from the study of macroscopic phenomena to that of microscopic nature, and the cross pollination between chemistry and material sciences has resulted in the emergence of cross-disciplines. For example, microscopic reaction kinetics, quantum electrochemistry, computational chemistry, surface structure chemistry, laser chemistry, solid material chemistry and physics, bioinorganic chemistry, quantum biochemistry, interstellar molecular spectroscopy and so on, all belong to this new frontier discipline.

Of course, beyond the field of chemistry, many important concepts related rare earth research also will be analyzed here within.

© Science Press 2023
C. Wang, *Theory and Application of Rare Earth Materials*,
https://doi.org/10.1007/978-981-19-4178-8_1

## 1.2 Quantum Chemical Basis

If a change in physical quantity is not continuous, but rather quantifiable by units, then that physical quantity is thus referred to as quantized, and the minimum unit is called the quantum of such a physical quantity. When a metal sheet is subjected to light, it releases electrons. This phenomenon is referred to as photoelectric effect. To explain the photoelectric effect, Einstein proposed the photon theory based on Planck's quantum hypothesis in 1905. The main points are as follows:

- The energy of radiation is not continuously varied but quantized, and the smallest unit of radiation energy is called the quantum or photon of light. A photon's energy $E$ is proportional to the frequency of radiation, that is:

$$E = hv, \tag{1.1}$$

where $h$ represents a Planck constant, whose value is:

$$h = 6.626176 \times 10^{-34} \, \text{J} \cdot \text{s}. \tag{1.2}$$

- Radiation is a flow of photons at the speed of light, whose intensity depends on the number of photons per unit volume.
- According to Einstein's law of mass and energy, the energy E of an object with a mass of m is:

$$E - mc^2, \tag{1.3}$$

And so, photons with energy $E$ must have mass m, that is:

$$m = \frac{E}{c^2} = \frac{hv}{c^2}. \tag{1.4}$$

The wavelength of light multiplied by the frequency $v$ equals to the light speed $c$:

$$AV = c, \tag{1.5}$$

- Photons possess momentum $p$, which is equal to the product of mass $m$ and velocity $c$, i.e.,

$$p = mc = \hbar/\lambda = hv/c. \tag{1.6}$$

- When photons collide with electrons, they must obey the laws of conservation of energy and momentum. Light has wave and particle duality, which is the unity of discontinuous particle and continuous volatility.

## 1.3 Schrödinger Equation

In the equations above, the left side of the equal sign indicates the properties of the particles, that is, the photon energy $E$, momentum $p$ and density $p$, while the right side of the equal sign indicates the fluctuation nature, that is, the frequency $v$ of light waves, wavelength $\lambda$ and field strength $\Psi$. The motion of a photon obeys a statistical law of the motion of a large number of photons, that is, the law of radiation.

In 1923 the French physicist de Broglie suggested that wave and particle duality is not only an optical phenomenon, but also suitable for physical particles such as electrons. For physical particles, the appropriate wavelength is equal to:

$$\lambda = \hbar/p = \hbar/mv, \tag{1.7}$$

from which, the obtained wavelength is called de Broglie wavelength with mass and velocity $v$. His hypothesis was confirmed by Davisson and Germer (1927) in the electron diffraction experiments.

## 1.3 Schrödinger Equation

During the 1870s, Maxwell established the electromagnetic theory and predicted its existence. Hertz soon proved the existence of electromagnetic waves by experiments, and proposed that light is an electromagnetic wave.

When a plane electromagnetic wave propagates in a vacuum, its wave equation is:

$$\nabla^2 \psi = \frac{1}{c^2} \frac{\partial^2 \psi}{\partial t^2}, \tag{1.8}$$

where $\Psi$ represents the electric field intensity $E$ or the magnetic field intensity $H$, $c$ is the light speed ($c = \lambda v$), $\nabla^2$ is called as Laplace operator:

$$\nabla^2 = \frac{\partial^2}{\partial x^2} + \frac{\partial^2}{\partial y^2} + \frac{\partial^2}{\partial z^2}. \tag{1.9}$$

When solving this partial differential equation by variation method, $\Psi$ can be written into the product of two functions, one of which $\Psi$ is a function of coordinates and the other $\varphi$ is a function of time, that is:

$$\Psi = (x, y, z, t) = \Psi(x, y, z)\varphi(t), \tag{1.10}$$

substitute Eq. (1.10) in Eq. (1.9):

$$\frac{\lambda^2}{\psi} \nabla^2 \psi = \frac{1}{v^2 \phi} \frac{d^2 \Phi}{dt^2}. \tag{1.11}$$

To the left of the equal sign is a function of coordinates, to the right of the equal sign is a function of time. To the left of the equal sign is a constant if the $t$ is changed and the $x$, $y$, $z$ is unchanged. If this constant is $-a^2$, then the right side of the equal sign must be equal to $-a^2$, thus Eq. (1.11) can be divided into the following two equations:

$$\frac{\lambda^2}{\psi} \nabla^2 \psi = -\alpha^2, \tag{1.12}$$

$$\frac{1}{v^2 \Phi} \frac{d^2 \Phi}{dt^2} = -\alpha^2. \tag{1.13}$$

Equation (1.14) is called the wave equation of standing wave:

$$\nabla^2 \psi + \left(\frac{2\pi}{\lambda}\right)^2 \psi = 0. \tag{1.14}$$

The above discussion is for light waves, while for the wave of matter fixed state of physical particulate movement (i.e., the state with certain energy); the equation should be associated with standing waves.

$$\nabla^2 \psi + \frac{8\pi^2 m}{h^2}(E - V)\psi = 0. \tag{1.15}$$

It is the First Equation of Schrödinger's equation, which is used to describe the fixed state of particle motion. This equation is the most important one in quantum mechanics. In many cases, the Schrodinger equation with time is required.

$$\left(-\frac{h^2}{8\pi^2 m} \nabla^2 + V\right)\psi = -\frac{h}{2\pi i} \frac{\partial \psi}{\partial t}. \tag{1.16}$$

Equation (1.16) is the Schrodinger equation with time, also known as the Second Equation of the Schrodinger equation.

To write the Schrodinger equation correctly, we must first find out the potential energy function, and the energy of the particle must be quantized. Usually the state with the lowest total energy is called the ground state and the different states are characterized by quantum numbers.

The structure of hydrogen atom or hydrogen-like ions is the simplest. The first step is to write the Schrodinger equation as if we are studying their structure.

## 1.4 Electron Configuration of Atoms

Xu and Wang (2010) discussed electron configuration of atoms. The work indicated that any atom consists of positively charged nuclei and negatively charged electrons. Electrons move around the nucleus in an orbital at different levels, and such levels from inside to outside are represented by $n = 1, 2, 3, 4, 5, \ldots$ The orbital at different levels are represented by symbols 1s, 2s, 2p, 3s, 3p, 3d, 4s, 4p, 4d, 4f, 5s, 5p, 5d, 5f, ..., in which the symbols s, p, d, f represent the different shapes: s is a spherical orbital without directionality and cannot be re-divided into a sub-orbital; p-orbital is dumbbell-shaped or two leaf-like clouds, which can be divided into three sub-orbitals in three different directions along the $x$, $y$ and $z$ axis; d-orbital can be divided into five sub-orbitals; and f-orbital can be divided into seven sub-orbitals. The s-, p-, d- and f-orbitals can be represented by an orbital quantum number $l$. The s-orbital $l = 0$, p-orbital $l = 1$, d-orbital $l = 2$, f-orbital $l = 3$ and so on, according to $2l = 1$, the calculated number of sub-orbitals s, p, d, f are 1, 3, 5, 7, respectively.

Electrons themselves have two kinds of spin motion and the spin orientation is opposite to each other, so the number of electrons filled in each orbital in s, p, d, and f is 2, 6, 10, and 14, respectively. The electrons in the atom are regularly filled in each sub-orbital by energy level. Through the periodic law and atomic spectrum, we know that the distribution of electrons outside the nucleus obeys three basic principles:

**Minimum Energy Principle**: The distribution of electrons outside the nucleus is based on the minimum energy principle.

**Pauli Exclusion Principle**: In one atom, it is impossible to have electrons with the exact same 4 quantum numbers (n, l, m, and s).

**Hund Rule**: Electrons distributed in equivalent orbitals, such as 3p orbitals or 5d orbitals or 7f orbitals, as far as possible occupy separately different orbitals and spin parallel. This electronic distribution minimizes energy. As a special case of the Hund rule, it is relatively stable for the state of the equivalent orbital with full, half-full, or empty. This can explain the particularity of some atomic electron distribution.

There are also the energy level interlacing phenomena. The order of electrons outside the nucleus is arranged according to their energy level, which depends on the nucleus to both electrons gravity and outer electrons repulsion. As the increase in the atomic number of elements, the electron energy level of the atom is 1s, 2s, 2p, 3s, 3p, 4s, 4p, 5s, 4d, 5p, 6s, 4f, 5d, 6p, 7s, 5f, and 6d from low to high.

The periodic table is a natural expression of the arrangement of electrons outside the nucleus. There are 17 rare earth elements including Sc, Y, and lanthanide.

## 1.5 Size and Valence State of Rare Earth Atoms and Ions

When the electrons of lanthanide atoms fill up the 4f orbitals of the inner layer, because one electron in the same 4f shell is not totally shielded by another electron, the effective nuclear charge number acting on each 4f electron increases as the atomic number increases, which causes the 4f shell and radius to shrink with the increase of atomic number in a phenomenon called "lanthanide contraction". When the coordinate number is 8, from $La^{3+}$ to $Lu^{3+}$, the ionic radius shrinks by 15.8%.

There's some regularity in the variation of the valence state of lanthanide ions. Lanthanum ($4f^0$) without 4f electrons, semi-filled gadolinium ($4f^7$) and fully filled lutetium ($4f^{14}$) has stable trivalent bonds. Adjacent lanthanide ions have the property of variable valence by losing or obtaining electrons, in order to tend to stable electron configurations $4f^0$, $4f^7$ and $4f^{14}$. The closer to lanthanide ions of lanthanum ($4f^0$), gadolinium ($4f^7$) and lutetium ($4f^{14}$), the greater variable valence tendency is. The right side ions of lanthanum and gadolinium tend to oxidize to high valences (e.g., $Ce^{4+}$ and $Tb^{4+}$), and the left side ones of gadolinium and lutetium tend to be reduced to low valences (e.g., $Eu^{2+}$ and $Yb^{2+}$).

The chemical properties of divalent and tetravalent lanthanide ions are obviously different from those of trivalent ones.

## 1.6 Specific Properties of the Lanthanides

The characteristics of the lanthanides depend on their specific 4f electron configurations. The different transitions of 4f electrons endow lanthanides with unique optical, magnetic, and electrical properties. Electrons can transition between different f–f and f–d levels, depending on the magnitude of the energy difference between the upper and lower levels, influencing the wavelength, and hence color, of the light absorbed or emitted. The larger the energy difference, the shorter the wavelength of absorption or emission. Since an f–f transition only involves electrons within 4f layers, the resulting spectra are less affected by external factors. They are characterized by long fluorescence lifetimes and narrow line shapes, and so the associated fluorescence is bright and suitable for displays and lighting. Some lanthanide ions emit in the near-infrared range, and the emissions can transmit in the atmosphere and optical fibers, making them widely applicable for ranging and optical communications.

In an f–d transition, 4f electrons are elevated to an outer layer, and the spectrum generated is broader in shape. For example, the trivalent cerium ion ($Ce^{3+}$) emits blue-purple light when it transitions from a 5d configuration with higher energy to an f configuration with lower energy. Because the d electrons are more exposed in the outer layer, the light absorption and emission observed are greatly affected by external factors. They are characterized by a wide spectral band, high intensity, and short fluorescence lifetimes, making them applicable in fields such as scintillation crystals and tunable lasers.

# 1.7 Valence Bond Theory

Considering the $4f^n$ configurations ($n = 1$–$13$) of 13 trivalent lanthanide ions (from $Ce^{3+}$ to $Yb^{3+}$) with unfilled 4f shells, there are 1639 energy levels, and the number of possible transitions between these is as high as 192,177. The lanthanides thus constitute a huge luminescence treasury. Only a few transitions are currently used for luminescent and laser materials, implying that the lanthanides still have important untapped potential as optical materials.

Lanthanides have as many as seven unpaired electrons in their 4f configurations. Their orbital motions and spin, strong spin-orbital coupling, and indirect exchange with the surrounding environment make the magnetic properties of lanthanides distinct from those of d-block transition elements such as Fe, Co, and Ni. They display high paramagnetic susceptibility, saturation magnetization, magnetic anisotropy, magnetostriction, magneto-optical rotation, and magnetic entropy effects, and are widely used in fields such as permanent magnets and magneto-restrictive, magneto-optic, and magneto-cooling materials.

Among the frameworks of layered structures composed of rare earths and d-block transition metal ions, the rare earths often provide stable structures conducive to electron transport. If defects are introduced, the valence state, spin state, and electron-transport properties of some d-block transition metal ions coexisting in the same compound will change, modifying conductivity and catalytic performance. This makes rare earths an important object in exploring new materials, such as semiconductors, electronic and ionic conductors, high-temperature superconductors, as well as new catalysts.

## 1.7 Valence Bond Theory

Even for the simple systems of $H_2$ and $H_2^+$, solution of the Schrödinger equation has proved difficult. For molecules in general, structures are much more complicated, and quantitative calculations become even more difficult. To circumvent this, the commonly used approximation methods are electron pairing and the molecular orbital method.

1. Electron pairing

Molecules consist of atoms, which contain unpaired electrons before they are combined. If the spin orientation of the unpaired electrons of two atoms is opposite, the pair of electrons can couple, and the coupling of each pair of electrons yields a covalent bond. Hence, the electron-pairing method is also called the valence-bond method. The hypothesis of valence-bond theory is consistent with the concept of covalent bonding developed earlier.

When two electrons compose a covalent bond, they can no longer form a covalent bond with a third atom. In other words, a covalent bond is saturated. The greater the electron cloud overlap, the more stable the resulting covalent bond.

8                                                          1   Quantum Chemical Basis of Chemical Bonds

According to the principle of maximum overlap of electron clouds, the formation of a covalent bond must take the direction of higher electron cloud density in the possible range; that is to say, a covalent bond has directivity.

2.   Molecular orbital method

Molecular orbital theory can be regarded as a generalization of the results of the linear variation method for processing $H_2^+$. The main points are as follows:

- It is assumed that the state of motion of each electron in a molecule can be described by the single-electron wave function $\Psi_\sigma$, of which the spatial part $\Psi$ is called a molecular orbital, $\Psi^2 \, d\tau$ represents the probability that this electron appears within the micro volume $d\tau$, $\Psi^2$ is the probability density (or electron cloud density), and $\sigma$ is the spin part.
- The molecular orbital $\Psi$ can be approximately represented by a linear combination of atomic orbitals.
- It is assumed that each molecular orbital $\Psi_i$ has a corresponding energy $E_i$, approximately representing the ionization energy of electrons in that molecular orbital. The electrons in the molecule are arranged in a molecular orbital, and conform to the Pauli law and the lowest-energy principle.
- In order to effectively compose a molecular orbital, the atomic orbitals must satisfy the following three conditions: energy proximity, symmetry matching, and maximum orbital overlap.

The motion of electrons in a molecule does not lead to their appearance at each point in space with equal probability. If an electron in a molecular orbital move most of the time near a nucleus or is controlled by it, then it is bound to be close to the atomic orbital.

The linear combination of two molecular orbitals $\Psi_1$ and $\Psi_2$ affords two molecular orbitals, of which one is the bonding molecular orbital and the other is the anti-bonding orbital; in two atomic orbitals, $E_1$ is lower than the lower energy orbitals and $E_2$ is higher.

3.   $\sigma$ orbital and $\sigma$ bond

Two molecular orbitals can be obtained by linear combination of H atom 1s orbitals $\Psi_a$ and $\Psi_b$:

$$\psi_\mathrm{I} = \psi_a + \psi_b = \frac{1}{\sqrt{\pi}}\left(e^{-ra}\right) + e^{-rb}$$

$$\psi_\mathrm{II} = \psi_a - \psi_b = \frac{1}{\sqrt{\pi}}\left(e^{-ra}\right) - e^{-rb}$$

(1.17)

The electron cloud distribution has cylindrical symmetry, and the symmetry axis is the straight line connecting the two nuclei, that is, the bond axis.

$\Psi_I$ in Eq. (1.17) is a centrosymmetric g-shaped orbital (from the German "gleich", meaning even); $\Psi_{II}$ is a centro-dissymmetric u-shaped orbital (from the German "ungerade", meaning odd). $\Psi_{II}$ has a nodal section of vertical bisection bond axis,

## 1.8 Structures of Multi-atomic Molecules

whereas $\Psi_I$ has no such nodal plane. All homonuclear diatomic molecules with anti-bonding and bonding orbitals share this property. Traditionally, an asterisk (*) is used to distinguish the anti-bonding molecular orbital from the bonding one.

According to molecular spectroscopy and quantum mechanical calculations, electrons in a molecular orbital such as $\Psi_I$ or $\Psi_{II}$, for which the component of angular momentum along the bond axis is equal to zero, constitute a $\sigma$ orbital. Assuming that $\Psi_I$ or $\Psi_{II}$ is generated by the 1s orbitals of two H atoms, $\Psi_I$ can be denoted as $\sigma_g 1s$ and $\Psi_{II}$ can be denoted as $\sigma_u^* 1s$. Here, the former is the ground state, whereas the latter is the repulsive state. The electron cloud with the bonding $\sigma_g 1s$ is the densest among the nuclei, with an interface like olives; the electron cloud with the anti-bonding $\sigma_u^* 1s$ is sparse between the nuclei, and its nodal surface is like a double-yolk-egg (Xu and Wang 2010).

The electrons in the $\sigma$ orbital, bonding orbital, and anti-bonding orbital are referred to as $\sigma$, $\sigma$ bonding, and $\sigma$ anti-bonding electrons, respectively. The bonding electrons make molecules stable, whereas the anti-bonding electrons impart a tendency for dissociation.

4. $\pi$ orbital and $\pi$ bond

Considering the z-axis as the bond axis, two $p_z$ orbitals can yield the $\sigma$ orbital, and two $p_y$ and two $p_x$ orbitals can constitute the molecular orbitals. For example, two $2p_{+1}$ (or $2p_{-1}$) orbitals generate molecular orbitals $\Psi_I$ and $\Psi_{II}$, respectively, which are not cylindrically symmetrical. The projection of the electron orbital angular momentum onto the bond axis at these two orbitals is $\hbar$, where $\hbar = \frac{h}{2\pi}$; a molecular orbital with a projection of the orbital angular momentum onto the bond axis of $\pm\hbar$ is called a $\pi$ orbital. Here, $\Psi_I$ is bonding and can be denoted as $\pi 2p_{+1}$; $\Psi_{II}$ has a facet perpendicular to the bond axis with anti-bonding character, and is denoted as $\pi^* 2p_{+1}$. For a homonuclear diatomic molecule, $Z_a = Z_b$. $2p_{-1}$ orbitals can compose $\pi_u 2p_{-1}$ and $\pi_g^* 2p_{-1}$ orbitals.

Two symmetry-matched d orbitals can also construct $\pi$ bonds. A $\pi$ bond composed of a symmetry-matched p orbital and a d orbital is called a p–d $\pi$ bond. These two $\pi$ bond types occur in complexes and hydrochloric acid.

## 1.8 Structures of Multi-atomic Molecules

In this section, we discuss the structures of multi-atomic molecules in terms of valence bond theory and molecular orbital theory.

1. Formation of $\sigma$ bonds and covalency of atoms

An atom A with $n$ unpaired electrons and n atoms B, each with an unpaired electron, can form a molecule $AB_n$. For example, an O atom has two unpaired electrons:

$$\left(1s\ \boxed{\uparrow\downarrow},\ 2s\ \boxed{\uparrow\downarrow},\ 2p\ \boxed{\uparrow\downarrow}\ \boxed{\uparrow}\ \boxed{\uparrow}\right)$$

An H atom has one unpaired electron $\left(1s\boxed{\uparrow}\right)$; thus, one O atom can combine with two H atoms to form an $H_2O$ molecule. Overall, two pairs of electrons are coupled to form two covalent single bonds ($\sigma$ bonds).

When one of the two atoms has a lone pair of electrons and the other has a free valence electron orbital, the pair produces a $\sigma$ bond. The combination of $NH_3$ and $BCl_3$ can be written as:

$$BCl_3 + NH_3 = BN + 3HCl.$$

Atoms that contain vacant orbitals in the valence electron layer are referred to as electron-deficient, such as Li, Be, B, Al, rare earths, and titanium atoms. Atoms that have more valence electrons than valence electron orbitals are referred to as multi-electron atoms, such as N, O, S, and Cl.

2. $\pi$ bond formation

After a $\sigma$ bond is formed between two adjacent atoms in a multi-atom molecule, if each has an unpaired p electron, these overlap in a transverse orientation to form a $\pi$ bond. This ($\sigma + \pi$) combination is referred to as a double bond, expressed as A = B, as in $H_2C = CH_2$. When there are two unpaired p electrons p electrons after forming a $\sigma$ bond (e.g., the bond axis is the z-axis, and the two p electrons are $P_x$ and $P_y$, respectively), two bonds can be generated, and this ($\sigma + \pi + \pi$) combination is called a triple-bond, expressed as A $\equiv$ B, as in HC $\equiv$ CH.

A $\pi$ bond formed by two overlapping p orbitals is called a p–p $\pi$ bond; a bond formed by two overlapping d orbitals is called a d–d $\pi$ bond; a bond formed by overlapping d and p orbitals is called a d–p $\pi$ bond; a bond formed by overlapping p and d orbitals is called a p–d $\pi$ bond.

3. $\delta$ bond formation

In the Re (III) complex anion $Re_2Cl_8^{2-}$, three characteristics of the bonding should be noted:

- The Re–Re bond energy is very large and the bond is very short at 224 pm. This is much shorter than the Re–Re distance in metallic Re (275 pm).
- The sets of four Cl atoms bound to the respective Re atoms in $Re_2Cl_8^{2-}$ adopt overlapping rather than staggered configurations. The Cl–Cl distance (332 pm) in this configuration is shorter than the sum of the van der Waals radii (360 pm).
- $Re_2Cl_8^{2-}$ is diamagnetic.

The Re–Re bond in $Re_2Cl_8^{2-}$ is a ($\sigma + \pi + \pi + \delta$) quadruple bond. There are many compounds containing quadruple bonds, such as $Mo_2Cl_8^{4-}$ and $Re_2(RCOO)_4X_2$.

## 1.9 Hybrid Orbitals

The central idea of valence bond theory is that bonding atoms form covalent bonds by electron pairing. Hybrid orbital theory was developed by Pauling and Slater in 1931. This theory can not only explain the tetrahedral structure of methane, but also the planar structure of the ethylene molecule, the linear structure of the acetylene molecule, and the geometric configurations of many other molecules. Pauling also rationalized more complex structures in terms of d orbital mixing to obtain s–p–d hybrid orbitals. Researchers in 1982 extended this principle to s–p–d–f hybrid orbitals according to a Group Theory method. They surmised that if the composition of s, p, d, and f is $\alpha$, $\beta$, $\gamma$, and $\delta$, the bonding ability is $f = \sqrt{\alpha} + \sqrt{3\beta} + \sqrt{5\gamma} + \sqrt{7\delta}$.

The atomic orbitals involved in hybridization should have similar energies. Hybridization can increase bonding ability, since the hybrid orbital has stronger directionality, which is beneficial for generating stable covalent bonds through overlap with the appropriate orbitals of another atom.

There are many types of hybrid orbitals. Some studies give the compositions and geometries of some examples, and the hybrid orbitals used by atoms with different valence states.

## 1.10 Cluster Compounds

The term "cluster" was defined by Cotton (1966). According to his original definition, a cluster is "… held together mainly or at least to a significant extent, by bonds directly between metal atoms, even though some non-metal atoms may also be intimately associated with the cluster", and "a cluster compound is a coordination compound containing three or more metal atoms bonded to each other …". By this definition, neither boranes nor carboranes can be counted as cluster compounds because they do not contain metal–metal bonds. However, boranes are clearly a class of cluster compounds if not named as such. To include boranes, carboranes, metal carboranes, and metal clusters, Xu et al. suggested that "clusters and cluster compounds are a cluster of atoms built by a polyhedron or cage formed by the direct bonding of three or more atoms to a peripheral atom or group". Lu JX in 1978 first translated cluster into "Yuanzicu" in China, after carrying out research in the United States as early as the 1930s and 1940s.

There are seven types of cluster compounds according to their elemental constituents (Lu 1981): Borane type; Carbon-borane type; Metal-borane type; Metal-carborane type; Transition-metal-cluster; Main-group-cluster; Pure-metal-cluster.

Considering that the number of bonding orbitals in a borane depends on the skeleton, Tang (1964) proposed the following rules for metal cluster compounds:

- Metal cluster compounds can be considered as metal atom polyhedral frameworks combined with coordination bodies.

- Each transition metal atom provides 9 valence orbitals (5d orbitals, 1s orbital, and 3p orbitals). If the number of metal atoms of the polyhedron skeleton is $n$, then the total number of valence orbitals provided is $9n$.
- Each edge of a polyhedron reflects the bond-forming interaction of two adjacent metal atoms. When two valence orbitals interact to produce a bonding skeleton orbital, an antibonding skeleton orbital must also be generated, so the number of antibonding skeleton orbitals is equal to the number of edges L of the polyhedron.
- When a metal atom polyhedron and any ligand each provide a bonding orbital, a new bonding orbital and an antibonding orbital are generated.
- The total number of bonding and non-bonding molecular orbitals of a metal cluster compound is $9n - L$, and the number of valence electrons of such a cluster should be $2 (9n - L)$.

In one Chinese study, a 48-atom-core rare earth cluster with a unique tubular structure was identified after introducing different template anions to induce slow hydrolysis at high temperature under weakly alkaline conditions. The cluster consists of two whorled 18-atom-core erbium clusters and an annular 12-atom-core erbium cluster. Previously, the same research group also synthesized two 36-atom-core rare earth clusters, of which a gadolinium cluster could potentially serve as a magnetic cooling material.

## 1.11 Molecular Spectroscopy

Molecules, like atoms, have characteristic energy levels, and hence corresponding characteristic spectra. Studied parameters include the relative vibration of each nucleus in the molecule, rotation of the whole molecule, the distance between the nuclei, the force constant of vibration, molecular moment of inertia, and so on. Molecular spectroscopy is one of the most important ways to study molecular structure and to conduct qualitative and quantitative analyses.

There are three quantized modes of movement within a molecule: the movement of electrons relative to the nuclei, the relative vibration of each nucleus, and the whole molecule rotation.

The energy differences between adjacent electron levels are generally of the order of 1–20 eV. The spectrum generated by electronic energy level transitions is located in the ultraviolet (UV, shorter wavelength than the visible, < 400 nm) and visible regions (wavelength 400–800 nm), and so is referred to as the electronic or UV/Vis spectrum.

A vibrational energy level transition inevitably accompanies an electronic energy level transition. The former interval is generally 0.05–1 eV. If it is 0.1 eV, then it is 2% of a 5 eV electronic level. The latter does not produce a line with a wavelength of 250 nm, but a series of lines with wavelength intervals of about 250 nm $\times$ 2% = 5 nm.

## 1.11 Molecular Spectroscopy

In fact, the spectra observed are much more complex because there are also rotational energy level transitions. The intervals between rotational energy levels are generally less than 0.05 eV, and can be as small as $10^{-4}$ eV. An electronic spectrum contains a number of band series, whereby different bands are equivalent to different electron energy levels. One spectral band series (i.e., relating to the same electron energy level transition, such as jumping from energy level A to B) contains several spectral bands with different vibrational energy level transitions.

Upon irradiating a molecule in the near-infrared ($\lambda = 1$–$2.5$ $\mu$m) or mid-infrared ($\lambda = 2.5$–$25$ $\mu$m) range, the energy of the radiation is insufficient to cause a transition of the electronic energy levels; it can only cause a transition of the vibrational and rotational energy levels. Such a spectrum is called a rotational or far-infrared one.

The rotational energy levels of some molecules need lower-energy microwaves (from $\lambda = 350$ $\mu$m to several cm), and so the transitions may be called a microwave spectrum. Some researchers summarized the spectral regions (Goto 1979; Xu and Wang 2010).

A common unit conversion relationship is:

If

$$E = \frac{1239.85}{\lambda/nm} \, eV,$$

then,

$$E = 1.23985 \times 10^{-4} \left( \tilde{v}/cm^{-1} \right) eV.$$

### Multi-atom molecular spectroscopy

Multi-atom molecular spectroscopy is an important tool for studying molecular structure, providing much of the knowledge about energy levels, electronic configurations, and geometrical structures. It is also used for analysis and identification. In recent years, various instruments for the determination of molecular spectra have undergone significant development. Instrument accuracy and resolution have been greatly improved, promoting development of the discipline and experimental applications. Such spectra, like the abovementioned two-atom molecular spectra, can also be divided into three kinds: electronic, vibrational–rotational, and rotational spectra.

### UV/Vis absorption spectra

The UV region and the visible region (200–800 nm) are the most convenient regions for studying spectra. To study the absorption spectra of molecules, the available instruments are ultraviolet and visible spectrophotometers. The UV/Vis absorption spectra of the molecule are produced by electron level transitions, accompanied by changes in the vibrational and rotational energy level. The transition of the electron level is mainly that of a valence electron; the internal electron energy level is low and not easy to excite under general conditions. According to their properties, various UV/Vis absorption spectra can be explored.

Type of transition

Valence electrons of ground-state organic compounds include bonding $\sigma$ electrons, bonding $\pi$ electrons, and nonbonded electrons. Under the action of light, these different types of valence electrons can undergo possible transitions to vacant orbitals of the molecule. The vacant orbitals of molecules include anti-bonding $\sigma^*$ and $\pi^*$ orbitals, so the possible transitions are $\sigma \rightarrow \sigma^*$, $n \rightarrow \pi^*$; for a nonbonded electron n, possible transitions are $n \rightarrow \sigma^*$, $n \rightarrow \pi^*$, and so on.

The types include:

*$\sigma \rightarrow \sigma^*$ Transition*;
*$n \rightarrow \sigma^*$ Transition*;
*$\pi \rightarrow \pi^*$ Transition*;
*$n \rightarrow \pi^*$ Transition*;
*$\pi \rightarrow \pi^*$ Transition of conjugated alkenes*;
*$\pi \rightarrow \pi^*$ Transition of aromatic hydrocarbons*;
*$p \rightarrow \pi$ Conjugation and auxochromes*.

The main external factors affecting the electronic spectral profiles of organic compounds are temperature and solvent.

The transition of the electron energy level is often accompanied by changes in vibrational and rotational energy levels. Instrument resolution is insufficient to separate energy levels that are very close, and so the electronic spectrum generally consists of broad peaks. Lower temperature can reduce the effects of vibration and rotation. The effect of solvent on the spectral shape is complex, mainly depending on the solvent properties.

Infrared spectra

The mid- and near-infrared spectra of multi-atom molecules are their vibrational–rotational spectra. At normal temperatures, most molecules are in the ground state, and the intensities of the double-frequency, sum-frequency, and difference-frequency bands are weak.

The infrared spectra of various compounds feature characteristic vibrational frequencies, and have become an important method for identifying functional groups and determining molecular structure, especially in the study of natural product structures and the quantitative analysis of carbon and nitrogen compounds. IR has been widely used in scientific research and industry.

The mid-infrared region is the most useful for identifying compounds; the 4000–1400 cm$^{-1}$ region is the group frequency zone, which is most valuable for the identification of functional groups; the 1400–650 cm$^{-1}$ region is the fingerprint region, which is the most sensitive to molecular structure changes and so is useful for compound identification.

The infrared spectra of purified compounds have been extensively studied, and an unknown structure can be identified and confirmed by comparison of its spectrum with that of a known standard; the near-infrared region of infrared spectra is best suited for quantitative analysis.

If an unknown cannot be found in the existing standard map, carefully analysis is required to elucidate the functional groups and chemical bonds of the compound. Raman spectroscopy, mass spectrometry, nuclear magnetic resonance, microwave spectroscopy, XRD analysis, chemical analysis, and so on, may also be applied as necessary.

Besides being widely used in organic synthesis and bioengineering, infrared spectroscopy has also been used to study the properties of new inorganic or functional materials, as in the study of perovskite high-temperature proton conductors. For example, there are O–H functional groups in $BaCe_{0.9}Y_{0.1}O_{3-\alpha}$ (Matsushita 2001).

# References

Cotton FA (1966) Q Rev Chem Soc 20:389

Davisson CJ, Germer LH (1927) Diffraction of electrons by a single crystal of nickel. Phys Rev 30:705

Gombas P (1963) The multiparticle problem in quantum mechanics. In: Pan ZC (trans). People's Education Press, Beijing

Goto KS (1979) Physical chemistry of oxides at high temperature for metallurgists, a course work. Tokyo Institute of Technology, Tokyo

Huang ZT (1994) Industrial catalysis. Chemical Industry Press, Beijing, pp 72–82

Jin SS (1955) Quantum chemistry application brief. China Science Books and Instruments Corporation, Beijing, pp 1–50, 119–156, 157–183

Lu JX (1981) Structural chemistry of atomic cluster compounds. In: Proceedings of the 1978 annual conference of China chemical society. Science Press, Beijing, pp 35–60

Matsushita E (2001) Tunneling mechanism on proton conduction in perovskite oxides. Solid State Ionics 145:445

Pauling LC (1939) The nature of the chemical bond, and the structure of molecules and crystals. Cornell University Press, Ithaca

Rare-earth element. Wikipedia. Last edited on 14 Apr 2021. https://en.wikipedia.org/wiki/Rare-earth_element

Tang YQ (1964) Statistical mechanics and its application in physical chemistry. Science Press, Beijing, pp 3–63

Xu GX, Wang XY (2010) Physical structure. Higher Education Press. Version 2. Science Press, Beijing, pp 11–46, 68–85, 144–322, 357–377

# Chapter 2
# Occurrence State and Resources of Rare Earth Elements

## 2.1 Situation of Rare Earth Elements in the Periodic Table

Rare earth elements, also known as rare earth metals, are located on the left side of Group III B in the Periodic Table, namely scandium (Sc) and yttrium (Y), plus the 15 lanthanide elements. This gives a total of 17 elements, which are commonly denoted as R or RE. From the discovery of the first rare earth element yttrium in 1794, to that of promethium (Pm) in 1945, it took 151 years to find them all from natural sources (Wang 2014).

Adjacent to the alkaline-earth metals, Res include scandium, with atomic number 21 and an outer electron arrangement of $3d^1 4s^2$, yttrium with atomic number 39 and an outer electron arrangement of $4d^1 5s^2$, and the lanthanides with atomic numbers 57–71.

The atomic numbers and outer electron arrangements for the lanthanides are 57 lanthanum (La), $5d^1 6s^2$; 58 cerium (Ce), $4f^1 5d^1 6s^2$; 59 praseodymium (Pr), $4f^3 6s^2$; 60 neodymium (Nd), $4f^4 6s^2$; 61 promethium (Pm), $4f^5 6s^2$; 62 samarium (Sm), $4f^6 6s^2$; 63 europium (Eu), $4f^7 6s^2$; 64 gadolinium (Gd), $4f^7 5d^1 6s^2$; 65 terbium (Tb), $4f^9 6s^2$; 66 dysprosium (Dy), $4f^{10} 6s^2$; 67 holmium (Ho), $4f^{11} 6s^2$; 68 erbium (Er), $4f^{12} 6s^2$; 69 thulium (Tm), $4f^{13} 6s^2$; 70 ytterbium (Yb), $4f^{14} 6s^2$; and 71 lutetium (Lu), $4f^{14} 5d^1 6s^2$.

While the lanthanides are commonly referred to as rare earth elements, scandium and yttrium are often considered separately. This is because the lanthanide elements have more similarity in electron layer arrangement. Except for lanthanum, the outermost electrons of the other lanthanide elements occupy the 4f sublayer. The electron layer arrangement of lanthanum can be ascribed to the attractive force on the final electron by the nucleus being less than the repulsive force of the inner electrons, such that the 4f sublayer cannot be filled even though it has vacancies. In the next element, cerium, the nuclear charge is increased by one, whereupon the attractive force of the nucleus becomes greater than the repulsion of the electrons in the same layer, and the final electrons begin to fill the 4f layer. In praseodymium, the last electron moves from the 5d to the 4f sublayer. This phenomenon of energy level interleaving has no effect on the outermost electrons; they are all $6s^2$ and so the elements have similar chemical properties.

© Science Press 2023
C. Wang, *Theory and Application of Rare Earth Materials*,
https://doi.org/10.1007/978-981-19-4178-8_2

How many rare earth elements should there be? In 1913, Moseley proposed a linear relationship of the form $\sqrt{v} = \alpha(Z - \sigma)$ between the square root of frequency $v$ of spectral lines in the element's X-ray spectrum and the atomic number $Z$, where $\alpha$ is a constant and $\sigma$ for La lines is about 7.4. This implies that only 14 lanthanide elements should exist between lanthanum (La) and hafnium (Hf), consistent with the conclusion that the 4f sublayer can only contain 14 electrons.

As regards the properties of the elements, the gloss of rare earth metals is between those of silver and iron, and their chemical reactivity is very high.

## 2.2 Discovery of the Rare Earth Elements

Rare earth elements have similar chemical properties and often co-exist in the same mineral, making them difficult to separate by chemical methods.

The path of discovery of the elements belonging to the group termed rare earth elements was particularly confused and chaotic. It started 200 years ago, in 1787, and it closed in 1947 with the discovery of promethium (Szabadvary 1988). To identify them, chemists have overcome these difficulties and corrected all kinds of errors, and it took 160 years to distinguish them all (Kaczmarek 1983; Liu 2009).

The first rare earth mineral to be discovered was gadolinite in the Swedish village of Ytterby by Arrhenius in 1787, after which many rare earth elements are named, from which the Finnish chemist Gadolin isolated a new oxide in 1794. Another Swedish chemist, Ekeberg confirmed the results in 1797. To commemorate Gadolin, this mineral was named gadolinite, and the oxide was named as yttria by using some letters from the name of the village. Subsequent research showed that yttria was not the oxide of a pure element, but contained yttrium mixed with other heavy rare earths.

The classification of rare earth elements is inconsistent between authors (Zepf 2013). Klaproth, German chemist discovered uranium (1789), zirconium (1789), and cerium (1803), who independently discovered cerium, a rare earth element, around the same time as Jöns Jacob Berzelius and Wilhelm Hisinger, in the winter of 1803.

Cerium was the first of the lanthanides to be discovered. In 1839 Carl Gustaf Mosander became the first to isolate the metal. Today, cerium and its compounds have a variety of uses. Cerium metal is used in ferrocerium lighters for its pyrophoric properties. Cerium-doped YAG phosphor is used in conjunction with blue light-emitting diodes to produce white light in most commercial white LED light sources. And, Pr, Nd, Sm, Eu, Gd, Tb, Dy, and La, were discovered. Ce, Pr, Nd, Sm, Eu, Gd, Tb, Dy are a new series of trivalent lanthanide.

The most common distinction between rare earth elements is made by atomic numbers (Rollinson 1993). Those with low atomic numbers are referred to as light rare earth elements, such as lanthanum, cerium, praseodymium, neodymium, promethium, samarium, europium, and gadolinium; those with high atomic numbers are the heavy rare earth elements, such as terbium, dysprosium, holmium, erbium, thulium, ytterbium, lutetium, scandium, and yttrium. Increasing atomic numbers between light and heavy rare earth elements and decreasing atomic radii throughout the series causes chemical variations.

## 2.3 Occurrence State of Rare Earth Minerals

Rare earth elements have obvious geochemical affinities for fluorine, oxygen, calcium, titanium, niobium, zirconium, carbonate, and phosphate.

Minerals containing rare earths include:

- Gadolinite ($Be_2FeY_2Si_2O_{16}$),
- Cerite ($CaCeSi_3O_{13}$),
- Yttrium phosphate ($YPO_4$),
- Cerium fluorocarbonate ($CeFCO_3$),
- Apatite $(CaCe)_5[(P, Si)O_4]_3(O, F)$,
- Chlorite $(nacace)_2Nb_2O_6F$,
- Monazite $(Ce, Y)PO_4$,
- Yttrium brown niobium ore $(Y, Ce, Th, Ca)(Nb, Ta)O_4$,
- Ytterbium ore $(Y, Ce, V, Ca)(Nb, Ta, Ti)_2O_6$,
- Black rare gold ore $(Y, Ca, Ce, V, Th)(Nb, Ta, Ti)O_6$,
- Limonite $(Ca, Ce, Th)_2(Al, Fe, Mn, Mg)_2(SiO_4)_3OH$,
- Yttrium arsenite $YAsO_4$,
- Hydrothium ore $(RE_2O_3 \cdot 3CO_2 \cdot 4H_2O)$.

With regard to the origin of these minerals, it is believed to have involved a large amount of gas originating from carbonate-rich substances in the upper mantle of the Earth's depths. The minerals were probably formed during natural processes of weathering and migration.

The two main phosphate-containing minerals are monazite and yttrium phosphate.

Rare earths occurring in minerals meet the following criteria: participation in mineral lattices; dispersal in suitable rock minerals; existence in minerals.

## 2.4 Rare Earth Resources and Optimization Strategies

### 1. Rare earth resources in China

After decades of hard work, the many Chinese geological workers have found that China is relatively rich in rare earth resources (Press Office of the State Council 2012), and that these are mainly distributed in 13 regions, such as Inner Mongolia, Jiangxi, and Guangdong (Xu and Wang 1990). The main rare earth mineral resources are as follows.

China produced 70% of world supply in 2018, down from 80.5% in 2017, and controls 36.6% of world reserves, according to the U.S. Geological Survey.

## Baiyun Obo mine in Inner Mongolia

Baiyun Obo mine is mainly an open-cast mine, and represents a large-scale comprehensive deposit of iron, rare earths, niobium, and other symbiotic minerals. It is a rare large deposit in the world, which was formed by alternating high-temperature hydrothermal solution after deposition, metamorphosis, and a magmatic period. Researchers found 14 rare earth minerals in the deposit, including fluorocarbonates, phosphates, and silicates. Rare earth minerals have fine particle size, and there is little difference between their processing requirements and those of other useful minerals and gangue minerals.

## Rare earth polymetallic deposit in Inner Mongolia

The orebody is an alkaline granitic porphyry of late Yanshanian origin in Mesozoic pyroclastic rocks. There is no obvious boundary between the orebody and the rock mass.

## Rare earth minerals in Weishan, Shandong

The vein originates in the quartz Neihui orthorhombic porphyry and its surrounding rock, black cloud oblique long gneiss. The number of veins in the region is large, and the extension direction is variable. The length of each vein is generally 100–400 m, with a width varying from a few cm to more than 10 m; the grade is 1–5%. The main industrial minerals are fluorocarbonate cerium ore and a small amount of calcium cerium fluorocarbonate ore. Rare earth elements are mainly found in independent minerals of large particle size, generally 0.4–0.5 mm.

## Mianning rare earth mine, Sichuan

At Mianning County, Sichuan Province, the mine is composed of three sections. Taking the Maoniuping rare earth ore section as an example, the estimated reserve is up to 1 million tonnes. The main minerals are fluorocarbonate cerium ore and a small amount of calcium cerium fluorocarbonate ore, along with silicon titanium cerium ore. The ore grade of the rare earth oxides is 0.5–5%. This mine is located at the junction of the Kangdian axis and the Ganzi geosyncline. The crustal movements are very strong.

## Southern ion adsorption rare earth ore

The deposits are found in Jiangxi, Guangdong, Guangxi, Fujian, Hunan, and other provinces, and are of the granite weathering type. The rare earth elements are mainly enriched in the fully weathered layer in the vertical direction, and their content in the horizontal direction is related to the thickness of the weathering layer. The developmental thickness of the weathered shell varies with the properties of the rock and the weathering conditions. Different degrees of weathering of such rock, ore structure, mineral composition, clay composition and content, and the type of mining area all influence the uniformity of the distribution of rare earth elements in minerals of different grades, and the technical parameters of the water separation method.

## 2.4 Rare Earth Resources and Optimization Strategies

### New type of tungsten manganese ore containing rare earths

According to a report in 2014, researchers at the Zhengzhou Institute of Comprehensive Utilization of Mineral Resources of the Chinese Academy of Geological Sciences found a new type of tungsten manganese ore containing rare earths in the course of a rock identification. The ore grade in terms of the total amount of rare earth oxides is 0.23%. Tungsten and the rare earth europium are enriched in barium-containing hard manganese, and the distribution is uniform, with an $Eu_2O_3$ content of 0.34–0.47%. This kind of industrial deposit containing rare earths comprises mainly endogenous fluorocarbonate cerium ore and monazite-type ore, an exogenous weathered shell-type ion adsorption layer, and monazite-type sedimentary placer ore.

### Seashore rare earth sand deposits

The coastal occurrence of rare earth industrial deposits is mainly concentrated on the western Guangdong coast, the western Taiwan coast, and the eastern Hainan coast. The rare earth minerals in seashore sand are mainly monazite and yttrium phosphate ore, and these are generally accompanied by ilmenite, zircon, and other useful minerals. Sea shore placer can be divided into sea-forming and sea-land mixed types. The former accounts for the main industrial ore deposit. The Nan Shan sea placer is a large seashore rare earth deposit, from which useful minerals such as zircon and titanium iron rutile can be recovered.

### Scandium resources of Gannan region

In September 1989, the "Achievement appraisal meeting of Nanling rare earth resources prospective survey" identified an evaluation criterion for scandium resources, that is, containing 20–50 ppm (ppm magnitude is $10^{-6}$) $Sc_2O_3$. Southern ion adsorbed rare earth ore is regarded as an associated scandium resource; ore with $Sc_2O_3 > 50$ ppm is regarded as an independent scandium resource (Xu 1990).

Gannan is a subtropical region with strong lithochemical weathering. The scandium orebody is mainly layered and occurs in the upper and middle parts of the weathering layer of the parent rock with a thickness of 1–9 m, which changes with geomorphology.

### 2. Further adjustment and optimization

China has a large production capacity for rare earth concentrate, amounting to more than 100,000 tonnes.

(1) Consumption of rare earth resources in China. China's rare earth reserves were claimed to account for 71.1% of global reserves, according to 1987 exploration results, but the white paper "China's Rare Earth Situation and Policy" released by the State Council Information Office in 2012 shows that this proportion is actually about 23%. Some analyses suppose that China has sold about 50% of its rare earth reserves.

(2) Since the introduction of an export quota management system for rare earths, new problems have arisen. The primary producers that buy rare earth raw materials are not subject to quota restrictions; some sole proprietorship and

joint venture companies simply process rare earth resources and then transport primary products abroad for further processing or storage (Xiong 2013).

(3) The environmental impact of rare earth mines is severe. Researchers at Beijing General Institute of Nonferrous Metals Research pointed out that the production of 1 ton of ion rare earth ore consumes 4–5 tonnes of ammonium sulfate, 1.7 tonnes of ammonium bicarbonate, 8–10 tonnes of hydrochloric acid, 6–8 tonnes of sodium hydroxide, and 1–1.2 tonnes of liquid ammonia. These chemical raw materials are ultimately converted into waste water, waste residues, or waste gas, and are not effectively recycled.

### 3. Foreign rare earth resources

Many countries around the world are now focusing on ensuring their own supplies of rare earth metals. Indeed, the possession of mineral deposits helps to increase national competitiveness.

**U.S.:** It was once the world's leading producer of rare earths, but production was phased out due to the difficulty of mining and environmental concerns. The largest ever rare earth mine (Mountain Pass) in the U.S. was closed in 2002, but production recommenced in 2018.

**Russia**: Its main source of rare earths is apatite ore. The state-owned Poctex and private enterprise ICT group have announced plans to exploit rare earth deposits in the Republic of Sakha (Yakut) and are preparing to invest billions of dollars in the Tom mine there. As one of the world's largest rare earth deposits, the mine has proven reserves of 154 million t, containing yttrium, scandium, terbium, and other rare earth elements.

**Canada**: In 2013, the Canadian Exploration Rare Minerals Corporation announced the discovery of rich rare earth deposits in northern Quebec. They predict an average annual rare earth oxide concentrate production of 13,650 tonnes, with the potential to become a global long-term supplier of heavy rare earths.

**Australia**: The country is a major producer of monazite, its mine occurs mainly in the western region; it has available rare earth resources, as well as uranium mining tailings in the Essa Mountains in central Queensland, and uranium gold deposits.

**India**: The country's original commercial source of rare earth materials was beach monazite from Kerala state. India is actively exploring new sources of rare earths.

**South Africa**: It has the world's only single-vein monazite mine, in Stingkamps Kral, Cape Province. In addition, rare earths are found in the coastal sands of Chazbe in the southeast, and monazite and fluorocarbonate cerium are found in Buffalo fluorite.

**Malaysia**: Its rare earth minerals, such as monazite, hussakite, and samarskite, are mainly recovered from tin tailings. It is one of the main suppliers of heavy rare earths and yttrium in the world (The discovery of heavy rare earth deposits in several states of Malaysia's Simma Peninsula 2014).

**Brazil**: It is the world's oldest rare earth production country, having exported monazite to Germany since 1884. The monazite resources are mainly concentrated on the eastern coast, from Rio de Janeiro in the south to Fortaleza in the north.

Additionally, Egypt, Afghanistan, Vietnam, North Korea, and Japan also operate rare earth mines.

## 2.5 Geochemistry of Non-mineral Rare Earth Occurrence States

Chen et al. (2007) of the Department of Geosciences, Zhongshan University, reviewed the Ce level abnormalities of surface geological bodies and their influencing factors in 2007. Besides redox conditions, microorganisms and organic matter are also important factors causing abnormal Ce levels in surface geological bodies. Geochemical studies of geological bodies have generated a lot of data, among which the differentiation of rare earth elements is most widely used to trace their origin and evolution.

Abnormal Ce distributions in water bodies and sediments are mostly attributed to environmental redox conditions. However, there are exceptions, such as positive Ce anomalies in aerobic alkaline natural water bodies such as Lake Van in Turkey and in oxidized surface water bodies such as the Sargasso Sea, which are likely related to the roles of organic matter and microorganisms. Characterized the effect of humic acid on the existence of rare earths in granite weathering, respectively. Haley et al. proposed that there are three sources of rare earths in the marine sediment pore, namely, iron-, organic coating layer-, and Ce oxide-source, and found the effect of organic coating layer on rare earth differentiation (including Ce anomalies). Wyndham et al. found that Ce abnormalities in corals have seasonal changes, mainly affected by the abundance of Ce oxidizing bacteria.

Researchers at the Institute of Oceanography, Chinese Academy of Sciences, and at the First Marine Institute of the State Oceanic Administration, studied the geochemistry of rare earth elements in cobalt-rich crusts in seamount in the central Pacific Ocean. The preponderance of $Ce^{4+}$ indicates an oxidative enrichment process. Cobalt-rich crusts are mostly distributed on seamounts, seafloor plateaus, and sea ridges. The total amount of rare earth elements within them is higher than that of normal deep-sea sediments and seawater. For the rare earth elements in such crusts, work has been carried out on their distribution characteristics and influencing factors, sedimentary environments, occurrence phases, and valence states, as well as material sources.

## 2.6 Distribution Characteristics in Beihai Peninsula

Ge et al. (2008) of the Guangzhou Institute of Geochemistry, Chinese Academy of Sciences, studied the geochemical characteristics of rare earth elements in the process of river and sea water mixing by selecting the estuary of the Nanliu River, the coastal area, and Weizhou island around Beihai peninsula in Guangxi.

The estuarine environment, as a buffer zone connecting rivers and oceans, has independent physical and chemical kinetic properties. Rare earth elements are valued in the field of environmental geochemistry because of their unique properties, which provide information concerning the source and reaction history of environmental substances in estuaries. In recent years, the wide application of rare earths has had an adverse impact on the offshore environment, especially the estuarine environment. Research is focused on the distribution of raw water, dissolved rare earth elements, their migration and transformation, material sources, and environmental behavior characteristics.

## 2.7 Distribution Characteristics in Xiamen West Sea

In 2010, Yao and Zhang of the School of Environment, Nanjing University of Technology, Hu of the Center for Modern Analysis, Nanjing University, and Liu of the School of Ocean and Environment, Xiamen University, have studied the distribution characteristics of rare earth elements in sediments (domestic scholars have studied the geochemistry of rare earth elements in seabed sediments of the Bohai Sea, Huang Hai, the Yangtze River, the East China Sea, the South China Sea, and the shallow beaches of Taiwan).

Xiamen West Sea area is a single half-closed bay, from the south of Gulangyu to Gaoji seawall, with a north–south extension and a narrow shape. Some suspended sediment is spreading to the west port area, which is the main feature of the West Sea area of Xiamen (Yao et al. 2010). Exploration of the regional geochemical characteristics of rare earth elements in soil sediments in this area is of great significance to further reveal their ecological and environmental impacts.

According to the characteristics of the West Sea area of Xiamen, eleven sampling points have been set up, namely four at the mouth of the Jiulong River and seven in the bay area. Rare earth elements are routinely determined by inductively coupled plasma mass spectrometry.

## 2.8 Tracing of Ancient Turquoise Origin Based on Rare Earth Elements

Turquoise is one of the "four famous jades" in China, and has been cherished by people of all nationalities since ancient times. From the early stage of various small ornaments to the late Neolithic to the bronze age, among the various materials applied in cultural relics and surface decoration mosaics, turquoise has been applied throughout the country.

Turquoise in China is mainly produced in Yunxian County and Zhushan, Northwest Hubei Province, and adjacent areas of the Baihe River, Shaanxi Province. In recent years, some scientific and technological archaeologists in China have also studied the ore source and function of turquoise. Most turquoise samples collected by She et al. (2009) are small fragments. When preparing the samples for the analysis of rare earth elements, the unweathered samples which have been treated without obvious inclusions, to form a combined sample and tested after grinding. All samples were subjected to XRD phase analysis and high-resolution inductively coupled plasma mass spectrometry.

Chen et al. (2014) of the China University of Mining and Technology discussed the characteristics of rare earth elements in groundwater in Wolong Lake in northern Anhui Province. Due to their unique geochemical behavior, rare earth elements have become a focus of geochemical research. Rare earth elemental analysis has been successfully applied in hydrogeochemical evolution, water–rock interaction, and groundwater run-off tracing. Results have shown that the rare earth elements of groundwater, as in surface water bodies, are affected by pH, redox conditions, dissolution–adsorption equilibria, and other processes. At the same time, the characteristics of rare earth elements in groundwater show similarities with their flow paths through surrounding rock, which implies that rare earths give a good indication of water–rock interactions. Therefore, there is a reliable theoretical basis to use the geochemical characteristics of rare earth elements in underground water bodies to assess the surrounding rock properties and groundwater sources of aquifers.

## 2.9 Rare Earth Content of Loess Plateau Surface Soil

The physical and chemical properties of various soils developed from different parent materials are different. The total amounts and compositions of rare earths vary accordingly. The special natural geographical environment of the Loess plateau in China makes the soil resources very complicated. Liu et al. (1995) of the Northwest Institute of Soil and Water Conservation, Chinese Academy of Sciences, and others studied the contents and regional distributions of rare earth elements in the soil of the Loess plateau.

A total of 64 typical soil profiles were selected across the Loess plateau, comprising 22 Loessal soils, 18 Heilu soils, 10 Lou soils, seven sierozem, and seven

26

gray cinnamonic soils. At the same time, surface soil samples from 40 sites were randomly collected. The samples were air-dried, crushed, ground, sieved (100 mesh), and homogenized. Portions (50–100 mg) were weighed, wrapped in clean high-purity aluminum foil, and set aside. The rare earth element content of each sample was analyzed by a neutron activation method, and the activation analysis procedure established by soil analysis was used to add international standard reference material to each batch of analytical samples to make quality control standard samples.

Although the content of rare earth elements in the Loess plateau was found to be similar to that of parent material, some differentiation among the various types of soils was identified due to the effects of region, biological environment, and human activities. The rare earth (La, Ce, Nd, Sm, Tb) contents in these soil types decrease in the order gray cinnamonic soil and Lou soil > sierozem > Heilu soil; the contents of Eu, Yb, Lu decrease in the order gray cinnamonic soil and Lou soil > Heilu soil and Loessal soil > sierozem.

## 2.10 Rare Earth Accumulation, Distribution, and Migration in Soil

There is increasing academic concern as to whether the long-term use of rare earths in agriculture and their accumulation in the soil have adverse effects on soil–plant environmental systems and groundwater.

Zhang et al. (1994) of Anhui Agricultural University and Chen of Nanjing Agricultural University studied Ce and Nd accumulation and migration in soil under conditions of simulated rare earth accumulation and annual rainfall. They predicted a migration to groundwater after long-term usage of rare earths.

Soils of nine representative types were collected from the national regions: chernozem (pH 8.2, organic matter 3.56%); Shajiang black soil (pH 8.0, organic matter 2.14%); sierozem (pH 8.1, organic matter 2.55%); Lou soil (pH 8.1, organic matter 1.64%); yellow tide soil (pH 8.2, organic matter 1.63%); chloasma (pH 6.0, organic matter 1.01%); paddy soil (pH 4.8, organic matter 2.03%); red soil (pH 5.0, organic matter 1.36%); and lateritel (pH 4.1, organic matter 2.17%). The rare earths were $Ce(NO_3)_3$ and $Nd(NO_3)_3$ solutions prepared at 50 mg/mL oxide concentration. The tracer Ce and Nd were made into working fluids, respectively.

## References

Chen BH, Wei HX, Huang ZT et al (2007) A review of Ce abnormalities and their influencing factors in surface geological bodies. Chin Rare Earths 28(4):80–83

Chen S, Gui HR, Sun LH et al (2014) Characteristics and significance of rare earth elements in multi-aquifer groundwater in Wolonghu coal mine, North Abhui Province. Chin Rare Earths 35(3):94–100

# References

Ge T, Tian Y, Wang Q (2008) Distribution characteristics of rare earth elements in water mixing process in the North Sea Peninsula. J Chin Soc Rare Earths 26(6):754–759

Kaczmarek J (1983) Rare earth discovery and industrial separation. Translated by Wei Y. Chin Rare Earths (For Rare Earth Appl Album) 60–73:135–166

Liu XZ (2009) Rare earth fine chemical chemistry. Chemical Industry Press, Beijing, pp 7–10

Liu PL, Tian JL, Li YQ et al (1995) Content of rare earth elements in surface soil of loess plateau and its effect on agriculture. J Chin Soc Rare Earths 13(2):155–158

NCEI, the National Centers for Environmental Information. https://www.ngdc.noaa.gov/

Press Office of the State Council (2012) White paper on rare earth status and policy in China, 20 June 2012. http://www.cre.net/show.php?contentid=102955

Rollinson HR (1993) Using geochemical data: evaluation, presentation, interpretation. Longman Scientific & Technical, Harlow, Essex

She LZ, Qin Y, Luo WG et al (2009) Using rare earth and other trace elements traces ancient turquoise origin in northwest Hubei. Chin Rare Earths 30(5):51–62

Szabadvary F (1988) The history of the discovery and separation of the rare earths. In: Handbook on the physics and chemistry of rare earths, vol 11, pp 33–80

The discovery of heavy rare earth deposits in several states of Malaysia's Simma Peninsula, 11 Dec 2014. http://www.cre.net/show.php?contentid=116924

Wang CZ (2014) Solid-state ion element sensing and applications. Metallurgical Industry Press, Beijing, p 334

Xiong XL (2013) Thoughts and suggestions on adjustment and optimization of China's national strategy for rare earth, 20 Dec 2013. http://www.cre.net/show.php?contentid=111498

Xu LF (1990) Analysis on geological characteristics and occurrence state of cormorant resources in south Jiangxi. In: Proceedings of the second annual meeting of China rare earth society, vol 1:10, pp 7–12

Xu GC, Wang JW (1990) General situation of rare earth resources and research progress of mineral processing technology in China. In: Proceedings of the second annual meeting of China rare earth society, vol 1:10, pp 1–6

Yao FZ, Hu X, Liu X et al (2010) Distribution characteristics of rare earth elements in surface sediments of Xiamen Sea area. J Chin Soc Rare Earths 28(4):495–499

Zepf V (2013) Rare earth elements: a new approach to the nexus of supply, demand and use: exemplified along the use of neodymium in permanent magnets. Springer, Berlin, London

Zhang JZ, Zhu WS, Zhang LG et al (1994) Study on accumulation, distribution and migration of rare earths in soil. In: Proceedings of the third annual conference of China rare earth society, vol 5, pp 202–204

# Chapter 3
# Properties, Compounds, and Complexes of Rare Earth Elements

## 3.1 Physical Properties of Rare Earth Elements

The rare earth elements comprise Sc, Y, and the lanthanide elements (La–Lu). Their physical properties and physicochemical behaviors are determined by the characteristics of their atomic electron layer arrangements (Gschneidner 1990). Several lists in this section give the characteristics of the atomic structure of rare earth elements and various physical properties. Among them, relative atomic masses are those in the 1987 standard and the values published officially in 1988; temperature data correspond to the International Temperature Standard (ITS-90) of 1990.

Part of the content has been previously published in "Physical Chemistry of Rare Earth Alkali Soil and Its Application in Materials" (Science Press), edited by Du et al. (1995). Other data have been taken from Gschneidner of the U.S. Rare Earth Intelligence Center. This part of the book was originally written by Wang and is incorporated here.

Goldschmidt (1978) and Delaeter (1988) summarized the atomic orders, relative atomic masses, outer electron configurations, and spectral ground states of $RE^{2+}$, $RE^{3+}$, and $RE^{4+}$ of the respective elements.

The crystal structures, metal ion radii, atomic volumes, and densities at 24 °C (297 K) or below, and the high-temperature crystal structures, metal ion radii, atomic volumes, and densities are summarized. The high-temperature transition points and melting points of the rare earth metals; their low-temperature transition points; the heat capacities, standard entropies, transition heats, and melting heats of the rare earth metals; their vapor pressures, boiling points, and sublimation heats, are listed into tables (Beaudry and Gschneidner 1978; Koskenmaki and Gschneidner 1978; Gschneidner and Calderwood 1986).

The magnetic properties of the rare earth metals, and their thermal expansion coefficients at room temperature, thermal conductivities, resistivities, and Hall coefficients are summarized into tables (Beaudry and Gschneidner 1978; McEwen 1978; Legvold 1980).

© Science Press 2023
C. Wang, *Theory and Application of Rare Earth Materials*,
https://doi.org/10.1007/978-981-19-4178-8_3

30          3 Properties, Compounds, and Complexes of Rare Earth Elements

Collocott et al. (1988) summarized the specific heat constants of electrons ($\gamma$), electron–electron coupling constants ($\mu^*$), electron–photon coupling constants ($\lambda$), Debye temperatures ($\theta_D$), and superconducting transition temperatures; Soott (1978) summarized the elastic moduli and mechanical properties at room temperature. Stretz and Bautista (1972), King et al. (1976), Baria et al. (1976), Van Zytveld (1989), and Li et al. (2009) listed liquid metal properties and effective ionic radii close to the melting points.

## 3.2 Chemical Properties of Rare Earth Metals

### (1) Chemical properties

Rare earth metals constitute group IIIB in the Periodic Table, adjacent to calcium, strontium, and barium in group IIA, and have similar chemical properties to these alkaline-earth metals. They are highly chemically reactive metals, with negative redox potentials and low ionization energies. Their first ionization energies are close to those of alkaline-earth metals, and their electronegativities are akin to that of calcium (Zinkevich et al. 2006).

### (2) Lanthanide contraction

Due to the characteristic electron arrangements of their atoms, the radii of the atoms and trivalent cations gradually decrease with increasing atomic number from La to Lu. This phenomenon is called "lanthanide contraction"

The electron distribution of lanthanide ions is $1s^2$, $2s^2$, $2p^6$, $3s^2$, $3p^6$, $3d^{10}$, $4s^2$, $4p^6$, $4d^{10}$, $4f^{14}$, $5s^2$, and $5p^6$. Nuclei and all electrons except $5s^2$ and $5p^6$ constitute the atomic core; $5s^2$ and $5p^6$ are the outer electrons. The effective nuclear charge Z* of the atomic core can be approximately expressed as:

$$Z^* = Z - \delta$$

where $Z$ is the nuclear charge of the atom and $\delta$ is the shielding constant.

For example, the nuclear charge of the $La^{3+}$ atomic core is 57, with 46 electrons; for the outer electrons, the contribution of each electron to $\delta$ in the inner layer is 1, and thus $Z^* = 57 - 46 = 11$ units of positive charge.

The $Ce^{3+}$ atomic core has 47 electrons, one of which is an f electron; for the s and p electrons in the outer layer, the contribution of each f electron to $\delta$ is 0.85, so $Z^* = 58 - 46 - 0.85 = 11.15$ units of positive charge.

With increasing positive charge of the nucleus, the shielding effect of the 4f electrons on the effective nuclear charge decreases (each 4f electron can only shield about 85% of the positive charge per unit). As a result, the electrostatic attractive effect of the effective nuclear charge on the outer electrons gradually increases, which causes contraction of the electron layer. The atomic radii of Eu and Yb are larger than those of the other rare earth elements, because their 4f electrons are in half-($4f^7$) and fully filled ($4f^{14}$) stable orbitals.

Regarding the ionic radii, the nuclear charge of each lanthanide element increases with increasing atomic number. Electrons are filled in the inner 4f sublayer, while the 5s and 5p of the outer layer remain unchanged. The 4f electrons cannot completely shield the increasing nuclear charge, so the electron cloud resides closer to the nucleus and the ionic radius gradually decreases.

With decreasing radii of the lanthanide cations, they are more readily coordinated by some ligands, and the pH at which hydroxide ion precipitation commences becomes lower. Rare earth metals are soluble in hydrochloric acid, sulfuric acid, and nitric acid. However, hydrofluoric acid and phosphoric acid form insoluble fluoride and phosphate protective films. The metals are unaffected by alkalis.

## 3.3 Reactions of Rare Earth Metals with Non-metals

**Reaction with oxygen**

(1) Oxides of rare earth elements

The oxides of rare earth elements can be broadly defined as binary compounds with oxygen. Compounds formed by a combination of oxygen and more electronegative fluorine are generally referred to as fluorides rather than oxides.

Rare earth metals react with oxygen in the air at room temperature, initially undergoing surface oxidation. The further extent of oxidation varies according to the structure and properties of the resulting oxide. For example, the oxidation rates of La, Ce, and Pr in air are fast, such that it is necessary to preserve these metals in oil. The oxidation rates of Nd, Sm, and heavy rare earth metals are slow, and they retain their metallic luster for a long time.

Burning the hydroxides of Ce, Pr, and Tb in air produces $CeO_2$, $Pr_6O_{11}(4PrO_2 \cdot Pr_2O_3)$, and $Tb_4O_7(2TbO_2 \cdot Tb_2O_3)$, respectively. $Pr_2O_3$ reacts with atomic oxygen at 450 °C, or reacts with molecular oxygen at 300 °C and $50.66 \times 10^5$ Pa to afford $PrO_2$. $TbO_2$ can also be obtained by the reaction of terbium oxides of different compositions with atomic oxygen.

The rare earth sesquioxides adopt three structures A, B, and C at up to 2000 °C. The structural form adopted mainly depends on the size of the metal cations and the formation temperature. Rare earth ions are seven-coordinated in A-type $RE_2O_3$; six oxygen atoms surround the metal atom in an octahedral arrangement, and the seventh resides on one surface of the octahedron. In the B-type structure, the metal is also seven-coordinated; six of the oxygen atoms are in an octahedral arrangement, and the seventh has a longer bond to the metal than the others. The C-type structure corresponds to the removal of 1/4 of the anions from the $CaF_2$ structure, and the metal ions are six coordinated. C-type $RE_2O_3$ readily forms mixed crystals with the oxides of high-valence Ce and Tb (all of which have $CaF_2$-type structures).

Below 2000 °C, the oxides of La to Nd are of A-type; those of Y and Ho to Lu are of C-type. The other rare earth oxides are of C-type at low temperature, but convert to B-type above a certain temperature. The structural forms of $RE_2O_3$ can be interconverted; for example, C-type $Eu_2O_3$ changes to the B-type structure at about 1100 °C, and the A-type at 2040 °C. There are two structural forms above 2000 °C, denoted as the H- and X-types.

Researchers at the Max-Planck Institut für Metallforschung in Germany, led by Zinkevich et al. (2006), studied correlations, crystal structures, defect structures, and thermodynamics in the Ce–O system (Гринберг 1956). $CeO_2$ is used for the deoxygenation and desulfurization of liquid steel, and as catalyst, and it is effective polishing compound and glass colorant and so on. $Ce_2O_3$ can be oxidized to $CeO_2$ by oxygen. $CeO_2$ and $Ce_2O_3$ are stoichiometric oxides, which gradually deoxidize from $CeO_2$ to $Ce_2O_3$.

The researchers have summarized the heats of generation of several rare earth metal oxides, and the relationship between standard free energy of generation and temperature of rare earth oxides.

(2) Preparation of rare earth oxides with specific physicochemical characteristics

The ever-increasing applications of rare earth oxides necessitate their production with different physicochemical properties. For example, rare earth oxides and composite oxide powders with different particle sizes, specific surface areas, or morphologies are required for use in superconducting materials, luminescent materials, functional ceramic materials, precision electroplating, catalysts, sensing materials, and high-melting-point and high-strength alloys.

The preparation methods are mainly based on various classical methods.

**Dissolution tendency of rare earth salts and dehydration of rare earth sulfates**

The $NO_3^-$, $ClO_4^-$, $CNS^-$, $BrO_3^-$, $CH_3COO^-$, $Cl^-$, $Br^-$, and $I^-$ salts of trivalent rare earth metals can be dissolved in water, but those of $F^-$, $OH^-$, $CO_3^{2-}$, $C_2O_4^{2-}$, $PO_4^{3-}$, and other anions are less soluble due to increased attraction between the ions. Rare earth oxides, hydroxides, and carbonates react with sulfuric acid to form the corresponding sulfates $RE_2(SO_4)_3 \cdot nH_2O$. Typically, La and Ce sulfates have $n = 9$ and the other rare earth sulfates have $n = 8$, but there are also cases of $n = 3, 5, 6$, etc.

The high-temperature dehydration of a hydrated sulfate or the thermal decomposition of an acid salt produces the corresponding anhydrous sulfate, which is further decomposed into oxygen-based sulfate on increasing the temperature, and finally the oxide is formed. The reaction proceeds as follows:

$$Ln_2(SO_4)_3 \cdot nH_2O \xrightarrow{155-260\,°C} Ln_2(SO_4)_3 + nH_2O \uparrow$$

$$Ln_2(SO_4)_3 \xrightarrow{800-850\,°C} Ln_2O_2SO_4 + 2SO_3 \uparrow$$

$$Ln_2O_2SO_4 \xrightarrow{1050-1150\,°C} Ln_2O_3 + SO_3 \uparrow$$

## 3.3 Reactions of Rare Earth Metals with Non-metals

Anhydrous rare earth sulfates readily absorb water and react exothermically when dissolved in water. The solubilities of these sulfates decrease with increasing temperature, making them easy to recrystallize and purify. Taking 20 and 40 °C as examples, it shows the solubilities of the rare earth sulfates decrease on going from Ce to Eu, and then increase on going from Gd to Lu.

Rare earth sulfates and alkali-metal sulfates can form $RE_2(SO_4)_3 \cdot M_2SO_4 \cdot nH_2O$ complex salts with $n = 0, 2, 8$, etc., according to the ternary aqueous salt system phase diagram for $RE_2(SO_4)_3 - M_2SO_4 - H_2O$. At temperatures higher than 90 °C, the anhydrous salt is formed. The solubilities of light rare earth sulfate complex salts are lower than those of salts incorporating the heavier rare earths. Therefore, on precipitating rare earth elements from mixed sulfate complex salts, light rare earths are preferentially precipitated, leaving much of the heavy rare earths in solution. Thus, light and heavy rare earth elements can be fractionated according to the solubilities of their sulfates.

Sulfates of the cerium group (La–Sm) are insoluble, those of the terbium group (Eu–Dy) are slightly soluble, and those of the yttrium group (Ho–Lu, Y) are freely soluble.

### Halides of rare earth elements

The main preparation methods of rare earth halides are as follows:

(1) Halogenation of the metals: in order to obtain pure trihalides, direct halogenation is favorable:

$$2RE + 3X_2 = 2REX_3$$

Such reactions are generally vigorous. Besides halogens, HX can also be used as the halide source, but this route has only been used to prepare fluorides and chlorides.

(2) Dehydration of hydrates: because hydrates undergo hydrolysis to form haloxides upon heating, so impurities are entrained in anhydrous halides.

$$REX_3 \cdot nH_2O \rightarrow REOX + 2HX + (n - 1)H_2O$$

Therefore, the dehydration of hydrated halides should be carried out in the presence of a dehydrating agent. The traditional laboratory method is dehydration in an HX atmosphere at 105–350 °C, whereby most of the water can be removed. This method is effective for chloride dehydration, but less so for bromides and iodides. The experimental set-up for the dehydration of rare earth chlorides is used schematically, including a tubular furnace and material pipe, that dry HCl enters pipe, through $LnCl_3$, $nH_2O$, to get HCl, $H_2O$.

Anhydrous $REBr_3$, except for that of Lu, can be obtained by carefully controlling the temperature under vacuum conditions. Anhydrous iodides can be prepared by heating the hydrates in an HI and $H_2$ atmosphere.

Adding excess ammonium halide to rare earth halides is a facile dehydration method. Typically, 6 molar equivalents of $NH_4Cl$ are added to 1 molar equivalent of $RECl_3$, or 12 molar equivalents of $NH_4I$ are added to 1 molar equivalent of $REI_3$.

Through a sequence of complete evaporation of excess water, slowly heating the product to 200 °C in vacuo to remove all moisture, then raising the temperature to 300 °C for sublimation of the excess ammonium halide, the anhydrous rare earth halide can be obtained (the added ammonium halide must be of high purity and free from impurities that might impair sublimation).

Taking the dehydration of $RECl_3 \cdot nH_2O$ as an example, the process can be represented as follows:

$$RECl_3 \cdot nH_2O \rightarrow REOCl + 2HCl \uparrow + (n - 1)H_2O \uparrow \tag{3.1}$$

If $NH_4Cl$ is added to assist dehydration, the reaction is:

$$REOCl + 2NH_4Cl \xrightarrow{\text{Heating}} RECl_3 + H_2O \uparrow + 2NH_3 \uparrow \tag{3.2}$$

Calculation of the equilibrium constant for Eq. (3.1) shows that the chemical stability of REOCl increases with increasing temperature and atomic number of the rare earth element. Therefore, the dehydration of heavy rare earth hydrated chlorides is more likely to produce chlorine oxides. Chlorine oxides are the main cause of oxygen and chlorine contamination and low metal recoveries in rare earth metal production. In industrial production, to prevent the formation of chlorine oxides, dehydration is performed under negative pressure and in the presence of dry HCl and $NH_4Cl$.

Anhydrous rare earth chlorides are produced by vacuum dehydration in the presence of $NH_4Cl$. Dehydration is performed in a horizontal drying kiln heated at the bottom.

In the industrial equipment for crystalline hydrate dehydration of mix rare earth chlorides, the hydrated rare earth chloride is mixed with 20 wt. % $NH_4Cl$ and loaded on enamel or ceramic plates placed in the frame of the feeder car. The thickness of the material layer is generally 20–30 mm. After the material car is pushed into the kiln, the kiln door is closed, and the pressure is reduced to less than 0.67 kPa by means of a mechanical vacuum pump as the sample is heated. Kiln heating programs often determine the quality of the dehydrated products. To obtain optimal products, the dehydration heating sequence is approximately room temperature $\rightarrow$ 100 °C $\rightarrow$ 120 °C $\rightarrow$ 155 °C $\rightarrow$ 200 °C (holding for the appropriate time at each temperature). The time required for the dehydration process depends on the amount of sample. Under the optimal process conditions, the yield of dehydrated product is more than 90%, the water-insoluble residue is less than 10%, and the residual moisture should be no more than 5%. Unqualified products can be reprocessed and refined; and equipment needs further improvement.

### Nitrogen compounds of rare earth elements

There are two common preparation methods for nitrogen compounds of rare earth elements.

(1) Direct combination of rare earth metals with nitrogen. RENs can be obtained by heating the metal to 800–1200 °C in a reaction furnace and injecting high-purity nitrogen. The reaction equation is as follows:

$$2RE + N_2 \xrightarrow{800-1200\,°C} 2REN$$

(2) Reaction of rare earth hydrides with nitrogen. The reaction temperature is 600–800 °C for La–Sm, 1000–1200 °C for Ga–Lu, and the equation is:

$$REH_3 + N_2 \rightarrow REN + NH_3 \uparrow$$

The nitrogen group comprises five elements: nitrogen, phosphorus, arsenic, antimony, and bismuth. Phosphides can be prepared by mixing two simple substances (the respective elements) in a sealable reaction tube, purging the air with argon, sealing, and gradually heating the mixture to 900 °C. The reaction vessel is then abruptly cooled. Excess phosphorus is removed by heating to 600 °C in vacuo.

Solutions of Eu and Yb in ammonia react with $PH_3$ to produce $RE^{2+}(PH_3)_3 \cdot 7NH_3$, thermal decomposition of which yields EuP and YbP. Heating rare earth metals and arsenic in sealed tubes generates REAs.

$$RE + As \xrightarrow{1000-1500\,°C.\ 10-15\,h} REAs$$

Rare earth antimonides can be obtained by mixing the rare earth metal and antimony in a 1:1 molar ratio and heating at 1000–1050 °C for several hours.

$$RE + Sb \xrightarrow{1000-1050\,°C} RESb$$

Rare earth bismuthides can be prepared by reaction of the rare earth metal with bismuth at a temperature below the bismuth melting point.

Most rare earth nitrides are semi-metallic conductors, such as ScN, GdN, and YbN, and show semiconductor properties. The melting points of rare earth nitrogen compounds are quite high, typically above for 2400 °C for RENs, 2200–2600 °C for REPs, and about 1800 °C for REB is. Certain iron ores from southern China contain As, Sb, and Bi, which are detrimental to steel and can be removed by adding rare earths to produce such high-melting-point compounds.

The enthalpies of formation ($\Delta H$) of rare earth nitrogen are summarized in a table (4.184 kJ/mol unit).

**Carbides of rare earth elements**

When rare earths are present in molten steel, in addition to the beneficial effects of deoxygenation and desulfurization, they may also cause problems by generating carbides in high-carbon steel. Rare earths produce three main types of carbides: $RE_3C$, $RE_2C_3$, and $REC_2$.

$RE_3C$ of Sm–Lu and Y have face-centered-cubic $Fe_4N$-type structures; $RE_2C_3$ of La–Ho and Y have body-centered-cubic structures; $REC_2$ of lanthanides and Y have body-centered-tetragonal $CaC_2$-type structures.

$RE_2C_3$ and $REC_2$ show metal-like conductivities. Except for those of Sm, Yb, and Eu, other rare earth elements adopt a trivalent state in $REC_2$. The melting points of $REC_2$ are generally higher than 2000 °C, and those of $La_2C_3$, $Pr_2C_3$, and $Nd_2C_3$ are 1430, 1557, and 1620 °C, respectively.

All rare earth carbides are hydrolyzed by water at room temperature to form the rare earth oxides and gaseous products. $RE_3C$ hydrolysis generates a mixture of methane and hydrogen. $RE_2C_3$ and $REC_2$ react with water to form acetylene accompanied by hydrogen and small amounts of hydrocarbons. This is reminiscent of the historical use of $CaC_2$ in acetylene lamps.

**Borides of rare earth elements**

Rare earth borides $REB_n$, with $n = 2$, 3, 4, 5, 6, or 12, have been synthesized. $REB_6$ is the most extensively studied. The simple preparation method involves direct combination of the component elements. Powders of the rare earth metal and boron are mixed in the required proportions and heated to 1300–2000 °C in vacuo or in a high-purity argon atmosphere.

The hexaborides of rare earth elements belong to the cubic crystal system. Metal atoms occupy each apex of the hexagonal body, while boron is located at the center of the cube. They have metal-type bonds and hence high electrical conductivity; their resistivities are similar to those of the rare earth metals.

Lanthanum hexaboride ($LaB_6$) can be used as a cathode emitting material. Compared with a tungsten cathode, it has advantages of low electron work function, high emission electron density, good ion-bombardment resistance, stable performance, and long service life. It has been used in high-precision instruments, such as plasma sources, scanning electron microscopes, Auger spectrometers, and electron probes.

## 3.4 Complexes of Rare Earth Elements

**Properties and factors**

Because of the strong coordination ability of oxygen, oxygen-containing ligands are most prevalent in rare earth complexes. Considering the coordination field effect, selected oxygen-containing ligands coordinate to light rare earths in the following order:

## 3.4 Complexes of Rare Earth Elements

$$H_2O < CH_3COO^- < CH_2(OH)COO^- < CH_2(OH)(CHOH)_4COO^-$$
$$< N(CH_2COO^-)_3 < DCTA < EDTA < DTPA$$

Because water has strong coordination ability towards rare earth cations, it has a great influence on the formation of complexes in aqueous solution. Ligands and water compete to coordinate with rare earth ions. Rare earth cations form complexes with inorganic ligands as follows.

(1) Trivalent rare earth cations do not interact with $NH_3$, $NO_2^-$, or $CN^-$ in aqueous solution, nor do they produce complex salts with $NO_2^-$; trivalent rare earth cations can only form $REX^{2+}$ complexes with halides, $ClO_4^-$, and $NO_3^-$, and these complexes are unstable.
(2) Some high- or low-**valent** rare earth cations can form complexes with certain ligands. Tetravalent rare earth cations ($Ce^{4+}$, $Pr^{4+}$, $Tb^{4+}$, and $Dy^{4+}$) have a stronger ability to form complexes than their trivalent counterparts.
(3) The stability order in which rare earth cations form complexes with inorganic ligands is as follows:

$$PO_4^{2-} > CO_3^{2-} > F^- > SO_4^{2-} \approx S_2O_3^{2-} > SCN^- > NO_3^- > Cl^-$$
$$> Br^- > I^- > ClO_4^-$$

The complexes of rare earth cations and phosphorus-containing ligands are typically highly stable chelates. The stability order of inorganic phosphorus-containing complexes of rare earths is:

$$PO_4^{3-} > P_2O_7^{4-} > P_3O_{10}^{4-} > P_4O_{12}^{4-} > P_3O_9^{3-} > H_2PO_4^-$$

Rare earth inorganic phosphorus-containing complexes bearing protons are less stable than those without protons, those with a cyclic structure are less stable than those with a straight chain, and the stability of the latter decreases with increasing chain length. The stability of light lanthanides is slightly higher than that of heavy lanthanides for complexes with low stability of $RENO_3^{2+}$ and $RECl^{2+}$.

### Coordination number and bond characteristics

The 4f electrons of lanthanide ions are in the inner layer of the atomic structure and are shielded by the full outer layer of $5s^2 5p^6$, so they are less affected by the ligand field. The stabilization energy of the coordination field is around $100\ cm^{-1}$, and due to the contraction of the 4f electron cloud, the 4f orbital shows little or no involvement in the formation of chemical bonds. Thus, the bonding in rare earth complexes is mainly ionic. However, with increasing atomic number of the lanthanide elements and the associated decrease in ionic radius, the covalent properties of the complexes gradually increase.

38 3 Properties, Compounds, and Complexes of Rare Earth Elements

Because of the large ionic radii of lanthanide elements compared to other common trivalent ions, their complexes can have a higher coordination number to accommodate the spatial requirements of ligands. The coordination numbers of lanthanide complexes range from 3 to 12, although 8 and 9 are the most common.

Although Sc and Y have no f orbitals, the radius of $Y^{3+}$ ions can be listed in the series of trivalent lanthanide ions, and where the ionic radius is the main influencing factor, the properties of Y complexes are similar to those of lanthanide complexes. However, where properties related to the 4f orbitals are the main influencing factor, Y and lanthanide complexes obviously have different properties.

The coordination properties of Sc differ more greatly from those of lanthanides, mainly in the following aspects:

(1) $Sc^{3+}$ has a small radius and belongs to the first transition row in the Periodic Table. Its d orbitals can participate in bonding, so its complexes have high covalent character. For example, the melting points of its complexes are lower than those of similar complexes of lanthanides and Y.

(2) $Sc^{3+}$ has a high ionic potential, making it easier to hydrolyze in aqueous solution. In alkaline or neutral solution, sufficient ligands are required to prevent its hydrolysis, whereas $Ln^{3+}$ and $Y^{3+}$, except when bearing particularly weak ligands, are generally not susceptible to hydrolysis. Weakly coordinated complexes of hydrated $Ln^{3+}$ and $Y^{3+}$ salts can be obtained in alcoholic solutions, but corresponding complexes of $Sc^{3+}$ are only formed with anhydrous ligands in anhydrous non-aqueous solvents.

For valence bond analysis of rare earth compositions, modern theories of coordination fields and molecular orbitals have been adopted (Liu 2007). Crystal field theory can still be effectively applied to lanthanide complexes of the f-block. Coordination fields and molecular orbital theory were introduced in the 1960s to illustrate the chemical bond characteristics between rare earth cations and ligands. Studies have shown that in lanthanide complexes, 4f orbitals shielded by the $5s^2$, $5p^6$ electrons can form weak covalent bonds with ligands. Overlap integration between the 4f orbitals of lanthanides and the $n$p, $n$s orbitals of the coordinating atoms is smaller than that between the unfilled orbitals of the outer shell.

Studies using modern physical testing methods, such as NMR and paramagnetic resonance spectra, show that the covalent bond effect of 4f orbitals is only slightly reflected. The physical and chemical properties of rare earth complexes also indicate that the valence bonds between the cations and coordinating atoms are mainly ionic in nature. Although metal–organic compounds involving σ and π bonds have a considerable degree of covalency, many phenomena indicate that they still have obvious ionic bond properties.

Xu GX et al. studied the molecular orbitals of lanthanide compounds, especially organometallic compounds. Their results showed that: (i) the combination of lanthanide atom and ligand atom is essentially a covalent bond with a varying degree of ionicity; (ii) the 4f orbitals of lanthanide atoms are basically localized, and the overlap integral with the electron cloud of the ligand is very small, reflecting little or no bonding; (iii) when 4f orbitals contain unpaired electrons, they can affect 5d orbitals through paramagnetic polarization, thus affecting the chemical properties of lanthanide elements. Ultrasensitive spectral bands due to f–f transitions can be used as a probe to assess the degree of covalency of bonds in rare earth complexes.

In $\pi$ complexes of lanthanides, besides the usual $\sigma$ bonds (electron cloud distribution symmetric about the axis) and the $\pi$ bond (electron cloud distribution with a symmetric band surface), there are distinct $\delta$ bonds (electron cloud distribution with two symmetric nodes) and a $\varphi$ bond (electron cloud distribution with three symmetric nodes).

**Complexes of rare earths and organic ligands**

There are many types of rare earth complexes with organic molecules as ligands, and with the development of modern synthesis and characterization techniques, new complexes are continuously being developed and applied. Several important types are described below.

1. Complexes of rare earths and oxygen-containing ligands

The coordination ability of oxygen towards rare earth cations is very strong, and the complexes formed by rare earths and oxygen-containing ligands constitute the most important category here. The main ligand types are alcohols and alcoholates, carboxylic acids, hydroxycarboxylic acids, $\beta$-diketones, carbonyl compounds, and macrocyclic polyethers.

(1) Rare earth alcoholates. Rare earths and alcohols form solvates and alcoholates. In solvates, the oxygen atom still bears a hydrogen atom in the alcohol group; in alcoholates, the rare earth atom replaces hydrogen in the alcohol group. The stability of an alcohol solvate is lower than that of a hydrate. Consequently, in a mixed water/alcohol medium, alcohol molecules in the solvation shells of rare earth cations are gradually replaced by water molecules as the amount of water is increased.

Anhydrous rare earth chlorides are soluble in alcohols and are solvated; slow evaporation of the volatiles from a saturated solution leads to precipitation of the crystalline solvate $RECl_3 \cdot nROH$. Longer or more branched carbon chains in the R group reduce the $n$ value. Anhydrous rare earth chlorides can form alcoholates $RE(OH)_3$ by exchange reaction with alkali-metal alcoholates in alcoholic media. Many solid rare earth alcoholates with methanol and $n$-butanol have been prepared. Aliphatic monoethanolates with $pK_2 > 16$ can only exist in non-aqueous solvents, decomposing to give rare earth hydroxide precipitates in water.

Polyphenols (for example, $C_{22}H_{18}O_{11}$) are more acidic than aliphatic alcohols and can react with rare earths to form alcohol compounds.

(2) Rare earth $\beta$-diketonates. Ketones and rare earths may form solvates, the most studied and applied of which are those of $\beta$-diketones, whereas those of monoketones have been little studied. $\beta$-Diketonates have found applications in the realms of extraction, co-extraction, luminescence, lasing, volatilization, and as displacement reagents.

β-Diketones are well known to tautomerize between ketone-type and enol-type structures.

β-diketones can be regarded as monobasic weak acids, and where appropriate, they can lose a proton to forma monovalent anion with two coordination points. A chelate can be formed with rare earth cations after deprotonation of the enol form.

Due to the formation of chelating rings and conjugated chains, complexes of β-diketones and rare earths are the most stable of those involving only oxygen ligands.

Acetoacetone, propionyl acetone, phenylacetone, and diphenylacylmethane can form stable complexes with rare earth cations such as $La^{3+}$, $Pr^{3+}$, $Nd^{3+}$, and $Y^{3+}$, and their stability constants are typically of the order of $10^{20}$–$20^{41}$. The stabilities of their complexes with rare earths increase in the order: phenylacylacetone > propionyl acetone > acetylacetone. Across the rare earth series, the stabilities of the complexes increase with increasing atomic number.

Many rare earth β-diketone complexes have good luminescence properties, such as those of phenylacylacetone, α-hydroxybenzophenone, hexafluoroacetone, and thiophene formoyl trifluoroacetone. These complexes have been used in laser technology. The organic ligands in rare earth β-diketone complexes can improve the photoluminescence efficiency of the cations (especially $Sm^{3+}$, $Eu^{3+}$, $Tm^{3+}$, $Dy^{3+}$) excited by mercury vapor lamps. Diketone complexes of rare earths, such as $Eu_3$, can be used as magnetic resonance imaging (MRI) displacement reagents. They are also the most volatile rare earth compounds characterized to date.

(3) Rare earth carboxylate complexes. Many organic acids can form stable complexes with rare earth cations. Among monocarboxylic acids, the complexes of acetate and rare earth ions are the most widely studied and used. Stability constants increase on going from La to Sm and Eu, and those of the heavier rare

## 3.4 Complexes of Rare Earth Elements

earths show little further change. There have also been many studies on rare earth complexes of propionic acid and isobutyric acid, the coordination abilities of which towards rare earth ions are similar to that of acetic acid, although they decrease in the order: acetic acid > propionic acid > isobutyric acid.

Among monocarboxylic acids, the most stable rare earth complexes are those with hydroxycarboxylic acids, because the hydroxy group in such carboxylic acids can contribute to the production of more stable chelate complexes.

The greater the positive charge on metal ion M, the stronger the interaction with the $\alpha$-hydroxyl oxygen atom, and the more stable the complex formed. $\alpha$-Hydroxyisobutyric acid, gluconic acid, citric acid, glyoxylic acid, lactic acid, malic acid, salicylic acid, and tartaric acid can form stable complexes with rare earth cations.

2. Complexes of rare earths with nitrogen-containing ligands

The coordination ability of N-ligands towards rare earth cations is weaker than that of O-ligands, making it difficult to produce rare earth complexes bound by N in aqueous solution. However, rare earth complexes with N-ligands can be prepared in appropriate non-aqueous polar solvents, and the coordination number can be as high as 8 or 9. Such complexes can be roughly divided into two categories.

(1) Complexes formed from rare earths and weakly basic nitrogen-containing ligands, such as phenanthroline, bipyridine, and phthalocyanine, which coordinate in appropriate solvents, for example, rare earth phthalocyanine complexes.

The compositions of rare earth complexes are related to the properties of inorganic anions. Complexes with two hydrogen-containing bidentate neutral ligands are obtained when the anion is $Cl^-$, $NO_3^-$, etc. With a coordination number greater than 6, anions participate in the coordination sphere, generating complexes with three nitrogen-containing bidentate neutral ligands with anions such as $SCN^-$. With weakly coordinating anions, such as $ClO_4^-$, complexes with four bidentate neutral N-ligands are generated.

(2) Complexes of rare earths with strongly basic N-ligands, such as amines and their derivatives, are a class of compounds containing RE–N bonds. N-ligands of rare earth organic amine compounds contain organic groups including $-NR_2$ and $-N(SiR_3)_2$. Such complexes can be prepared as powders by treating anhydrous rare earth nitrides with polydentate ligands, such as ethylenediamine, propanediamine, diethylenetriamine, or triethylenetriamine, in acetonitrile. These complexes have considerable thermal stability, but are easily hydrolyzed by atmospheric moisture. In the last decade, many new rare earth complexes with organic amines have been synthesized, most notably trimethylsilylamino complexes, which not only show good solubility in organic solvents, but are also versatile reaction precursors. They can be converted into pure alkoxy rare earth metal compound which can be used in the preparation of electronic materials, etc.

Furthermore, it has been found that Re–N bonds can undergo many chemical reactions, such as the formation of cyclopentadiene rare earth amides, which are effective catalysts. They can not only catalyze organic reactions, such as amination and cyclization, but also the polymerization of some polar and non polar monomers.

3. Complexes of rare earths with N,O-ligands

There are at least two potentially coordinating atoms in organic N,O-ligands, so they can form stable chelates with rare earth ions. Such N,O-ligands include amino carboxylic acids, pyridinic dicarboxylic acids, and is ohydroxamic acids.

Rare earth complexes with amino carboxylic acids have many chelate rings and good stability, and find wide application. The complexes mainly incorporate linear or cyclic polyamines, and polycarboxylic acid ligands, including aminotriacetic acid (NTA), ethylenediamine tetraacetic acid (EDTA), imino $N,N$-diacetic acid (IMDA), and diethylenetriamine pentacetic acid (DTPA). The order of their coordination abilities is: DTPA > DCTA > EDTA > HEDTA > NTA > IMDA.

The 1:1 or 1:2 complexes formed with rare earth ions can be crystallized from aqueous ethanolic solution due to their strongly coordinating carboxyl groups. They have important applications in nuclear magnetic resonance (NMR) techniques and are also commonly used in the separation and analysis of rare earths. For example, EDTA has been widely used in ion-exchange separation and analysis techniques. In addition to complexes of organic N,O-ligands, there are also complexes involving coordination through sulfur, phosphorus, arsenic, and so on, but these are less stable.

4. Multi-component and bi- or polynuclear complexes of rare earths

### 3.4 Complexes of Rare Earth Elements

Rare earth complexes can incorporate two or more ligands. These ligands can have various coordination modes, being either neutral or anionic. They can coordinate through oxygen, nitrogen, or other atoms. As a result, there are many kinds of multi-component complexes, in which the coordination number at the rare earth cation can vary in the range 6–12. There are also ion-association-type rare earth complexes, as exemplified by $[RE (PyO)_6]^{3+}[Cr (NCS)_6]^{3-}$, where PyO denotes pyridine oxide.

5. Rare earth metal–organic compounds

The term rare earth metal–organic compounds refer to compounds containing rare earth metal–carbon bonds. They have many important chemical and physical properties, which can be exploited in organic synthesis and polymer synthesis.

Rare earth complexes commonly involve cyclic ligands bound through their $\pi$-electron systems. Cyclopentadiene $(C_5H_5^-)$ is the most common organic $\pi$ ligand, while carborane $(C_2B_3H_{11}^{2-})$ is the most common inorganic ligand. Such ligands have similar molecular orbitals and can form $\pi$ bonds with the central metal cations. Many studies have been carried out on rare earth metal carborane complexes, including those of Sc and Y. These complexes are sensitive to the reaction conditions, especially in the $C_2B_4$ system. As light change in the reaction conditions can lead to products with very different structures.

### Main uses of rare earth complexes in the materials field

In addition to their important applications in rare earth hydrometallurgy and analytical chemistry, rare earth complexes have been widely used in modern industry and high-tech fields.

1. Application of rare earth complexes in luminescent materials

The photoluminescence phenomenon of some rare earth complexes has long been observed. During the 1960s and 1970s, with the development of laser technology, researchers systematically examined photo luminescent rare earth complexes in order to find working materials for lasers, and made many important observations (Crawford et al. 2020). The study of rare earth complexes progressed from solution to solid-state chemistry, and a large number of complexes have been characterized. Rare earth $\beta$-diketone complexes show favorable properties, and they have been used as working materials in laser output.

It has been demonstrated that the quantum efficiencies of $RE^{3+}$ complexes are strongly enhanced when they are photo reduced in alcohols. For example, an ethanolic solution of an $Eu^{3+}$ crown complex emits strong blue light under UV irradiation, endowing it with good application prospects.

2. Application of rare earth complexes in nuclear magnetic resonance (NMR)

Rare earth complexes are applied in NMR spectroscopy as displacement reagents, for example, widely used Eu (fod)$_3$. Here, the chemical formula of ligand H (fod) is 1, 1, 1, 2, 2, 3, 3–7F-7, 7-dimethyl-acetyl caproyl $[CH_3CF_2COCH_2\text{-}COC(CH_3)_2CH_3]$.

In recent years, the technology of nuclear magnetic imaging developed for medical diagnosis has relied on rare earth complexes as contrast agents. These are a class of paramagnetic substances that facilitate the relaxation of protons in water, thus accelerating the acquisition of images by increasing the signal intensity contrast. The resulting enhanced images permit the differentiation of normal or abnormal tissues and organs for diagnosis. The most promising of these contrast agents are complexes of rare earths with triamine derivatives such as DTPA, most notably $Gd^{3+}$ with linear (e.g., DTPA) or large cyclic (e.g., DOTA) aminocarboxylic ligands, as in $(NMG)_2[Gd(DTPA)(H_2O)]$, here, NMG is n-methyl glucosamine.

3. Application of rare earth complexes in organic synthesis

(1) Activation of carbonyl groups

Transition metal organic compounds containing d electrons are capable of activating carbonyl groups. Studies of the reactivity of complexes containing RE–C bonds have revealed that RE–C σ-bonds also have this characteristic. CO can be inserted to obtain rare earth acyl complexes.

(2) Activation of saturated hydrocarbon bonds

Activation of saturated hydrocarbon bonds (C–H) is an issue that has not been well resolved in homogeneous catalytic reactions. Watson first found that both $(C_5Me_5)LnCH_3$ and $(C_5Me_5)LnH$ complexes are capable of activating saturated C–H bonds, and the reaction is mild and high-yielding.

4. Application of rare earth complexes as catalytic materials

Some rare earth metal–organic complexes, such as those of Er, Yb, and Sm, can be used as homogeneous catalysts to hydrogenate alkenes or alkynes at room temperature and atmospheric pressure to afford the corresponding alkanes or alkenes. Effective catalysts include $[(C_5Me_5)Sm(THF)_2]$ and $[(C_5Me_5)_2SmH_2]$, of which the latter shows the highest catalytic activity. Such Sm complexes can hydrogenate ethylene to ethane with conversions of up to 120,000 $h^{-1}$ at an $H_2$ pressure of 1 Pa at 25 °C.

Because the efficacy of the hydrogenation reaction is related to the properties of the rare earth cation and the ligands, as well as the coordination mode and the solvent, it is necessary to carry out detailed and rigorous experimental studies under theoretical guidance. Rare earth complexes also have important applications in homogeneous polymerization, as in the production of synthetic rubber.

Effective utilization of the optical, electrical, and magnetic properties of rare earth elements and their properties related to energy storage and energy conversion is central to the design of new rare earth materials (Shun 2019). Through the interaction of rare earth ions with ligands, these properties can be changed, modified, and enhanced to a large extent, so the development of rare earth complexes has broad prospects.

# References

Baria DN, King TS, Bautista RG (1976) The normal spectral emittance of yttrium, lanthanum, cerium, praseodymium and neodymium above 1000 K. Metall Mater Trans B 7:577–580

Beaudry BJ, Gschneidner KA Jr (1978) Preparation and basic properties of the rare earth metals. In: Gschneidner KA Jr, Eyring L (eds) Handbook on the physics and chemistry of rare earths, vol 1. North-Holland Physics Publishing, Amsterdam, pp 173–232

Collocot SJ, Hill RW, Stewart AM (1988) Physical properties of the rare earth metals. J Phys F 18:L223

Crawford SE, Ohodnicki PR, Baltrus JP (2020) Materials for the photoluminescent sensing of rare earth elements: challenges and opportunities. J Mater Chem C (24)

Delaeter JR (1988) Atomic weights of the elements 1987. J Phys Chem Ref Data 17:1791

Du T, Han QY, Wang CZ (1995) Physical chemistry of rare earth alkali soil and its application in materials. Science Press, Beijing

Goldschmidt ZB (1978) Atomic properties (free atom). In: Gschneidner KA Jr, Eyring L (eds) Handbook on the physics and chemistry of rare earths, vol 1. North-Holland Physics Publishing, Amsterdam, pp 1–171

Гринберг AA (1956) Chemical summary of complexes. In: Shen PW, Wang JZ, Ma W (trans). Higher Education Press, Beijing

Gschneidner KA Jr (1990) Bulletin of alloy phase diagrams. Ames Lab 11(3):216

Gschneidner KA Jr, Calderwood FW (1986) Intra rare earth binary alloys: phase relationships, lattice parameters and systematic. In: Gschneidner KA Jr, Eyring L (eds) Handbook on the physics and chemistry of rare earths, vol 8. North-Holland Physics Publishing, Amsterdam, pp 1–161

King TS, Baria DN, Bautista RG (1976) The normal spectral emittance of erbium, dysprosium and samarium above 1000 K. Metall Mater Trans B 7:411–415

Koskenmaki DC, Gschneidner KA Jr (1978) Cerium. In: Gschneidner KA Jr, Eyring L (eds) Handbook on the physics and chemistry of rare earths, vol 1. North-Holland Physics Publishing, Amsterdam, pp 337–377

Legvold S (1980) Rare earth metals and alloys. In: Wohlfarth EP (ed) Handbook of ferromagnetic materials, vol 1. North-Holland Physics Publishing, Amsterdam, pp 183–295

Li M, Liu ZG, Wu JX et al (2009) Rare earth elements and analytical chemistry. Chemical Industry Press, Beijing

Liu GH (2007) Rare earth materials. Chemical Industry Press, Beijing

McEwen KA (1978) Magnetic and transport properties of the rare earths. In: Gschneidner KA Jr, Eyring L (eds) Handbook on the physics and chemistry of rare earths, vol 1. North-Holland Physics Publishing, Amsterdam, pp 411–488

Shun O (2019) Energy transfer processes in polynuclear lanthanide complexes. Springer, Singapore

Soott T (1978) Elastic and mechanical properties. In: Gschneidner KA Jr, Eyring L (eds) Handbook on the physics and chemistry of rare earths, vol 1. North-Holland Physics Publishing, Amsterdam, pp 591–705

Stretz LA, Bautista RG (1972) Temperature, its measurement and control in science and industry. Part 1, vol 7. In: Plumb HH (ed). Instrument Society of America, Pittsburgh, p 489

Van Zytveld J (1989) Liquid metals and alloys. In: Gschneidner KA Jr, Eyring L (eds) Handbook on the physics and chemistry of rare earths, vol 12. North-Holland Physics Publishing, Amsterdam, pp 357–407

Zinkevich M, Djurovic D, Aldinger F (2006) Thermodynamic modelling of the ceria-oxygen system. Solid State Ionics 177:989–1001

# Chapter 4
# Rare Earth Invar Alloys, Intermetallic Compounds, and Composite Oxides

## 4.1 Invar Alloys

Invar alloys are a special class of alloys showing little or no expansivity with temperature in the normal working range. The term "Invar" originated from a study of Guillaume (1897) on opposite-centered cubic FeNi alloys in the early nineteenth century. He discovered that the thermal expansion of such alloys as a function of temperature was anomalous; that is to say, the length of a sample remained constant, corresponding to a thermal expansion coefficient of zero or near-zero, over a wide temperature range. This discovery was of very high practical importance, since this alloy could be applied in precision instruments, integrated circuit matrices, and seismograph components. Furthermore, Guillaume found that the thermal expansion of FeNiCr alloy was temperature-independent over a wide range, allowing this alloy to be used as a pendulum material. He won the Nobel Prize in Physics in 1920 for this research.

The pioneering work of Guillaume attracted worldwide interest, and through a series of studies it was found that Invar anomalies are not limited to FeNi and FeNiCr ferromagnetic alloys with a face-centered cubic structure (Hidnert et al. 1988). Indeed, many other ferromagnetic and antiferromagnetic binary, ternary, and multiple alloy systems show Invar properties. Invar systems can have face-centered cubic, body-centered cubic, hexagonal, or other structures, or even be amorphous. Among alloys formed by rare earth (RE) metals and transition group metals (TM), $La(Fe_xAl_{1-x})_{13}$ and $La(Fe_xSi_{1-x})_{13}$ with cubic structures (Palstra et al. 1983; de Boer et al. 1987), and $REFe_{14}B$ and $REFe_{12-x}(TM)_x$ with tetragonal crystal systems, also show obvious Invar properties (Buschow 1988). It is believed that a necessary condition for Invar behavior in alloys is that they should contain at least one 3d transition metal.

To date, many anomalies in important physical properties of alloys have been considered as Invar properties. There are many theoretical models for describing Invar properties, such as that of intrinsic antiferromagnetism in the domain magnetic moment model, the differential short-range ordered domain model, the localized

© Science Press 2023

C. Wang, *Theory and Application of Rare Earth Materials*,

https://doi.org/10.1007/978-981-19-4178-8_4

48 4 Rare Earth Invar Alloys, Intermetallic Compounds, and Composite Oxides

surrounding model, the in homogeneity model, the $2\gamma$ and weak nonlocal ferromagnetism in the energy-band model, the weak nonlocal ferromagnetic spin fluctuation model, the finite temperature rigid performance band model, the finite temperature electron–phonon interaction model, the finite temperature localized environment model, and so on.

Of these models, the most important and widely discussed is the Weiss $2\gamma$ state model (Weiss 1963). In 1963, Weiss proposed that $\gamma$-Fe has two different magnetic states, the $\gamma_1$ state with low spin and the $\gamma_2$ state with high spin. The $\gamma_1$ state has a smaller magnetic moment and volume, whereas those of the $\gamma_2$ state are larger. For Fe–Ni alloys, Weiss assumed that the change in energy between the two states is a function of the Ni concentration, such that the $\gamma_2$ state in $Fe_{65}Ni_{35}$ becomes the ground state and the $\gamma_1$ state becomes the excited state. As the temperature is increased, the total number of $\gamma_1$ excited states occupies a smaller volume, thus compensating for expansion of the lattice. The Weiss model has since been improved and generalized (Bendick and Pepperhoff 1979; Chikarumi 1980). The modern energy-band calculation method confirms the validity of the Weiss $2\gamma$ model to some extent. Carbone et al. (1987) used spin polarization angle-splitting photoemission techniques to investigate ordered $Fe_3Pt$ single-crystal planes. Hasegawa (1985) confirmed the ordered $Fe_3Pt$ splitting band structure through calculations. By studying the dependence of reflection intensity on temperature for ordered $Fe_3Pt$ single crystals, as well as measuring the specific heat capacities of Invar alloys, they confirmed the high-low spin-state transition and temperature-induced changes in electronic structure (Bendick and Pepperhoff 1979, 1981; Bendick et al. 1978, 1979). Thus, anomalies in Invar properties can be attributed to changes in the thermal activation process and thermodynamic stability.

As Invar theory has developed, new types of Invar alloys have been discovered, including those with rare earth components.

$La(Fe_xSi_{1-x})_{13}$ and $La(Fe_xAl_{1-x})_{13}$ show distinct Invar properties in terms of thermal expansion. There are three different magnetic regions in the $La(Fe_xAl_{1-x})_{13}$ system in the range $0.46 \leq x \leq 0.92$. It is believed that the presence of exceptional antiferromagnetic regions is caused by special crystal structures. It allows a large coordination number of up to 12 between Fe–Fe at a very small distance. Girord et al. found that Invar-type anomalies also arise in $RE_2Fe_{14}B$-type compounds. As permanent magnet materials, such Invar alloys are of great significance both theoretically and in applications such as electric drive devices. Measurements of the magnetic properties of $REFe_{12-x}(TM)_x$ alloys show a high Curie transition temperature. The lattice parameter shows the range of compositions that can stably exist.

Xu et al. (1991, 1992a, b, 1994) studied the thermodynamics of the $RE_2Fe_{14}B$ and $REFe_{12-x}(TM)_x$ systems. At the same time, they also explored effective methods for the preparation and treatment of alloys (Xu et al. 1991, 1992a, b, 1994; Yu et al. 1995), as elaborated in the following.

Using $CaF_2$ as a solid electrolyte, the electromotive force (EMF) method was applied to determine the activities of rare earth components in Invar alloys and other thermodynamic properties, using the pure RE metal and $REF_3$ as a reference electrode and the RE Invar alloy and $REF_3$ as the electrode to be tested. The assembled batteries

## 4.1 Invar Alloys

were of the form:

$$Mo \mid RE, REF_3 \mid CaF_2 \text{ (single crystal)} \mid REF_3, RE_2Fe_{14}B \mid Mo$$

$$Mo \mid RE, REF_3 \mid CaF_2 \text{ (single crystal)} \mid REF_3, REFe_{12-x}V_x \mid Mo$$

The electrode reactions can be expressed as:

$$\text{Reference electrode } RE + 3F^- = REF_3 + 3e$$

$$\text{Tested electrode } REF_3 + 3e = [RE]_{alloy} + 3F^-$$

The battery reaction is:

$$RE = [RE]_{alloy}$$

where $[RE]_{alloy}$ denotes the rare earth in the alloy, and its relative partial molar free energy $\Delta\overline{G}_{[RE]}$ is equal to the free energy of the battery $\Delta G$, that is:

$$\Delta\overline{G}_{[RE]} = \Delta G = -3FE$$

Here, $E$ is the EMF of the battery and $F$ is the Faraday constant equal to 96,485 C/mol, assuming solid pure rare earths as the standard state with an activity of 1.

$$\Delta\overline{G}_{[RE]} = RT \ln \alpha_{[RE]}, \, \alpha_{[RE]} = \exp(-3FE/RT)$$

From the above formula, we can evaluate the EMF of the battery at a certain temperature, as well as the activity value and other thermodynamic parameters of the rare earth alloy at that temperature.

The rare earth alloys prepared for the experiment included five $RE_2Fe_{14}B$ rare earth alloys, and three $REFe_{12-x}V_x$ rare earth alloys with five kinds of ratios, a total of 20 samples. In these preparation materials, irons are high purity, secondary electron bombardment, and deoxygenation treatment ones, and vanadium of purity $> 99.95\%$, boron of purity $> 99.99\%$, and rare earths La, Ce, Nd, Sm, and Y, each of purity $\geq 99.5\%$. The powders were pressed and submitted to degassing in a tungsten wire vacuum furnace.

Each alloy was prepared in a miniature magnetically controlled vacuum casting arc furnace, with six water-cooled copper crucibles per furnace, one of which contained $Zr_{84}Al_{16}$ alloy as an aspirate deoxidizer. After freezing, the samples were subjected to vacuum annealing, $RE_2Fe_{14}B$ for 500 h at 900 °C, $REFe_{12-x}V_x$ for 500 h at 850 °C.

Each alloy was analyzed by XRD and observed by scanning electron microscopy. Most of the samples were seen to consist of a single phase, with only a few showing a small amount of an accessory phase.

In order to further observe the morphologies of the alloy surfaces and the uniformities of their compositions, several samples were analyzed by electron probe. Except for high rare earth contents found in individual alloys at their grain boundaries, the compositions of most prepared alloys were in accordance with the stoichiometric ratio, indicating that they met the requirements for our research.

Commercially available $REF_3$ does not meet the purity requirements of this study, so we prepared samples in our laboratory. Each $RE_2O_3$ (99.5%) was fluorinated by the hydroamination method, using dry $NH_4HF_2$ of high purity as the fluorination agent. The reaction equation is:

$$RE_2O_{3(s)} + 6NH_4HF_{2(s)} \xrightarrow{300\,°C,10h} 2REF_{3(s)} + 6NH_4F_{(g)} \uparrow +3H_2O_{(g)} \uparrow$$

Volatilized $NH_4F$ was present in the product, which can inhibit the hydrolysis of $REF_3$, and the resulting product was of purity > 99%. This reaction can be carried out in a completely closed system, and the gas can be absorbed in a solvent. The product was confirmed as $REF_3$ by XRD analysis.

The $REF_3$ and $RE_{powder}$ were intimately mixed in a box filled with Ar in a mass ratio of 2:8 and then pressed into a circular sheet of $\varnothing 10$ mm $\times$ 3 mm, one side of which was pressed into a section of $\varnothing 0.5$ mm molybdenum wire to ensure good contact between them. The same method was used to prepare the $REF_3$–RE alloy electrode sheet to be tested. To eliminate the influence of contact resistance, that is, the internal factors of the battery, on the EMF measurement, each electrode slice was wrapped with molybdenum foil, placed in a quartz tube, and purged with Ar three times. Finally, under Ar atmosphere, the quartz tube was sealed with a hydrogen–oxygen flame and the contents were annealed at 600 °C.

To measure the electromotive force, the electrode sheet and $CaF_2$ single-crystal electrolyte were assembled into a battery and placed in a quartz tube (Yang and Wang 2004). A small amount of La metal was placed at one end of the tube to remove the residual oxygen, and then the tube was evacuated, filled with Ar, and sealed (Fig. 4.1). The experimental temperature was 607–747 °C.

The battery was placed in a double-wound Fe–Cr–Al resistance furnace with earthed and shield sleeve, a constant temperature band $677 \pm 0.5$ °C, and length 30 mm, and measured with a pure Pt thermocouple. A high-internal-resistance ($10^{14}$ $\Omega$) solid-state potentiometer connected to a Keithley 192 digital voltmeter (internal resistance $10^{12}$ $\Omega$) was used to measure the battery EMF, and an $X$–$Y$ function recorder was used to acquire the EMF polarization curve to determine whether the battery reaction had reached equilibrium. The EMF change was $\leq \pm 1$ mV. After the experiment, the electrode and electrolyte surfaces were examined; no reaction products or cracks were found.

## 4.1 Invar Alloys

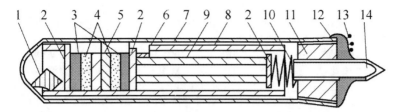

**Fig. 4.1** Battery assembly diagram. *Note 1* Lanthanum block (used to deoxidize); *2* corundum sheet; *3* tested electrode; *4* single crystal CaF$_2$; *5* reference electrode; *6* internal electrode lead (Pt or Ag); *7* quartz tube; *8* corundum tube; *9* corundum top rod; *10* spring; *11* stopper; *12* seal adhesive; *13* electrode lead; *14* fused glass tube

YFe$_{12-x}$V$_x$ is taken as an example to illustrate the experimental results, as shown in Figs. 4.2 and 4.3 (Figs. 4.1, 4.2 and 4.3 are kind permission from Metallurgical Industry Press).

The figures show that the thermodynamic stability of the alloy is poor, the activity of Y is large, and the trend in the activity of Y is similar to that in the partial

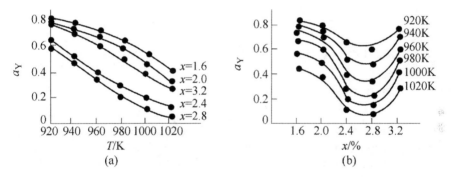

**Fig. 4.2** Relationship between activity-temperature (**a**), activity-mole fraction (**b**)

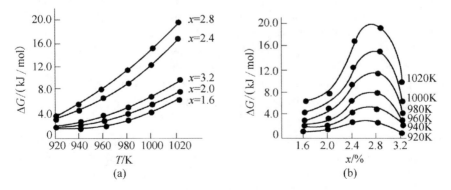

**Fig. 4.3** Relationship between free energy-temperature (**a**), free energy-molar fraction (**b**)

52      4   Rare Earth Invar Alloys, Intermetallic Compounds, and Composite Oxides

molar free energy. The most stable composition corresponds to the minimum of the composition–activity curve and the maximum of the $\Delta G$ versus $x$ plot.

Results for Invar alloys incorporating other rare earth metals show that the activity values of Sm in $SmFe_{12-x}V_x$ are two to three orders of magnitude smaller than those of Y in $YFe_{12-x}V_x$; that is, the chemical stability of Sm-containing alloys is higher than that of those containing Y, which may be due to strong interactions between the Sm ($4f^6$) electrons and the V and Fe electron orbitals therein.

Besides thermodynamic properties, we also determined the Curie temperatures of five $RE_2Fe_{14}B$ and $NdFe_{12-x}V_x$ alloys. The measured values were the same or similar to those reported in the literature. Thermal expansion measurements of $RE_2Fe_{14}B$ and $REFe_{12-x}V_x$ demonstrated variable levels of Invar features at around room temperature, with those of $RE_2Fe_{14}B$ alloys being stronger. $Nd_2Fe_{14}B$, $Sm_2Fe_{14}B$, and $Y_2Fe_{14}B$ show almost no thermal expansion in the range from room temperature to 200 °C. The properties of the $REFe_{12-x}V_x$ alloys are greatly influenced by the V/Fe ratio. The thermal expansions of $NdFe_{10}V_2$, $SmFe_{10}V_2$, and $YFe_{10}V_2$ are very small in the range from room temperature to 300 °C.

Above all, when selecting Invar alloys, the physical and thermodynamic properties should be considered comprehensively.

## 4.2   Thermodynamics of Intermetallic Compounds

(1)   Thermodynamic study of intermetallic compounds

The thermodynamic properties of intermetallic compounds can be determined indirectly or directly by the EMF method with a solid electrolyte. Researchers have found that $CaF_2$ single-crystal solid electrolyte readily reacts with the electrode material at high temperatures, so the measurements are usually carried out below 827 °C. Rezukhina et al. measured the standard free energy of generation of $LaNi_5$ in the temperature range 831–921 °C. Since $LaNi_5$ is an important oxygen-storage material for which abundant thermodynamic data are available over a wide temperature range. In 1994, Xiao et al. (1993, 1994a, b) synthesized polycrystalline $0.95LaF_3 \cdot 0.05CaF_2$ for use as a solid electrolyte, and assembled the following batteries to measure the standard free energy of generation of $LaNi_5$, as well as that of $CeNi_5$. The battery forms were:

(1)   $Mo \mid La, LaF_3 \mid 0.95LaF_3 \cdot 0.05CaF_2 \mid LaNi_5, Ni, LaF_3 \mid Mo \quad 630 - 900\,°C$
(2)   $Mo \mid La \mid 0.95LaF_3 \cdot 0.05CaF_2 \mid LaNi_5, Ni \mid Mo \quad 630 - 900\,°C$
(3)   $Mo \mid Ce, CeF_3 \mid 0.95LaF_3 \cdot 0.05CaF_2 \mid CeNi_5, Ni, CeF_3 \mid Mo \quad 600 - 750\,°C$

The difference between (1) and (2) is that the reference electrode of (2) does not include $LaF_3$. Instead, the $LaF_3$ in the solid electrolyte participates in the electrode reaction to establish the corresponding electrode potential.

When expressed in terms of La and Ce, the battery reactions of (1) and (3) are:

Reference pole: $3F^- - 3e = \frac{3}{2}F_2$

## 4.2 Thermodynamics of Intermetallic Compounds

$$\frac{3}{2}F_2 + RE = REF_3$$

Pole to be tested: $REF_3 + 5Ni = \frac{3}{2}F_2 + RENi_5$

$$\frac{3}{2}F_2 + 3e = 3F^-$$

General response: $RE + 5Ni = RENi_5$.

For (2), the corresponding cell reaction equation can be obtained by substituting $REF_3$ into a solid solution $LaF_3$. The corresponding free energy change is:

$$\Delta G = \Delta G^{\ominus}_{RENi_5} = -3FE$$

For all of the above reactions, pure solid matter is regarded as the standard state.

For examples, $LaF_3$ and $CeF_3$ were prepared by the abovementioned hydroamination method; $LaNi_5$ and $CeNi_5$ synthesis, annealing, phase analysis, battery assembly, EMF measurement, were all the same as the above thermodynamic study of rare earth Invar alloys.

Experimental results showed the EMFs of both cell reactions fluctuated in the range $\pm 0.15$ mV, with very good reproducibility.

Map temperature $T$ by EMF ($E$), it is a linear relation. According to:

$$\Delta G^{\ominus}_{LaNi_5} = -3FE$$

and

$$\Delta G^{\ominus}_{CeNi_5} = -3FE$$

get:

$$\Delta G^{\ominus}_{LaNi_5} = -152,590 + 13.143T \text{ (error} \pm 150\,J)$$

this is the average of batteries (1) and (2).

$$\Delta G^{\ominus}_{CeNi_5} = -157,600 + 25.514T \text{ (error} \pm 150\,J),$$

indicating that $LaNi_5$ and $CeNi_5$ are stable.

Pratt collected thermodynamic data on solid and liquid alloys studied by scientists using different solid electrolytes up to 1990, including $CaF_2$ as a solid electrolyte, and rare earth alloys such as Co–La, Ni–La, Pd–Ce, Pd–Gd, Pd–Y, Y–Co, Y–Fe, and Y–Ni.

(2) Thermodynamic study of rare earth intermetallic compounds

54      4 Rare Earth Invar Alloys, Intermetallic Compounds, and Composite Oxides

Alloy has different components and contents, so the material has varied properties. Irrespective of whether optical, electrical, magnetic, or thermal properties are to be exploited, all require the material to have thermodynamic stability at the operating temperature.

The solid-electrolyte EMF method has been widely used in the thermodynamic study of alloys. For thermodynamic study of binary intermetallic compounds (Subbarao 1980; Kubaschewski and Alcock 1977; Rapp 1970a, b; Pratt 1990), a typical battery form is:

$$M \mid A, AX_n \mid X^{x-} \mid [A]_{\text{alloy}}, AX_n \mid M$$

A solid-oxide electrolyte is used for research purposes through the establishment of an oxygen concentration cell. It takes the form:

$$M \mid A, AO \mid O^{2-} \mid [A]_{\text{alloy}}, AO \mid M$$

where A is the more reactive metal with respect to B in the A–B alloy; the opposite arrangement cannot be a reversible battery. To study the thermodynamics of particularly reactive metals, in order to avoid reaction with the solid-oxide electrolyte, it is advisable to choose single-crystal $CaF_2$ for this purpose.

The alloys formed by the addition of rare earth metals to precious metals have special properties. Paasch et al. studied the thermodynamic properties of Pd–RE (RE = Gd, Y, Ce) alloys. Each alloy was melted in a small induction furnace and then annealed for several days at 750 °C. Phase analysis was carried out by XRD. The activities of the rare earth elements were determined by the EMF method, taking the pure material as the standard state.

The experimental results showed the activities of the rare earth elements in the Pd–Y and Pd–Gd alloys to be very small; at 750 °C, $a_Y$ and $a_{Gd}$ were both of the order of $10^{-18}$. The phase boundary can be accurately delineated from the activity value for a constant composition, and the accuracy can be up to 0.1% (mole fraction).

Partial molar excess free energy, that is, the deviation from an ideal solution, can be expressed as:

$$\Delta G_i^E = RT \ln \alpha_i - RT \ln x_i$$

The relationship between partial molar excess free energy and content $x_i$ of rare earth elements in Pd alloys shows a large negative value, indicating the stability of the rare earth metals therein. For Pd–Gd and Pd–Y alloys, the most stable compositions are those with rare earth metal contents of $x_i = 0.05$. Regarding the Pd–Ce system, although there are only three data points, it is evident that the partial molar excess free energy of Ce in this alloy is more negative, that is, it is more stable. Paasch and Schaller (1983) surmised that the thermodynamic properties of the system are mainly regulated by the electronic effect, and that the excess-function has a minimum value with the change of composition. It indicates that the atomic size also has an effect.

4.3 Use of the Gibbs–Duhem Equation to Calculate Activity 55

In order to distinguish the two effects, partial molar excess free energy is divided into two parts: expansion effect and electronic effect.

$$\Delta G_i^E = \Delta G_i^{Ed} + \Delta G_i^{Ee}$$

In the formula, $\Delta G_i^{Ed}$ is the partial molar excess free energy change caused by the expansion effect, and $\Delta G_i^{Ee}$ is that due to the electronic effect. Paasch et al. evaluated the alloy expansion effect and the electronic effect partial molar excess free energy change ($x_i \to 0$) at 750 °C, and summarized into a table, in which, the values of $\Delta G_i^{Ed}$ are positive, indicating that when rare earth elements are dissolved in Pd, energy is required to expand the lattice. Thus, $\Delta G_i^{Ee}$ due to the electronic effect should be more negative than the value of $\Delta G_i^E$.

The thermodynamic study of ternary alloys is the same as that of two-phase alloys in terms of experimental methods and techniques, but the design of test points has distinct characteristics. After the activity of one active component is obtained, the equal activity line is drawn, and then the equal activity lines of other components can be calculated and drawn according to the ternary system Gibbs–Duhem equation.

## 4.3 Use of the Gibbs–Duhem Equation to Calculate Activity

After the activity of one component of a binary or ternary solid phase or melt is obtained experimentally, the activity of another component can be obtained by using the Gibbs–Duhem equation (Darken 1950; Wei 1964).

The following relationships hold between the partial molar properties of each component, which constitute the Gibbs–Duhem equation (Wei 1964; Darken 1950):

$$n_1 dG_1 + n_2 dG_2 + \cdots = \Sigma n_i dG_i = 0$$

$$x_1 dG_1 + x_2 dG_2 + \cdots = \Sigma x_i dG_i = 0$$

(1) Application of the Gibbs–Duhem equation to binary systems:

$$x_1 dG_1 + x_2 dG_2 = 0$$

Because $G_1 = G^\ominus + RT \ln \alpha_1$; $G_2 = G^\ominus + RT \ln \alpha_2$

$$dG_1 = RTd \ln \alpha_1; \quad dG_2 = RTd \ln \alpha_2$$

So, $x_1 d \ln \alpha_1 + x_2 d \ln \alpha_2 = 0$

$$\int_{x_1=1}^{x_1} d\ln\alpha_1 = -\int_{x_1=1}^{x_1} \frac{x_2}{x_1} d\ln\alpha_2$$

$$\ln\alpha_1 = -\int_{x_1=1}^{x_1} \frac{x_2}{x_1} d\ln\alpha_2$$

The above formula can be replaced:

$$\ln\alpha_1 = -\int_{x_2=0}^{x_2} \frac{x_2}{x_1} d\ln\alpha_2$$

However, the area integral method is limited because the curve cannot intersect with the transverse coordinate axis, and accurate results cannot be obtained.

Expressed in terms of the activity coefficient, the Gibbs–Duhem equation is:

$$x_1 d\ln\gamma_1 + x_2 d\ln\gamma_2 = 0$$

$$\ln\gamma_1 = -\int_{x_1=1}^{x_1} \frac{x_2}{x_1} d\ln\gamma_2$$

or

$$\ln\gamma_2 = -\int_{x_2=0}^{x_2} \frac{x_1}{x_2} d\ln\gamma_1$$

This formula is again limited, and cannot yield an accurate value. Therefore, the function to solve this problem is:

$$\alpha_i = \frac{\ln\gamma_i}{(1-x_i)^2}$$

Reduction to $x_1 d\ln\gamma_1 + x_2 d\ln\gamma_2 = 0$ is as follows:

$$d\ln\gamma_1 = -\frac{x_2}{1-x_2} d\ln\gamma_2$$

$$\int_{x_1=1}^{x_1} d\ln\gamma_1 = -\int_{x_1=1}^{x_1} \frac{x_2}{1-x_2} d\ln\gamma_2$$

### 4.3 Use of the Gibbs–Duhem Equation to Calculate Activity

$$\ln \gamma_1 = - \int_{x_2=0}^{x_2} \frac{x_2}{1-x_2} d\ln \gamma_2$$

Using the stepwise integration method:

$$\ln \gamma_1 = -\frac{x_2}{1-x_2}\ln \gamma_2 + \int_{x_2=0}^{x_2} \ln \gamma_2 d\left(\frac{x_2}{1-x_2}\right)$$

$$\ln \gamma_1 = -\frac{x_2}{(1-x_2)^2}(1-x_2)\ln \gamma_2 + \int_{x_2=0}^{x_2} \ln \gamma_2 \frac{dx_2}{(1-x_2)^2}$$

we obtain

$$\ln \gamma_1 = -\alpha_2 x_1 x_2 + \int_{x_2=0}^{x_2} \alpha_2 dx_2$$

Because $dx_2 = -dx_1$, it follows that

$$\ln \gamma_1 = -\alpha_2 x_1 x_2 - \int_{x_1=0}^{x_2} \alpha_2 dx_1$$

or

$$\ln \gamma_1 = -\alpha_2 x_1 x_2 - \int_{x_1=1}^{x_1} \alpha_2 dx_1$$

Therefore, knowing the $\alpha$ group function of one component, we can calculate $\gamma$ of the other component and thus calculate the activity according to the relationship $\alpha = xy$.

(2) Application of the Gibbs–Duhem equation to ternary systems

For application of the Gibbs–Duhem equation to ternary systems, the Darken (1950), Wagner (1952), Schuhmann, and Zhou methods are available, among others.

If $G$ denotes the molar thermodynamic function of a solution of a ternary system or solid solution, and $G_i$ denotes the corresponding molar function of component $i$, we can write:

$$G = x_1 G_1 + x_2 G_2 + x_3 G_3$$

so,

$$\left(\frac{\partial G}{\partial x_2}\right)_{\frac{x_1}{x_3}} = x_1\left(\frac{\partial G_1}{\partial x_2}\right)_{\frac{x_1}{x_3}} + G_1\left(\frac{\partial x_1}{\partial x_2}\right)_{\frac{x_1}{x_3}} + x_2\left(\frac{\partial G_2}{\partial x_2}\right)_{\frac{x_1}{x_3}} + G_2$$

$$+ x_3\left(\frac{\partial G_3}{\partial x_2}\right)_{\frac{x_1}{x_3}} + G_3\left(\frac{\partial x_3}{\partial x_2}\right)_{\frac{x_1}{x_3}}$$

Let $R = (x+a)^n = \sum_{k=0}^{n}\binom{n}{k}x^k a^{n-k}$   $R = \frac{x_1}{x_3}$.

Because $x_1 = 1 - x_2 - x_3$, $x_1 = 1 - x_2 - \frac{x_1}{R}$.

Or $x_1 = \frac{k}{1+k}(1 - x_2)$.

Then $\left(\frac{\partial x_1}{\partial x_2}\right)_k = -\frac{k}{1+k}$.

Or $\left(\frac{\partial x_1}{\partial x_2}\right)_{\frac{x_1}{x_3}} = -\frac{x_1}{1-x_2}$.

Analogously:

$$\left(\frac{\partial x_3}{\partial x_2}\right)_{\frac{x_1}{x_3}} = -\frac{x_3}{1-x_2}$$

So

$$\left(\frac{\partial G}{\partial x_2}\right)_{\frac{x_1}{x_3}} = -x_1\left(\frac{\partial G_1}{\partial x_2}\right)_{\frac{x_1}{x_3}} - \frac{x_1}{1-x_2}G_1 + x_2\left(\frac{\partial G_2}{\partial x_2}\right)_{\frac{x_1}{x_3}} + G_2$$

$$+ x_3\left(\frac{\partial G_3}{\partial x_2}\right)_{\frac{x_1}{x_3}} + \frac{x_3}{1-x_2}G_3$$

Because

$$x_1\partial G_1 + x_2\partial G_2 + x_3\partial G_3 = 0$$

it follows that:

$$\left(\frac{\partial G}{\partial x_2}\right)_{\frac{x_1}{x_3}} = -\frac{x_1}{1-x_2}G_1 + G_2 - \frac{x_3}{1-x_2}G_3$$

$$(1-x_2)\left(\frac{\partial G}{\partial x_2}\right)_{\frac{x_1}{x_2}} = -x_1 G_1 + G_2 - x_2 G_2 - x_3 G_3 = G_2 - G$$

while

$$G_2 = G + (1-x_2)\left(\frac{\partial G}{\partial x_2}\right)_{\frac{x_1}{x_3}}$$

### 4.3 Use of the Gibbs–Duhem Equation to Calculate Activity

Dividing the above expression by $(1 - x_2)^2$ gives:

$$\frac{G_2}{(1 - x_2)^2} = \frac{G}{(1 - x_2)^2} + \frac{1}{1 - x_2}\left(\frac{\partial G}{\partial x_2}\right)_{\frac{x_1}{x_3}} = \left[\frac{\partial}{\partial x_2}\left(\frac{G}{1 - x_2}\right)\right]_{\frac{x_1}{x_3}}$$

so

$$\partial\left(\frac{G}{1 - x_2}\right) = \frac{G_2}{(1 - x_2)^2}\partial x_2 \left(\text{when } \frac{x_1}{x_3} \text{ does not change}\right)$$

There are two methods for integrating the expression.
The first approach:

$$\int_1^{x_2} \partial\left(\frac{G}{1 - x_2}\right) = \int_1^{x_2} \frac{G_2}{(1 - x_2)^2}\partial x_2 \left(\text{when } \frac{x_1}{x_3} \text{ does not change}\right)$$

so

$$G = (1 - x_2)\left[\int_1^{x_2} \frac{G_2}{(1 - x_2)^2}dx_2\right]_{\frac{x_1}{x_3}} - x_1\left[\int_1^0 \frac{G_2}{(1 - x_2)^2}\partial x_2\right]_{x_3=0}$$

$$- x_3\left[\int_1^0 \frac{G_2}{(1 - x_2)^2}\partial x_2\right]_{x_1=0}$$

Integral condition for deriving the formula: when $x_2 \to 1$, $G_2 = 0$.
The second approach:

$$\int_0^{x_2} \partial\left(\frac{G}{1 - x_2}\right) = \int_0^{x_2} \frac{G_2}{(1 - x_2)^2}\partial x_2 \left(\text{when } \frac{x_1}{x_2} \text{ does not change}\right)$$

so $G = (1 - x_2)\left[G_{x_2=0} + \int_0^{x_2} \frac{G_2}{(1-x_2)^2}\partial x_2\right]_{\frac{x_1}{x_3}}$.

The above two methods are suitable for thermodynamic functions with pure matter as the standard state.

In order to obtain the activity of the ternary system, the activity of several ternary elements with a certain $x_1/x_3$ ratio must be first known.

## 4.4 Thermodynamic Study of Rare Earth Compounds

Metallurgists and materials scientists need to know the thermodynamic stabilities of substances involved in reactions. The traditional methods for obtaining thermodynamic data are calorimetry and chemical equilibrium methods, but there are many influencing factors, so it is difficult to obtain accurate data. Kiukkola and Wagner first determined thermodynamic data for compounds by a solid-electrolyte EMF method in 1957. This method has great advantages, since the free energy change of the battery reaction is directly related to the EMF of the battery:

$$\Delta G = -nFE$$

The EMF of a battery, $E$, can be accurately determined by means of a high-resistance ($\geq 10^{10}$ $\Omega$) digital voltmeter.

With an appropriate solid electrolyte and reference electrode, a battery free from side reactions, no mixing potential etc., the measured thermodynamic data should be accurate. The measurable $p_{O_2}$ range of $ZrO_2$ groups at 870 °C is $10^{-17}$ to $4 \times 10^5$ Pa, while $ThO_2(Y_2O_3)$ solid electrolyte can be measured down to $10^{-25}$ Pa. $ZrO_2^-$ based materials are suitable for solid electrolytes at high $p_{O_2}$; $ThO_2(Y_2O_3)$ is suitable for low $p_{O_2}$.

When selecting a reference electrode, in addition to requiring accurate thermodynamic data, it is also desirable that the $p_{O_2}$ values of the selected reference and the tested material are similar. Moreover, the general EMF of the battery should be no greater than 500 mV, to avoid any influence of internal current or the diffusion of oxygen molecules under a large chemical potential difference. The change of battery electronic error should be $\pm$ (0.2–0.5) mV, or $\pm$ 1 mV, for accurate measurement. This needs to be based on EMF values.

Reference electrodes can be divided into two categories: gas reference electrodes and co-existing phase reference electrodes.

(1) Gas reference electrode

Available air and mixed gas with a certain gas-phase partial pressure, such as $H_2/H_2O_{(g)}$ or $CO/CO_2$, can be used to set a certain $p_{O_2}$ value in the range $10^{-17}$–$10^5$ Pa. Similarly, $H_2/H_2S$ can be used to set the gas-phase content of sulfur, $H_2/NH_3$ for the gas-phase content of nitrogen, and $H_2/HCl$ for the gas-phase content of chlorine, and so on.

The reason for using a mixed gas as the reference electrode is that thermodynamic data can be reliably calculated from the spectral data of the gas molecules, and that it can be easily applied at close to atmospheric pressure.

Relationships between standard free energy of generation and temperature for some gases (e.g. $O_2$, $S_2$, $N_2$, $H_2$, $P_2$, $Cl_2$) have been shown in a detailed table, for example, when $C + O_2 = CO_2$, $\Delta G^{\ominus}/(J/mol)$ is $- 394,130 - 0.84T$, error ($\pm$)/(kJ/mol) is 0.42, and temperature range/K is 298–2473; when $H_2 + 1/2S_2 = H_2S$, $\Delta G^{\ominus}/(J/mol)$ is $- 90,165 + 49.29T$, error ($\pm$)/(kJ/mol) is 0.63, and temperature range/K is 717–2473. In the reaction formulas, except $C$ is solid, the rest is gas.

## 4.4 Thermodynamic Study of Rare Earth Compounds

### (2) Co-existing phase reference electrode

A mixture of a metal and its low-valent or high-valent oxide can be used as a reference electrode, such as Cr, $Cr_2O_3$ or $Cu_2O$, $CuO$. The reactions are:

$$\frac{4}{3}Cr_{(s)} + O_2 = \frac{2}{3}Cr_2O_{3(s)} \quad \Delta G^{\ominus}_{Cr_2O_3} = -RT \ln \frac{1}{p_{O_2}}$$

$$Cu_2O_{(s)} + \frac{1}{2}O_2 = 2CuO_{(s)}$$

$$\Delta G^{\ominus}_{CuO} = -RT \ln \frac{1}{p_{O_2}^{\frac{1}{2}}}$$

Most d sub-group metals and rare earth metal oxides have a variety of valence states, as well as non-stoichiometric forms, such as a series of stoichiometric or non-stoichiometric oxides in the $Ti-TiO_2$, $Ce-CeO_2$, and $V-V_2O_5$. For example, $Ce-CeO_2$ has dozens of intermediate oxides (Matvei et al. 2006); $V-V_2O_5$ intermediate oxides include $VO$, $V_2O_3$, $V_3O_5$, $V_4O_7$, $V_5O_9$, $V_6O_{11}$, and $V_7O_{13}$.

Hence, although there are comparison expressions, such as $Ce+O_2 = CeO_2$, $\frac{4}{5}V+O_2 = \frac{2}{5}V_2O_2$, and the corresponding relationship between $\Delta G^{\ominus}$ and temperature, this only indicates the matter and corresponding energy changes at initial-and-end states, while does not mean that there is an equilibrium relationship between Ce and $CeO_2$ or V and $V_2O_5$.

Only the $\Delta G^{\ominus}$ of the equilibrium reaction has a $\Delta G^{\ominus} = -RT \ln K$ relationship, otherwise the mixed $p_{O_2}$ value or the $p_{O_2}$ value of an unknown reaction is obtained. An exception is that Ta and $Ta_2O_5$ co-exist in equilibrium, owing to favorable atomic and crystal structures, such that quantitative change causes qualitative change.

Accurate thermodynamic data are required for a reference electrode. Rapp (1970a, b) recommends 36 pairs of co-existing systems as a reference electrode selection, and summarized into a table, in which $p_{O_2}$ values are listed in decreasing order. For example, when temperature range/K is 900–1154, coexistence response is $Pd + 1/2O_2 = PdO$, standard free energy change/J $mol^{-1}$ is $- 114,220 + 100.0T$, $\lg p_{O_2}/1000$ °C is $+ 6.09$; when temperature range/K is 884–1126, coexistence response is $2Mn_3O_4 + 1/2O_2 = 3Mn_2O_3$, standard free energy change/J $mol^{-1}$ is $- 113,390 + 92.05T \pm 795$, $\lg p_{O_2}/1000$ °C is $+ 5.31$.

As regards commonly used reference electrodes, there are currently 17 salient literature reports. For example, the Japan Society for the Promotion of Science (JSPS) analyzed related articles, and found the results in three articles to be broadly similar. The average of these three data sets is recommended as follows:

$$\Delta G^{\ominus}_{Cr_2O_3} = - 1,115,747 + 250.45T \pm 1255 \text{ J/mol}$$

The temperature range of the three studies is 800–1327 °C. In the absence of a phase transition, thermodynamic data can be extrapolated to 1600 °C, corresponding to the steel-making temperature. All the materials used by the researchers were of high purity, so users should also ensure the purity of reference electrode materials. The metal phase powder generally requires a purity of 99.99% and the oxide should be a spectroscopically pure reagent. The mass ratio of metal to oxide is generally 90:10, and the metal phase can serve as a good contact between solid electrolyte, electrode lead, and gas phase. The powder should be fully mixed and ground to a particle size of 135–250 $\mu$m.

Wagner et al. studied the thermodynamic properties of oxides by a solid-electrolyte EMF method in 1957. In the following 15 years or so, scholars studied almost all single oxides and composite oxides, as well as some other compounds of theoretical and practical significance. Relevant theories and methods are summarized in the following sections.

## 4.5 Thermodynamic Study of Single Oxides

This requires a battery consisting of a solid-oxide electrolyte and two oxide systems, one of which is a reference electrode for which accurate thermodynamic data are known, and the other is the electrode to be tested (Japan Society for the Promotion of Science, JSPS 1955). The battery takes the form:

$$M \mid A, AO \mid Solid\ Electrolyte \mid BO, B \mid M$$

$$p_{O_2}^{I}\ p_{O_2}^{II}$$

If $p_{O_2}^{II} > p_{O_2}^{I}$, then

$$E = \frac{RT}{4F} \ln \frac{p_{O_2}^{II}}{p_{O_2}^{I}} = \frac{G_{BO}^{\ominus} - G_{AO}^{\ominus}}{2F}$$

The equilibrium oxygen partial pressure of the reference electrode may be greater than or less than that of the electrode to be tested. If the oxygen partial pressure of the reference electrode is known, then the equilibrium oxygen partial pressure of the compound to be tested can be obtained according to the EMF of the battery. Thus, the standard free energy of the compound to be tested can be obtained.

Equilibrium gas-phase oxygen partial pressure is a characteristic quantity to determine the stability of a single metal oxide or compound. The formation reaction of a single metal oxide $M_xO_y$ can be expressed as:

$$X M_{(s)} + \frac{y}{2} O_2 = M_x O_{y(s)}$$

With pure metal and pure oxide as the standard state, the activity is 1, then:

$$\Delta G^{\ominus}_{M_x O_y} = \frac{y}{2} RT \ln p_{O_2}$$

According to the thermodynamic relations:

$$\Delta G^{\ominus}_{M_x O_y} = \Delta H^{\ominus}_{M_x O_y} - T \Delta S^{\ominus}_{M_x O_y}$$

For metallurgy and materials research, the expressions

$$\Delta H^{\ominus}_{M_x O_y} = A, \text{ and } \Delta S^{\ominus} = B$$

are commonly used as an approximation, using $\Delta G^{\ominus} = A + BT$ (J/mol) to characterize the relationship between $\Delta G^{\ominus}$ and $T$.

## 4.6 Thermodynamic Study of Composite Oxides

(1) Composite oxide profile

Composite oxides include compounds such as rock structures, spinel phases, and silicates. For AO, $B_2O_3$ composite oxides, metallurgists and materials researchers are accustomed to writing $AO \cdot B_2O_3$, but chemically and structurally it is more appropriate to write $AB_2O_4$. Composite oxides can be prepared by solid-phase direct synthesis. They have non-stoichiometric ratios, often forming a solid-solution phase with a limited range, often containing lattice defects. The defect form is related to component concentration, temperature, and oxygen partial pressure n the gas phase. The premise of battery design is that the studied system is an equilibrium phase; the metal phase should exist only in the form of pure matter.

One of the battery forms for studying the thermodynamics of complex oxides can be expressed as:

$$\text{A, AO | Oxide Electrolyte | A, } B_2O_3, AB_2O_4$$

Battery reaction:

$$AO + B_2O_3 = AB_2O_4$$

For thermodynamic study of composite oxides formed by active metal oxides and $Cu_2O$, $NiO$, $Cr_2O_3$, etc., $ZrO_2$ groups and $ThO_2(Y_2O_3)$ solid electrolytes can be employed. For example, Tretyakov et al. (1976) used air as a reference electrode to form the following battery:

$$\text{Pt | } O_2 \text{ (air) | } ZrO_2(Y_2O_3) \text{ | } CuLn_2O_4, Cu_2O, La_2O_3 \text{ | Pt}$$

$$p_{O_2} = 0.21 \times 10^5 \, \text{Pa}$$

They also investigated the thermodynamic properties of a series of composite oxides containing lanthanide (Ln) element oxides. The composite oxides were $CuLn_2O_3$ (Ln = La, Nd, Sm, Eu, and Gd) and $Cu_2RE_2O_3$ (RE = Tb, Dy, Er, Yb, and Y), and the experimental temperatures were 950–1150 °C.

Their results showed that the stability of $CuLn_2O_4$ compounds generated by simple oxides decreases in the order from La to Gd, while the stability of $Cu_2Ln_2O_5$ compounds increases in the order from Tb to Yb.

Sreedharan et al. (1976) and Sreedharan and Chandrasekharaiah (1986) used $ThO_2(Y_2O_3)$ as a solid electrolyte to form the following battery:

$$Pt \,|\, Ni, NiO \,|\, ThO_2(Y_2O_3) \,|\, La_2NiO_4, La_2O_3, Ni \,|\, Pt$$

They measured the relationship between $\Delta G^{\ominus}$ (J/mol) and temperature for $NiO_{(s)} + La_2O_{3(s)} = La_2NiO_{4(s)}$ in the range 850–1100 °C:

$$\Delta G^{\ominus}_{La_2NiO_{4(s)}} = 25{,}568 - 30.18T \pm 190 \, (\text{J/mol})$$

Wang et al. (1985) used $ZrO_2(MgO)$ as a solid electrolyte to assemble the following battery:

$$Mo \,|\, Cr, Cr_2O_3 \,|\, ZrO_2(MgO) \,|\, YCrO_3, Y_2O_3, Cr \,|\, Mo$$

The battery reaction is

$$Y_2O_{3(s)} + Cr_2O_{3(s)} = 2YCrO_3$$

They also measured the relationship between $\Delta G^{\ominus}$ and temperature for $YCrO_3$ in the range 909–1113 °C to modify the electronic conductivity of the solid electrolyte. It was shown that:

$$\Delta G^{\ominus}_{\substack{2YCrO_3 \\ (Y_2O_3 - Cr_1O_2)}} = 183{,}900 - 146.3T \pm 1450 \, (\text{J/mol})$$

Composite oxides formed by oxides of alkali metals, alkaline-earth metals, rare earth metals, and B, Al, Ga, Ti, Zr, Hf, Si, and Nb, for which equilibrium oxygen partial pressures are very low, will increase the electronic conductivity as $ZrO_2$-based solid electrolytes operate at a very low oxygen partial pressure. Hence, single-crystal $CaF_2$ should be used as a solid electrolyte for thermodynamic study. The following reactions will occur at the electrolyte/electrode interface:

$$CaF_{2(s)} + \frac{1}{2}O_2 = CaO_{(s)} + F_2 \text{ or } 2F^- + \frac{1}{2}O_2 = O^{2-} + F_2$$

## 4.6 Thermodynamic Study of Composite Oxides

The equilibrium constant of the reaction is:

$$K = \frac{p_{F_2}}{p_{O_2}^{\frac{1}{2}}}$$

At a certain temperature, $K$ is a fixed value. If the $p_{O_2}$ values on both sides of the solid electrolyte are equal, the battery EMF depends on the fluorine potential difference on either side of the electrolyte.

(2) Relevant experiments and issues

Beng and Wagner (1961) studied solid $CaO\text{-}SiO_2$ thermodynamics by using $CaO\text{-}SiO_2$ phase. Taylor and Schmalzried (1964) used $CaF_2$ electrolyte batteries to study the $\Delta G^{\ominus}$ of $CaTiO_3$ and $Ca_4Ti_3O_{10}$ in the $CaO\text{-}TiO_2$ system according to Wagner's working principle. They also used a Sr electrolyte battery to study $\Delta G^{\ominus}$ of $SrTiO_3$ and $Sr_4Ti_3O_{10}$ in the $SrO\text{-}TiO_2$ system, and an $MgF_2$ electrolyte to study $\Delta G^{\ominus}$ of $MgAl_2O_4$ in the $MgO\text{-}Al_2O_3$ system. Some researchers have carried out similar work on the thermodynamics of other systems.

Wang et al. (1987) used single-crystal $CaF_2$ as a solid electrolyte to study the standard free energy of generation of $LaAlO_3$, whereby the experimental temperature was 823–950 °C.

The materials used in these experiments were $La_2O_3$ 99.99%, calcined at 900 °C to remove the $H_2O$ present in $La(OH)_3$, $Al_2O_3$ 99.999%, calcined at 1400 °C to remove water in the form of hydrates or hydroxide ions, and $LaF_3$ prepared by hydrofluoric acid precipitation. LaOF was prepared from appropriate amounts of $La_2O_3$ and $LaF_3$, and fired at 1400 °C for 30 h. Its identity was confirmed by XRD analysis. Single-crystal $CaF_2$ (15 mm × 5 mm) was purchased from the Changchun Institute of Optical Precision Machinery and Physics, Chinese Academy of Sciences.

The electrode consisted of $La_2O_3$ and LaOF, pressed at 14.8 t/cm$^2$ to become 15 mm × 5 mm sheets and calcined at 1000 °C for 24 h. LaOF, $LaAlO_3$, and $Al_2O_3$ were similarly mixed and pressed into a sheet, and calcined at 1000 °C for 30 h. The identities of the products were confirmed by XRD analysis. The reference electrode sheet and the electrode sheet to be tested were flattened. A $CaF_2$ single-crystal wafer, the reference electrode sheet, the electrode sheet to be tested, and a welded Pt wire or Pt sheet were then assembled on a corundum tube.

This study required pure $O_2$. Hence, $O_2$ from a commercial cylinder was sequentially passed through a KOH column, two silica gel columns, and two $P_2O_3$ columns to remove $CO_2$ and $H_2O$. The pure $O_2$ was and then divided into two channels, and passed through a capillary flowmeter and a stainless steel micro-needle valve into the furnace tube; flow through the two poles of the battery, to ensure the $p_{O_2}$ equality at two poles, flow is kept low so as not to produce a cooling effect.

After the experiment, no reaction of the electrolyte and electrode material was evident. The results show certain regularity of the relationship between the change of standard free energy and temperature when $LaAlO_3$ is formed by oxide:

$$\Delta G^{\ominus}_{2LaAlO_3(La_2O_3 \cdot Al_2O_3)} = 34,160 - 37.81T \pm 180 \,(\text{J/mol}) \,(823 - 950\,^{\circ}\text{C})$$

Kallarackel et al. (2013) used the $ZrO_2(Y_2O_3)$ solid-electrolyte EMF method to determine the relationship between free energy and temperature for $LaCrO_4$, $La_2CrO_6$, and $La_2Cr_3O_{12}$. The battery forms were:

$$Pt, LaCrO_4 + LaCrO_3 \mid ZrO_2(Y_2O_3) \mid O_2(0.1\,MPa), RuO_2, Pt$$

$$Pt, La_2CrO_6 + LaCrO_3 + La_2O_3 \mid ZrO_2(Y_2O_3) \mid O_2(0.1\,MPa), RuO_2, Pt$$

$$Pt, La_2Cr_3O_{12} + LaCrO_4 + Cr_2O_3 \mid ZrO_2(Y_2O_3) \mid O_2(0.1\,MPa), RuO_2, Pt$$

(Note: $RuO_2$ as catalyst).

Researchers collected standard free energy of generation data for composite oxides (including spinel-type oxides) by solid-electrolyte EMF methods published prior to 1980, and listed into a detailed table, for example, when reaction $CaO_{(s)} + Mo_{(s)} + O_2 = CaMoO_{3(s)}$, temperature range/K is $\approx$ 1273, $\Delta G^{\ominus}/(J/mol)$ is $-605,772 + 154.05T$; when reaction $3Dy_2O_{3(s)} + W_{(s)} + 3/2O_2 = Dy_6WO_{12(s)}$, temperature range/K is 1200–1400, $\Delta G^{\ominus}/(J/mol)$ is $-935,460 + 227.27T$.

Pratt collected research data on composite compounds in 1990. The contents of the rare earth oxides were as follows.

(1) Using stable $ZrO_2$-based solid electrolytes:
$La_2O_3$-$\beta$-$Al_2O_3$, $Ce_{1-x}Ca_yO_{2-y-x}$, $Dy_2O_3{\cdot}Bi_2O_3$, $(Er_2O_3)_{0.455}{\cdot}(Bi_2O_3)_{0.545}$, $Y_3Fe_5O_9$, (La, Pr, Eu, Y)$FeO_3$, (Sm, Eu, Gd, Tb, Dy, Ho)$CoO_3$, $La_4Co_3O_{10}$, $La_4Ni_3O_{10}$, La(Mn, Fe, Co, Ni)$O_3$, $La_2$(Mn, Co, Ni, Cu)$O_4$, $LaVO_3$, $LaVO_4$, (La, Pr, Nd)–(Co, Cu, Ni)–O, (La, Pr, Nd, Sm, Eu, Gd, Tb, Dy, Y)$_2O_2S$, (La, Pr, Nd, Sm, Eu, Gd, Tb, Dy)$_2O_2SO_4$, (Nd, Pr)$_2CuO_4$, (Nd, Pr)$CuO_2$, $PrCoO_3$, $Pr_4Co_3O_{10}$, $YBa_2Cu_3O_x$, etc.
(2) Using $ThO_2(Y_2O_3)$ solid electrolyte:
Dy–W–O, $Er_6WO_{12}$, $Eu_6WO_{1.2}$, $La_2NiO_4$, $Sm_6WO_{12}$, $VO_2$–$ZrO_2$–(La, Ce, Nd)$O_{2-x}$.
(3) Using $CaF_2$ solid electrolyte: BaO–$CeO_2$, $BaCeO_3$, $Y_3Fe_5O_{12}$.
(4) Using $SrF_2$ solid electrolyte: SrO–$CeO_2$.

## 4.7 Thermodynamic Study of Non-stoichiometric Compounds

Oxygen defects and metal defects in the lattice are often present in single oxides and composite oxides, forming non-stoichiometric compounds. These defects are related to the partial pressure of oxygen in the gas phase, and during use it is essential to know the relationship between the defects and oxygen partial pressure.

The non-stoichiometric thermogravimetric method by coulometric titration of solid-electrolyte batteries is more accurate than the thermogravimetric method, but the air chamber must be well sealed.

## 4.7 Thermodynamic Study of Non-stoichiometric Compounds

The principle of the method is to pass a given amount of oxygen through a solid electrolyte into a gas chamber containing the sample to be tested by means of direct current (Subbarao 1980; Fischer and Sank 1991). The relationship between the amount of oxygen entering and the power-on time is:

$$n_{O_2} = \frac{It}{4F}$$

where $n_{O_2}$ is the amount of material (mol), $I$ is the titration current (A), $t$ is the titration time (s), and $F$ is the Faraday constant (96,485 C/mol).

After the oxygen enters the gas chamber, it exceeds the equilibrium partial pressure of the non-stoichiometric sample. Hence, it is gradually absorbed by the sample, establishing a new equilibrium relationship. When a certain amount of oxygen is introduced and absorbed, a new equilibrium relationship is established. The defect degree of a specimen has a definite value under a certain oxygen partial pressure; the change ($\Delta x$) is obtained from the titrated amount of $O_2$ ($\Delta n_{O_2}$) according to the following relationship, taking FeO as an example:

$$\Delta x = \frac{2M_{\text{FeO}}}{m_{\text{FeO}}} \Delta n_{O_2}$$

where $M$ is the molar mass (g/mol) and $m$ is the mass (g).

See the relationship between iron defect $x$ at 1200 °C, $Fe_{1-x}O$, and oxygen partial pressure in double logarithmic coordinates, the following equations are followed in the range $0.10 - x$:

$$\frac{1}{2}O_2 + Fe_{Fe} = V''_{Fe} + 2h' + FeO$$

$$[V''_{Fe}] = x = Kp_{O_2}^{\frac{1}{6}}$$

The slope of the curve $d\lg x/d\ln p_{O_2} = 1/6$. This value decreases as $x$ is further increased. Near the $Fe_{1-x}O/Fe_3O_4$ equilibrium ($x = 0.145$), it is 1/8. It corresponds to the $Fe-Fe_{1-x}O$ equilibrium at 1200 °C, $x = 0.05$.

The compound $Fe_3O_4$ is also Fe-defective. It can be expressed as $Fe_{3-y}O_4$, and the non-stoichiometric phase above 900 °C has a narrow concentration range.

Coulometric titration has been used to study the non-stoichiometric formulations of a series of oxides. The batteries used refer to the coulometric titration cell form used by Rapp et al. in 1969, the setting includes Pt/Pt–Rh thermocouple wire, coated Pt electrode, $ZrO_2(CaO)$ electrolyte, pyrex glass seal, Pt wire, oxide samples (in Pt net), and sealed $Al_2O_3$ tubes. Ito et al. (1987) studied $V_2O_{5-x}$, they used a tubular solid electrolyte to circumvent the problem of sealing a sheet electrolyte, the setting includes Pt, Pt·Rh (13%), A sample wrapped in Pt foil, Corundum, $Al_2O_3$ ceramics, $B_2O_3 \cdot Al_2O_3$ ceramics, $B_2O_3$, and $ZrO_2(Y_2O_3)$ electrolyte tubes.

Non-stoichiometric $SnO_{2-x}$ finds multiple applications. Yang et al. (1992) used $ZrO_2(Y_2O_3)$ tubes as solid electrolytes to study its composition. The battery form was:

$$Pt \mid air\ (clear)\ or\ Ni, NiO \mid ZrO_2(Y_2O_3) \mid SnO_{2-x} \mid Pt$$

The study used a double-bend resistance wire furnace, a precision galvanostatic potentiometer, high-resistance ($10^{11}$ Ω) precision digital voltmeters to determine battery EMF, and a stopwatch to record the titration time, and, validated it in turn. The value of $x$ in $SnO_{2-x}$ changed after each titration, and the corresponding $p_{O_2}$ value was calculated according to the relationship:

$$E = \frac{RT}{nF} \ln \frac{p_{O_2}^{II}}{p_{O_2}^{I}}$$

These experimental results are summarized in Figs. 4.4 and 4.5, respectively. Figure 4.4 is diagram of $\ln p_{O_2}$ changes with $x$ in the $SnO_{2-x}$, and Fig. 4.5 is relationship between $\ln p_{O_2}$ and $\ln x$ in $SnO_{2-x}$ (notes: 1—990 K (717 °C); 2—885 K (612 °C); 3—783 K (510 °C); 4—746 K (473 °C); 5—720 K (447 °C)).

The former shows that, at the studied temperature, and in the range $x < 0.022$ for $SnO_{2-x}$, the slope of the straight line was between $-1/5.7$ and $-1/8.5$. The negative slope indicates that $SnO_{2-x}$ exhibits n-type conduction behavior. Its relationship with gas-phase oxygen is as follows:

$$O_O = \frac{1}{2}O_2 + V_{\ddot{O}} + 2e'$$

The equilibrium constant of the reaction is:

$$K_O = [V_{\ddot{O}}][e']^2 p_{O_2}^{\frac{1}{2}} / [O_O]$$

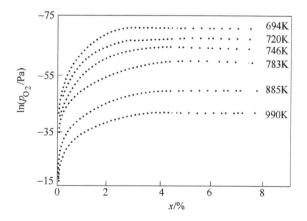

**Fig. 4.4** Diagram of $\ln p_{O_2}$ changes with $x$ in the $SnO_{2-x}$

## 4.7 Thermodynamic Study of Non-stoichiometric Compounds

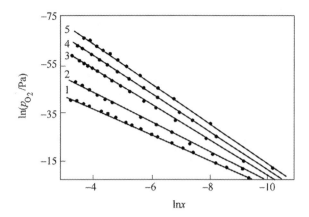

**Fig. 4.5** The relationship between $\ln p_{O_2}$ and $\ln x$ in $SnO_{2-x}$

It can be seen from Fig. 4.4 that the gas-phase $p_{O_2}$ decreased continuously with increasing $x$ defect concentration and then remained constant after a certain $x$ value was attained. This corresponds to entering a two-phase co-existence zone. XRD analysis proved the co-existence of Sn and $SnO_2$; the equilibrium $p_{O_2}$ values at different temperatures are plotted against $x$ in Fig. 4.5, which summarizes good regularity.

All $p_{O_2}$ values are the equilibrium values of the reaction:

$$Sn_{(s)} + O_2 = SnO_{2(s)}$$

$$\ln p_{O_2} = 24.3 - \frac{69,080}{T}$$

from which

$$\Delta G^{\ominus}_{SnO_2} = -574,330 + 202T$$

Coulometric titrations have revealed intermediate oxides in a series of metal oxide systems. For example, study of the Ti–O system showed that $TiO_{2-x}$ sequentially involves intermediate oxides such as $Ti_{10}O_{19}$, $Ti_9O_{17}$, $Ti_8O_{15}$, $Ti_7O_{13}$, $Ti_6O_{11}$, $Ti_5O_9$, $Ti_4O_7$, and $Ti_3O_5$ as $x$ is increased from 0.1 to 0.35. The corresponding phase equilibrium relationships involve $Ti_{10}O_{19}/Ti_9O_{17}$, $Ti_9O_{17}/Ti_8O_{15}$, $Ti_8O_{15}/Ti_7O_{13}$, $Ti_7O_{13}/Ti_6O_{11}$, $Ti_6O_{11}/Ti_5O_9$, $Ti_5O_9/Ti_4O_7$, and so on. For each of the two-phase co-existing regions on the composition EMF relationship curve, the value is constant, as shown by a horizontal line segment. In this method, the system thermodynamics is studied, but phase equilibrium relationships are also revealed, which can supplement and verify the phase diagram.

Ce has dozens of non-stoichiometric oxides due to the versatile deployment of its 4f electrons.

## 4.8 Thermodynamic Study of Non-oxide Systems

In principle, provided that there is a reversible chemical reaction associated with oxygen, a battery can be designed to meet the needs of non-oxide thermodynamics research by electrochemical methods.

(1) Thermodynamic study of sulfides and sulfates

On the basis of the M-S-O system phase diagram, for example, there is a balance between MS and MO boundaries:

$$MS_{(s)} + \frac{3}{2}O_2 = MO_{(s)} + SO_2$$

The standard free energy of formation of MnS can be determined as follows:

$$(-) \, Pt \, | \, SO_2, \, MnO, \, MnS \, | \, ZrO_2(CaO) \, | \, O_{2(air)} \, | \, Pt \, (+)$$

Reference pole (+): $\frac{3}{2}O_{2(air)} + 6e = 3O^{2-}$.
Tested pole (−): $3O^{2-} - 6e = \frac{3}{2}O_2$ (tested).
Battery reaction: $\frac{3}{2}O_2$ (air) $+ MnS = MnO + SO_2$ (ordinary pressure).
Accordingly:

$$\Delta G = \Delta G^{\ominus} + RT \ln \frac{P_{SO_2}}{P_{O_2}^{\frac{3}{2}}}$$

$$\Delta G = -6FE$$

and

$$\Delta G^{\ominus} = \Delta G^{\ominus}_{SO_2} + \Delta G^{\ominus}_{MnO} - \Delta G^{\ominus}_{MnS}$$

(using pure oxygen at 101,325 Pa as the standard state, such that $\Delta G^{\ominus} = 0$ and there is no $\Delta G^{\ominus}_{O_2}$ item in the above formula).
It follows that:

$$-6FE = \Delta G^{\ominus}_{SO_2} + \Delta G^{\ominus}_{MnO} - \Delta G^{\ominus}_{MnS} + RT \ln \frac{p_{SO_2}}{p_{O_2}^{\frac{3}{2}}}$$

and

$$\Delta G^{\ominus}_{MnS} = 6FE + \Delta G^{\ominus}_{SO_2} + \Delta G^{\ominus}_{MnO} - \frac{3}{2}RT \ln p_{O_2}$$

Here, $\Delta G^{\ominus}$ of $\Delta G^{\ominus}_{MnO}$ and the oxides are known, $\Delta G^{\ominus}_{SO_2}$ is also known from molecular spectroscopy, as well as $p_{O_2}$ is known, so only a certain temperature of

## 4.8 Thermodynamic Study of Non-oxide Systems

the battery electromotive force value is obtained, can $\Delta G^{\ominus}_{MnS}$ values be obtained and relationship between $\Delta G^{\ominus}_{MnS}$ and temperature.

For determination of the standard free energy of generation according to the M-S-O system phase diagram, the reaction producing sulfate is:

$$MO_{(s)} + SO_2 + \frac{1}{2}O_2 = MSO_{4(s)}$$

When MO and $MSO_4$ co-exist, for a certain $p_{SO_2}$ value, the $p_{O_2}$ value is fixed and four phases co-exist. The battery form is:

$$(-)\, Pt \,|\, SO_{2\,(ordinary\ pressure)},\ MO,\ MSO_4 \,|\, ZrO_{2\,(base\ electrolyte)} \,|\, O_2 \,|\, Pt\ (+)$$

### (2) Thermodynamic study of non-oxides by fluoride solid electrolyte

Fluoride solid electrolyte with $t_i \approx 1$ has a broad $p_{F_2}$ and $t$ relation range. It can be used in thermodynamic studies of carbides, sulfides, borides, and phosphides. We take the case of carbides as an illustrative example.

To determine the thermodynamic properties of carbides, fluorine-containing compounds were added on either side of a $CaF_2$ solid electrolyte to form a fluorine concentration difference battery. For example, Mn, $MnF_2$ was selected as the reference electrode, and the tested electrode was composed of $MnF_2$, C, and $MnC_x$ mixed powders pressed into a table (Tanaka et al. 1973). At a certain temperature, the battery EMF is determined by the fluorine partial pressures on either side of the $CaF_2$ solid electrolyte, that is:

$$E = \frac{RT}{nF} \ln \frac{p_{F_2(MnF_2-MnC_x-C)}}{p_{F_2(MnF_2-Mn)}}$$

The standard free energies of generation of $Mn_7C_3$, $Mn_8C_3$, and $Mn_{23}C_6$ were measured. The thermodynamic properties of a compound can be rationalized by reference to the Mn-C phase diagram, and the thermodynamics of other carbides can be assessed by the equilibrium relationship.

The above battery reactions are:

Battery 1
Reference pole $(-)$: $7Mn + 14F^- = 7MnF_2 + 14e$
Tested pole $(+)$: $7MnF_2 + 3C + 14e = Mn_7C_3 + 14F^-$
Battery reaction: $7Mn + 3C = Mn_7C_3$.
Taking the activities of Mn and C as 1, then:

$$\Delta G^{\ominus}_{Mn_7C_3} = 14FE$$

Battery 2
Reference pole $(-)$: $Mn + 2F^- - 2e = MnF_2$
Tested pole $(+)$: $MnF_2 + Mn_7C_3 + 2e = Mn_8C_3 + 2F^-$

Battery reaction: $Mn + Mn_7C_3 = Mn_8C_3$

$$\Delta G^{\ominus}_{Mn_8C_3} = \Delta G^{\ominus}_{Mn_7C_3} - 2FE$$

Battery 3
Reference pole $(-)$: $7Mn + 14F^- - 14e = 7MnF_2$
Tested pole $(+)$: $7MnF_2 + 2Mn_8C_3 + 14e = Mn_{23}C_6 + 14F^-$
Battery reaction: $7Mn + 2Mn_8C_3 = Mn_{23}C_6$

$$\Delta G^{\ominus}_{Mn_{23}C_6} = 2\Delta G^{\ominus}_{Mn_8C_3} - 14FE$$

Regarding the thermodynamic study of sulfide, boride, and phosphide systems, the battery design principles are similar to those for carbides. Carbides of rare earth elements are less stable, akin to $CaC_2$. That is to say, they release acetylene upon contact with water, and are flammable.

For comparison of the stabilities of rare earth compounds with other compounds, the reader can refer to Wang's book "Methods for the Study of Metallurgical Physical Chemistry" (Beijing: Metallurgical Industry Press, 2013).

# References

Bendick W, Pepperhoff W (1979) J Phys F Met Phys 9:2185
Bendick W, Pepperhoff W (1981) J Phys F Met Phys 11:57
Bendick W, Ettwig HH, Pepperhoff W (1978) J Phys F Met Phys 8:2535
Bendick W, Ettwig HH, Pepperhoff W (1979) Thermally excited electron transitions in FCC iron alloys. J Magn Magn Mater 10:214–216
Beng R, Wagner C (1961) Shock waves in chemical kinetics: the hydrogen-bromine reaction. J Phys Chem 65:1308
Boer FRD, Huang YK, Mooij DBD et al. (1987) Magnetic properties of a series of novel ternary intermetallics ($RFe_{10}V_2$). J Less Common Met 135:199–204
Buschow KHJ (1988) J Less Common Met 144:65
Carbone C, Kisker E, Walker KH et al (1987) Phys Rev B 35:7760
Chikarumi S (1980) J Magn Magn Mater 15–18:1130–1135
Darken LS (1950) Application of the Gibbs–Duhem equation to ternary and multicomponent systems. J Am Chem Soc 72:2909
Fischer WA, Sank D (1991) Metallurgical electrochemistry. In: Wu XF (trans). Northeast Institute of Technology Press, Shenyang
Guillaume CE (1897) Comp Rend Acad Sci Ser 125:235
Hasegawa A (1985) J Phys Soc Jpn 54:1477
Hidnert P, Souder B, Buschow KHJ (1988) J Less Common Met 136:207
Ito Y, Maruyama T, Saito Y (1987) Nonstoichiometry in vanadium pentoxide by coulometric titration. Solid State Ionics 25:199–205
Japan Society for the Promotion of Science. https://dl.acm.org/institution/60015540
Kallarackel TJ, Gupta S, Singh P (2013) Thermodynamic properties of $LaCrO_4$, $La_2CrO_6$, and $La_2Cr_3O_{12}$, and subsolidus phase relations in the system lanthanum–chromium–oxygen. J Am Ceram Soc 96(12):3933–3938

# References

Kubaschewski O, Alcock CB (1977) Metallurgical thermo-chemistry, 5th edn. Pergamon Press, Oxford, New York, Toronto, Sydney, Paris, Frankfurt

Matvei Z, Dejan D, Fritz A (2006) Thermodynamic modeling of the cerium-oxygen system. Solid State Ionics 177:989

Paasch S, Schaller HJ (1983) Thermodynamic properties of Pd-X-alloys, with X = Gd, Y, Ce. Ber Bunsenges Phys Chem 8:812

Palstra TTM, Mydosh JA, Nieuwenhuys GJ (1983) J Magn Magn Mater 36:290

Pratt JN (1990) Applications of solid electrolytes in thermodynamic studies of materials: a review. Metall Trans A 21:1223

Rapp RA (1970a) Physicochemical measurements in metals research. Interscience Publishers, New York, London, Sydney, Toronto

Rapp RA (1970b) Physicochemical measurements in metals research, part 2. Inter-Science Publishers, New York, London, Sydney, Toronto

Report of the Japan Small Commission for Steel Sensors. New development of steel sensors—focusing on solid electrolyte sensors. Japan Academic Renewal Association (JAC) 19 Member Council (JAC) Sensor Committee (JAC), Tokyo

Schuhmann R (1955) Application of Gibbs–Duhem equations to ternary systems. Acta Metall 3:219

Sreedharan OM, Chandrasekharaiah MS (1986) Standard Gibbs' energy of formation of $LaFeO_3$ and comparison of stability of $LaMO_3$ (M = Mn, Fe, Co or Ni) compounds. J Mater Sci 21(7):2581

Sreedharan OM, Chandrasekharaiah MS, Karkhanavala MD (1976) Free energy of formation of lanthanum nickelate. High Temp Sci 8:179

Subbarao EC (1980) Solid electrolytes and their applications. Plenum Press, New York, London

Tanaka H, Kishida Y, Kotani T, Moriyama J (1973) Thermodynamic study of the Mn-C system by electro-motive force measurements. Jpn Acad Met 37(5):568

Taylor RW, Schmalzried H (1964) The free energy of formation of some titanates, silicates, and magnesium aluminate from measurements made with galvanic cells involving solid electrolytes. J Phys Chem 68(9):2444

Tretyakov YD, Kaul AR, Makukhin NV (1976) An electrochemical study of high-temperature stability of compounds between the rare earths and copper oxide. J Solid State Chem 17:183

Wagner C (1952) Thermodynamics of alloys. Addison-Wesley, London

Wang CZ, Ye SQ, Zhang X (1985) Thermodynamic properties of $Y_2O_3 \cdot Cr_2O_3$ composite oxides. Acta Phys Sin 8:1017–1026

Wang CZ, Xu XG, Man HY (1987) Thermodynamic properties of lanthanum aluminate, $LaAlO_3$. Inorg Chim Acta 140:181–182

Wei SK (1964) Activity in metallurgical physicochemical applications. China Industrial Press, Beijing

Weiss RJ (1963) Proc Phys Soc (Lond) 82:281–288

Xiao LS, Yu HL, Wang CZ et al (1993) A study on the standard free energy of formation of $LaNi_5$ and $CeNi_5$. J Rare Earths 11(1):28

Xiao LS, Li GQ, Sui ZT, Wang CZ (1994a) Impedance spectroscopy study of solid electrolyte $La_{1-x}Ca_xF_{3-x}$. J Chin Ceram Soc 22:6

Xiao LS, Yu HL, Wang CZ et al (1994b) $LaNi_5$ and $CeNi_5$ criteria to generate free energy. J Chin Soc Rare Earths 1:15

Xu XG, Li GG, Wang CZ (1991) J Less Common Met 175:271

Xu JM, Sui ZT, Wang CZ (1992a) J Alloy Compd 190:L9–L12

Xu JM, Wang CZ, Sui ZT (1992b) J Alloy Compd 190:L5–L7

Xu JM, Wang CZ, Sui ZT (1994) Thermodynamic properties of rare earth invar alloys. J Phys Chem 10(3):276

Yang LZ, Sui ZT, Wang CZ (1992) A study on the nonstoichiometry of tin oxides by coulometric titration. Solid State Ionics 50:203–208

Yang YJ, Wang CZ (2004) Y1-XCaXF3-XYOF impedance spectroscopy of mixed-phase solid electrolytes. 2004 National Conference on Metallurgical Physical Chemistry, 308–311

Yu HL, Xu JM, Wang CZ (1995) Thermochemical properties of rare earth invar alloys. J Phys Chem 11(6):564

# Chapter 5
# Rare Earth Nanomaterials

## 5.1 On Nanomaterials

Nanometer (nm) is a measure of scale: 1 nm $= 10^{-9}$ m or 1 nm $= 10^{-3}$ $\mu$m. Generally, the term "nanoparticles" is used to refer to particles ranging in the size from 1 to 100 nm. At the end of 1974, Japan applied the term "nano" to technology, but it only started to be applied to materials during the 1980s. As early as 1861, with the establishment of colloid chemistry, researchers began to study colloidal systems (1–100 nm). Platinum black is made up of platinum nanoparticles. In 1967, Kubo et al. proposed the quantum confinement or limited domain theory of ultramicroparticles for the study of metal particles, thus facilitating the exploration of nanoscale particles by experimental physicists. In 1984, the Gleiter group in Germany first prepared nanocrystalline Pd, Cu, and Fe with clean surface interfaces by inert gas evaporation and in situ pressurization, and then found that $CaF_2$ nanocrystals and $TiO_2$ nanoceramics have good toughness at room temperature, offering a new approach to toughening ceramics.

In July 1990, the first international academic conference on nanoscience and technology was held in Baltimore, USA, and nanomaterials science was officially published as a new branch of materials science, with unifying concepts. Nanomaterials, nanobiology, nanoelectronics, and nanomechanics were then formally introduced, and official academic journals were published: *Nanostructured Materials*, *Nanobiology*, and *Nanotechnology*. Since then, nanomaterials have attracted global attention in the realms of materials, physics, and chemistry. In 1999, the concept of nanomaterials engineering was officially presented at an autumn conference, held in Boston, USA, which encompassed research on nanomaterials through nano-synthesis, nano-addition, the development of new nanomaterials, and nano-addition to modify new materials, thereby expanding the application of nanotechnology (Cardoso et al. 2019).

Further attention is paid to nanoscale particles and clusters, nanofilaments, nanorods, nanotubes, nanocables, and nano-assembly systems, which are assembled into systems with nanostructures in one-, two-, or three-dimensional space. These

© Science Press 2023

C. Wang, *Theory and Application of Rare Earth Materials*,

https://doi.org/10.1007/978-981-19-4178-8_5

76 5 Rare Earth Nanomaterials

contain solid components of the nano-units, but the matrix supporting them is also of nanoscale dimensions.

## 5.2 Basic Features and Characteristics of Nanoparticles

Nanoparticles range in size from 1 to 100 nm. They can be broadly classified according to their dimensions: large nanoparticles are 10–100 nm, medium nanoparticles are 2–10 nm, and small nanoparticles are 1–2 nm; those of less than 1 nm are called clusters. The basic characteristics of nanoparticles are their surface and volume effects (Liu 2009).

(1) Surface effect: stems from the fact that the ratio of the number of surface atoms to the total number of atoms in nanoparticles increases greatly with decreasing particle size. The surface energy and surface tension of nanoparticles also increase, which causes a change in their properties. (2) Volume effect: small size effect; quantum size effect; quantum tunneling effect.

Nanoparticles have some special properties, as briefly described below.

(1) Thermal properties: the melting point, onset of sintering, and crystallization temperatures of nanoparticles are much lower than those of conventional powders. Their surface energies are large, and the number of surface atoms is greater. These surface atoms have incomplete coordination and high activity, which results in a sharp decrease in the melting point of nanoparticles.
(2) Conductivity: decreases due to the quantum tunneling effect. As particle size is decreased, the critical temperature of superconductivity $T_c$ increases. As the particle size decreases, the low-frequency lattice vibration will be eliminated, which increases the $T_c$ value by a few percent.
(3) Magnetic properties: distinct from those of conventional granular materials: superparamagnetism; coercivity.
(4) Optical properties: macroscopic metal materials, such as gold, silver, and copper, have high reflectivity to sunlight. When the surface is smooth, the reflectivity is almost 100%, especially in the long band region.
(5) Adsorption: adsorption is a binding phenomenon between different phases in contact. Nanoparticles show strong adsorption due to their large specific surface areas and incomplete coordination of their surface atoms.

## 5.3 Structure of Nanoparticles

Nanoparticles can adopt many different morphologies, such as spherical, globular, flake, rod, reticulate, tubular, chain, and sponge-like. The main factors affecting morphology are dimensionality, composition, impurities, and the preparation method. Nanoparticles can be classified into nano-grain and nano-non-grain according to their aggregation state. Generally, nanocrystalline grains are single crystals. With increasing number of grains, the probability of producing defects increases. This may lead to metastable phases, which become clusters at grain sizes $\leq 1$ nm. Clusters are aggregates composed of several to hundreds of atoms. The shapes of clusters are various; they have not yet formed regular crystals. In addition to some lattice forms of macroscopic single crystals, the configurations of clusters can also have geometries with five-fold symmetry axes, such as icosahedra or pentagonal decahedron.

A cluster is particularly stable when it contains a certain number of atoms, known as the magic number. If the number of atoms is 4, a tetrahedron can be formed, and the order of magic numbers formed by closely packed tetrahedra is 8, 12, 16, 20, 24, 36, 52, 84, and so on. This sequence of magic numbers also holds for antimony clusters. The known structures of clusters include linear, layered, tubular, onion-like, skeleton-like, and globular.

As noted above, the proportion of atoms on the surface of nanoparticles is very large. Thus, nanoparticles are composed of a surface layer and an inner layer, and the arrangement of atoms in the surface layer shows neither long-range nor short-range order. Due to the high activity of surface atoms, gas molecules are readily adsorbed to form an adsorption layer thereon. The adsorption layer also affects the structure of the surface layer. The atoms in the inner layer of a single crystal are arranged periodically and have a complete lattice structure with long-range order. A nanocrystalline structure is characterized by a quasi-crystal configuration with five-fold symmetry axes. The distribution of atoms in a quasi-crystal shows long-range ordering, but the atom positions are not periodic.

## 5.4 Nano Block Materials

The term "nano block" refers to a nano solid bulk. A block composed of nanoparticles as structural units may be regarded as a nano block. Compared with a general bulk, a nano bulk has three main features: a nanoscale effect, high concentration interfacial effect, and interaction between nanostructured units. The thermal, optical, electrical, magnetic, and mechanical properties are distinct from those of general crystals and amorphous materials. The structure of nanocrystals can be modulated by applying different preparation methods.

(1) Nano-scale effect: the structural units in nanocubes are of nanometer scale, similar to nanoparticles, and the nanoscale effect is manifested through the small size effect, quantum size effect, and macroscopic quantum tunneling effect. Because of these effects, the atomic arrangement cannot be treated as having infinite long-range order, and the electron continuous band splits into discrete energy levels. That is, the energy is quantized and can cause obvious changes in macroscopic properties.

(2) High concentration interfacial effect: the adjacent nanoparticles in nanocubes form grain boundaries due to their different orientations. For instance, for 5 nm particles, the interfacial concentration is $10^{14}/cm^2$; the grain boundary thickness is 1–2 nm, and the grain boundary volume accounts for about 50% of the total volume. Thus, the atoms in the nanoparticles are in two situations, either within the particles or within the grain boundaries. Atom arrangements in the grain boundary are different from those in the grain, having neither long-range nor short-range order, and are constantly changing.

(3) Interactions between nanostructured units: the structural units of a nano-block differ from individual nanoparticles in the way they interact. A nanocrystalline bulk structure can be divided into nanocrystals and nano-sized amorphous crystals. Nanocrystalline microcrystals can be divided into two components, namely the grain component and the interface component. In the former, all atoms are located at lattice points within the grains; in the latter, all atoms reside within the interface between the grains.

## 5.5 Nanocomposites

(1) Characteristics of nanocomposites

The term "nanocomposite" refers to any composite material with components of nanometer dimensions. Nanocomposites typically show superior properties to those of general composites. In general, for maximum effect, the nanoscale component should be the active component in a nanocomposite. The following five parameters are used to characterize nanocomposites.

**Composition**

Composition is the volume fraction or mass fraction of each component in the composite. Clearly, the properties of nanocomposites will depend on the content of each component.

**Connectivity**

Connectivity is the way in which each component is interconnected in three-dimensional space, usually expressed using the representation method proposed by Newnham, taking 0 to represent micropowder, 1 to represent a fiber, 2 to represent a film, and 3 to represent the spatial network formed by three-dimensional interconnection. For example, type 0–3 represents a composite of active micro powder dispersed in a continuous matrix; it is different from type 3–0. A three-component composite has 20 connection types, and the mode has a great influence on the coupling characteristics of each component of the composite.

**Symmetry**

The symmetry of a composite is a characteristic of the internal structure of each component in the material and its spatial geometrical configuration, which can be characterized by the crystallographic point group. When the symmetry of a composite cannot be expressed by a crystallographic point group, it can be expressed by a Curie group or black-white group.

**Scale**

The scale of a composite mainly refers to the dimensions of the active components therein. When the active component is of nanoscale dimensions, it will produce a nanoscale effect, a high concentration interface effect, and a coupling effect between the components. These have a great influence on the properties of a composite.

**Periodicity**

The periodicity of a composite refers to the periodicity of the geometric distribution of components therein. This has an effect on the properties of the material, especially those related to excitation waves. Because the nanoscale dimensions are much smaller than the excitation wavelength, nanocomposites can be regarded as a continuous medium.

(2) Types of nanocomposites

Nanocomposites are mainly classified according to their matrix or dispersed phase. The dispersed phase is generally the active component, as exemplified by rare earth nanoluminescent materials, rare earth nanomagnetic materials, rare earth nanocatalysts, and rare earth nano-hydrogen-storage alloys.

## 5.6 Preparation of Rare Earth Nano-Oxide Materials

Rare earth nano-oxides have special physical and chemical properties, so methods for their preparation have become a research hotspot. The preparation methods are classified according to the state of the starting material, namely solid-phase, liquid-phase, and gas-phase methods, of which the first two are most commonly applied.

1. Solid-phase method

The solid-phase method is a traditional powder preparation method. It is inexpensive, high-yielding, and involves a simple preparation encompassing two processes: mechanical crushing and solid-state reaction. Bondioli et al. used $(NH_4)_2Ce(NO_3)_6$ as a raw material, which they dispersed in 1:1 mixtures of KOH/NaOH, $NaNO_3/KNO_3$, or LiCl/KCl. After the reaction, the product was washed with distilled water to remove impurity ions and dried at about 100 °C. Three kinds of nano-$CeO_2$ were thereby obtained, with particle sizes in the range 10–20 nm (Jia et al. 2008). The nano-$CeO_2$ generated in LiCl/KCl showed a superior particle distribution.

2. Liquid-phase methods

Liquid-phase methods include precipitation, microemulsion formation, sol–gel synthesis, hydrothermal synthesis, low-temperature combustion, and electrochemical methods.

## (1) Precipitation method

This generally involves the addition of a precipitant to a solution containing metal ions to form an insoluble compound, such as a hydroxide, carbonate, oxalate, or other precursor, which is then thermally decomposed to afford a nano-oxide powder. A variety of rare earth oxides have been prepared by this method, such as $CeO_2$, $La_2O_3$, $Y_2O_3$, $Tm_2O_3$, and $Er_2O_3$.

Li et al. used a surfactant as a protective agent to prepare $CeO_2$ nanoparticles with a uniform particle size distribution and a monodisperse particle size of only 2–4 nm. Zhu et al. prepared $Y_2O_3$ nanometer powder by a urea decomposition method and an alternating quenching–fast heating technique. The resulting regular spherical particles showed good dispersion and a narrow size distribution, with a minimum particle size of 65 nm.

The basic principle of the alternating quenching–fast heating technique can be rationalized as follows. In terms of the phase precipitation process, the key to the formation of nanoscale uniformly dispersed colloidal particles is to control the uniformity of the nucleation process and to minimize the rate of nuclear growth. The nuclear growth rate can be easily reduced by lowering the temperature. Therefore, a solution that has already nucleated should be quenched. By repeated quenching–fast heating, increasing the number of nucleations, nanoscale dispersed particles are ultimately obtained.

## (2) Microemulsion method

In recent years, the method of preparing superfine particles from reverse micelles or water/oil (W/O) microemulsions has gradually developed. A microemulsion used to prepare nanoparticles is usually a W/O-type system and generally consists of four components: reactants, surfactant, organic solvent, and aqueous solution. Each water core in the W/O microemulsion serves as a microreactor, and is surrounded by a single molecular layer interface composed of the reactants and surfactant, the size of which can be controlled between tens and hundreds of angstroms (Å). The preparation of nanoparticles by the microemulsion method often involves dissolution of the two reactants in two microemulsions with exactly the same composition. These are then mixed under the requisite conditions, whereupon the reactants generate an insoluble precipitate through material exchange. Subsequent thermal treatment of the precipitate yields nanoparticles.

## (3) Sol–gel method

The sol–gel method is commonly used to prepare nanomaterials. The principle is that a metal alcoholate or inorganic salt is hydrolyzed or polymerized to obtain a sol, which is further condensed into a gel, and then the gel is thermally dehydrated to obtain an ultrafine powder. Dong et al. prepared $Yb_2O_3$ nanocrystalline grains by the sol–gel method. The particles were spherical and uniform in size, with an average size of less than 100 nm. $Yb_2O_3$ solution and citric acid were mixed in the requisite proportions and the mixture was slowly concentrated at 70 °C to form first a sol and then a gel, which was dried and calcined to obtain $Yb_2O_3$ nanocrystals.

### 5.6 Preparation of Rare Earth Nano-Oxide Materials

(4) Hydrothermal synthesis

Hydrothermal synthesis is performed in a specially designed closed reactor (auto-clave), using an aqueous solution as the reaction medium. Heating the reactor produces high-temperature, high-pressure reaction conditions therein, which forces normally insoluble substances to dissolve. Their recrystallization produces dispersed nanocrystalline nuclei. In the hydrothermal process, water is involved in the reaction as a chemical component. It serves both as a solvent and as a promoter of mineralization, while also acting as a medium for pressure transfer. A number of non-aqueous solvents can serve the same function, such as ethanol, phenylethylenediamine, and $CCl_4$, allowing the non-aqueous solvothermal preparation of nano-powders.

(5) Low-temperature combustion method

The low-temperature combustion method utilizes the heat released by the redox reaction of the mixed raw materials composed of oxidant and reducing agent (as fuel), so that the reaction is carried out in the form of self-spreading combustion at 1000–1600 °C. During the combustion reaction, large amounts of gas and heat are released, enabling the product to be fully dispersed and sintered at the atomic scale, affording a nano-powder of uniform composition and fine size.

Wang et al. synthesized $Y_2O_3$, $CeO_2$, $Nd_2O_3$, $Pr_6O_{11}$, and $Sm_2O_3$ nano-powders by employing a nitrate-glycine system. The precursors used in the preparations were hydrated rare earth nitrates, which were dissolved in distilled water in a certain proportion with glycine in a beaker. Most of the water was evaporated by heating in an electric furnace. The residues became gelatinous and started to bubble. They were further heated until combustion commenced, whereupon rare earth nano-oxide powders were obtained.

(6) Electrochemical method

This method can be used to produce many ultrafine metal powders that cannot be prepared or are difficult to prepare by other methods. Li et al. prepared nanocrystalline $CeO_2$ powder of different grain sizes by precipitation, sol–gel, and electrochemical methods, respectively. The specific operation of the electrochemical method involved the preparation of $Ce(NO_3)_3 \cdot 6H_2O$ and $NH_4NO_3$ solutions at certain concentrations, mixing them in the requisite proportions, and placing the mixture in an electrolytic cell. After adjusting the pH, the solution was electrolyzed under magnetic stirring. It gradually turned turbid as a rust-colored flocculent precipitate was formed. After the electrolysis, the electrolyte was filtered, and the collected solid was washed, dried, and ground to obtain light-yellow nanometer $CeO_2$ powder. Comparing the samples prepared by the three methods, the authors found that the average particle size of the $CeO_2$ prepared by the electrochemical method was the smallest and the dispersion was the best, but the yield was low.

3. According to pressure

Escudero et al. (2017) think that the synthesis methods of RE-based nanoparticles can be divided into two general groups, according to the pressure at which the reaction takes place: reactions carried out at atmospheric pressure and hydrothermal or

82                                                          5  Rare Earth Nanomaterials

solvothermal methods. The examples of synthesis conditions for some RE-based nanoparticles have been summarized in a table, and the factors include composition, solvent, additives temperature ($^\circ$C), time, and morphology.

## 5.7   Preparation of Multi-dimensional Rare Earth Materials

With the development of nanotechnology, research work has gradually progressed from zero-dimensional nanoparticles to one-dimensional, two-dimensional, and three-dimensional nanomaterials. Hong (2006) discussed the preparation and assembly of multi-dimensional rare earth nanomaterials. Research on rare earth nanoparticles has progressed from an early focus on their preparation to more in-depth study of their formation and growth mechanisms, as well as their characteristics and uses.

(1)  Zero-dimensional rare earth nanoparticles

Zhou et al. (2008) prepared a series of rare earth nanohydroxides and oxides by a precipitation method with ethanol as a dispersant and protectant. They found that, with increasing atomic number from La to Lu, the particle sizes of cubic $Ln_2O_3$ showed lanthanide shrinkage, the lattice distortion increased, and the diffraction peak intensity showed a characteristic bimodal inverted "W"-shaped variation. They also prepared $CeO_2$ nanocrystals by a sol–gel method, and investigated the photovoltaic properties and quantum size effects of such materials of different particle sizes by surface photovoltaic spectroscopy and other means.

Tsai studied the formation and growth process of nano-$CeO_2$ and found that temperature, seed concentration, and particle morphology were correlated. The German company Siemens uses mechanical alloying and solid-state reaction to prepare nanocrystalline rare earth permanent magnet materials, such as NdFeB and SmFeN. Of these, NdFeB powders are formed by alloying in a planetary ball mill under hydrogen atmosphere. For example, hard magnetic $Nd_2Fe_{11}B$ was formed after 30 h of ball-milling and conducting heat treatment (600 $^\circ$C, 1 h); the optimum coercivity can be obtained after annealing at 700 $^\circ$C for 15–30 min. The particle size of $Nd_2Fe_{11}B$ is about 50 nm.

The raw materials were mixed under a hydrogen atmosphere prior to ball-milling to avoid their oxidation, but cannot be done in anhydrous ethanol, because it contains trace combined or adsorbed water. The applicability of other organic solvents requires further study.

(2)  One-dimensional rare earth nanowires, nanorods, and nanotubes

In recent years, research on one-dimensional rare earth nanomaterials has developed rapidly, and almost all one-dimensional materials of rare earth oxides and hydroxides have been studied. The main synthesis routes are hydrothermal and template methods. Template methods can involve soft templates, such as surfactants and

## 5.7 Preparation of Multi-dimensional Rare Earth Materials

carbon nanotubes, or hard templates, such as anodic alumina templates with display holes (AAO).

Fang et al. synthesized a series of lanthanide orthophosphate ($LnPO_4$) single-crystal nanowires with different crystalline phases by a hydrothermal method. Cao et al. synthesized $LaPO_4$ and $CePO_4$ nanowires and nanorods of diameters 20–60 nm and lengths of hundreds of nm to several $\mu$m by a hydrothermal-microemulsion method. Yan et al. synthesized hexagonal phase $LnPO_4 \cdot nH_2O$ (Ln = La, Ce, Pr, Nd, Sm, Eu, Gd) and orthorhombic phase $LnPO_4 \cdot nH_2O$ (Ln = Tb, Dy) one-dimensional nanomaterials by a hydrothermal method, and $LaNiO_3$ nanowires by a sol–gel method. Yin and Hong (2005) synthesized $La(OH)_3$ nanorods by a hydrothermal–microemulsion method, controlling the synthesis by changing the reactant concentrations, time, and temperature. They also synthesized various rare earth hydroxide nanorods by a dissolution–crystallization process.

Yada et al. synthesized rare earth (Er, Tm, Yb, and Lu) oxide nanotubes with an inner diameter of 3 nm by urea homogeneous precipitation using sodium dodecyl sulfate as a template. Wu also used sodium dodecyl sulfate as a template to synthesize $Y_2O_3$:Eu nanotubes with diameters of 20–30 nm, and studied spectral differences between the tubular and bulk materials. Tang synthesized $Tb(OH)_3$ nanotubes by a hydrothermal method in the presence of a surfactant, and obtained $Tb_2O_3$ nanotubes after burning treatment. It was found that the absorption spectra shifted towards higher energy and showed peak broadening, whereas the fluorescence spectra showed more precise spectral lines than in the case of the bulk material, which might be attributed to a surface effect. Researchers at the Brookhaven Laboratory, USA, have also prepared $CeO_2$ nanotubes for use as a catalyst.

(3)  Two-dimensional rare earth nanofilms

Two-dimensional (2D) rare earth nanomaterials are mainly thin-film materials, which can be divided into two categories: complex and oxide nanofilms. Rare earth oxide nanofilms can be prepared by physical or chemical methods. In the physical method, the corresponding rare earth oxide or pure metal serves as the precursor. Through a process of electron-beam evaporation or electron-beam bombardment, the precursor is deposited on a prepared substrate to obtain the required oxide nanofilm, which has high chemical stability. Chemical methods include spray pyrolysis, chemical vapor deposition, and sol–gel methods, which are inexpensive, easy to operate, and widely used. Jenouvier used a sol–gel method to synthesize rare earth erbium- or thulium-doped yttrium titanate ($Y_2Ti_2O_7$) thin films, and studied the relationships between up-conversion emission and crystallinity and rare earth ion-doping concentration and the effect of ytterbium co-doping.

Rare earth organic complexes have excellent luminescence properties, but poor photostability and thermal stability. The introduction of rare earth complexes into organic–inorganic exchange networks by sol–gel methods not only solves the stability issue of nanoparticles, but also produces films with good process ability and favorable functional properties. Cong et al. used micelles formed from polystyrene (PS) and poly (4-vinyl pyrrolidone) (PVP) in the presence of 1,10-phenanthroline

(Phen) in a mixed solvent of ethanol and $N,N$-dimethylformamide (DMF) as nanoreactors. The resulting Eu(III)-(PS-b-P4VP)-Phen·5DMF complexes were spin-coated on quartz or glass substrates to form nanoscale ordered Eu(III)block-copolymer composite luminescent films. The films had strong Eu(III) characteristics and long lifetimes.

(4)  3D rare earth nanomaterials

Three-dimensional (3D) rare earth nanomaterials are mainly rare earth nanoceramics. Nano $ZrO_2$-$Y_2O_3$ ceramics have high strength and toughness, and can be used as cutting tools, wear-resistant parts, and ceramic engine parts. The Japan Science and Technology Agency has developed some stable $ZrO_2$ nanoceramic materials that are mass-produced by a hydrothermal method. Using $YCl_3$ as a raw material and adding urea as a precipitant, nanoceramic powder with a purity of more than 99.9% and an average particle size of 30 nm is routinely prepared by hydrothermal synthesis in an autoclave. The material obtained by sintering this powder has high strength and toughness, and can be used to make cutting tools and molds.

## 5.8  Composites and Assembly of Rare Earth Nanomaterials

Rare earth nanocomposites can not only modify materials and endow them with new properties, but also prevent the agglomeration of nanoparticles, which has wide application prospects. Louis et al. combined Au nanoparticles with a rare earth oxide (Tb: $Gd_2O_3$) to form a nanocomposite. The Au nanoparticles in this complex play a role of absorbing energy and transferring it to (Tb: $Gd_2O_3$), thereby improving the luminescence efficiency of the latter. This composite can be used as a fluorescent probe for biological detection and so on.

Core–shell structured materials can be obtained by surface-coating treatment, such as Eu(TTA)$_3$·$2H_2O$ coated on the surface of Ag colloidal nanoparticles, which renders such nanoparticles more luminescent. Rare earth nanocomposites are introduced into colloidal $SiO_2$, giving a core–shell structure with a smooth surface. The product displays the characteristic emission of $Eu^{3+}$ ions, but with much higher thermal stability than that of the pure rare earth complex.

Coating relatively expensive rare earth materials on inexpensive non-rare earth materials to obtain core–shell structured composites can greatly reduce the amount of rare earth element needed, while also permitting shape control of the rare earth nanoparticles.

Cao et al. coated a layer of $Y_2O_3$ doped with different rare earth ions on carbon nanotubes by a chemical precipitation method, and recovered the nanotubes after a burning treatment. The available assembly techniques for materials are principally plate printing, scanning probe manipulation assembly, molecular beam epitaxy, and self-assembly.

Tsyetkov et al. loaded the rare earth erbium into an inverse opal structure composed of $SiO_2$ nanospheres and formed a composite with porous anodic alumina by a self-assembly technique. They studied the elemental composition, the concentration of rare earth ions, the composition of the medium, and the luminescence properties of the product. This kind of luminous complex may be used in the storage and transmission of optical information.

## 5.9 $CeO_2$ Nanofibers

There are two main forms of nanofibers, namely ultrafine fibers with diameters less than 100 nm, and fibers filled with nanoparticles to modify their properties, that is, nano-modified fibers. Many methods are available for preparing nanofibers, among which electrospinning has emerged as an important method for preparing a variety of inorganic oxide nanofibers because they are much finer than those prepared by traditional methods (with a minimum diameter as low as 1 nm).

Yang prepared $CeO_2$ nanofibers with diameters of 50–100 nm by electrospinning. Dong et al. synthesized $PVP/Ce(NO_3)_3$ composite fibers by this method; after calcination at 600 and 800 °C, the fibers showed a hollow structure; the surface was porous and the average diameter was about 600 nm. Such nanomaterials with large surface area and small pore size can be applied as catalysts and hydrogen storage materials. Gu et al. prepared $CeO_2$ nanofibers by a reverse micelle method without the need for a hard template; by adjusting the aging temperature, they succeeded in preparing $CeO_2$ nanobelts and nanorods.

(1) Specific $CeO_2$ shapes

Besides $CeO_2$ nanospheres, nanorods, nanotubes, and nanofibers, other specific morphologies have appeared in recent years, such as dendritic, flower-like, and dumbbell-like. These morphologies broaden the applicability of $CeO_2$. Zhang et al. (2012) used amino acids as novel crystal growth modifiers to control the morphology of $CeO_2$ under hydrothermal conditions. Different amino acids can influence the morphology and size of $CeO_2$ precursors, and hence the morphology of the final product.

(2) Doped nano-$CeO_2$ UV-shielding materials

The potential harm of ultraviolet (UV) radiation to humans has attracted wide attention, and research on various UV-shielding materials has become a hot topic. UV-shielding materials can be divided into inorganic and organic categories. The inorganic materials are widely used, and include nano-sized $TiO_2$, nano-$ZnO$, and nano-$CeO_2$. Of these, nano-$CeO_2$ shows strong visible light transmittance and represents an ideal UV-shielding material. However, pure nanometer $CeO_2$ powder is yellow and shows a strong catalytic activity towards oxidation, which limits its practical application. Recently, the Ce–O octa-coordination structure in $CeO_2$ has been stabilized by cationic doping or surface modification treatment. Oxygen defects in the lattice are caused by charge balance, which reduces the catalytic activity towards oxidation to some extent, but does not completely overcome its shortcomings.

Dai et al. (2011) adopted $F^-$ with a smaller radius than that of $O^{2-}$, and $Sr^{2+}$ or $Ca^{2+}$ with radii larger than that of $Ce^{4+}$. They modified the lattice constant of nano-$CeO_2$, making it closer to the ideal value for a stable octa-coordination structure at 0.732 by anion-cation co-doping, also studied the change of lattice constant and UV-shielding performance caused by doping.

The spectral scanning range was 200–500 nm, and the band-gap width of doped $CeO_2$ crystals was calculated from the formula $E_g$ (eV) $= 1240/\lambda_g$, X-ray diffraction analysis with Phillips PANalytical X′-Pert Pro MPD X-ray diffractometer. Such luminous complexes may also be used in the storage and transmission of optical information.

(3) Preparation of powders

A certain concentration of cerous carbonate solution was prepared by dissolving cerium phosphate in concentrated nitric acid. The sample preparation includes: preparation of cerium oxide co-doped with strontium fluoride, preparation of calcium fluoride co-doped with cerium oxide. The experimental results are discussed in several aspects such as structure and composition, morphology and particles, and UV-shielding performance.

Other researchers, such as Liu et al. (2001, 2011), also independently discussed the modification of nano-powders.

## 5.10 Study of Low-Dimensional Nanostructured $Gd_2O_3$

Because cerium is the most abundant of the light rare earth metals and $CeO_2$ finds many uses, research on nanometer $CeO_2$ is the most extensive. Nevertheless, Zhang et al. (2012) studied the synthesis and morphology control of $Gd_2O_3$ low-dimensional nanostructures. The outer layer of the Gd atom has the electron arrangement $4f^7 5d^1 6s^2$. That is to say, the $Gd^{3+}$ electron configuration is $4f^7$, a half-filled f-sublayer. The development of synthetic methods for obtaining low-dimensional $Gd_2O_3$ nanostructures with controlled morphology is meaningful from both theoretical and practical perspectives.

For the preparation of $Gd_2O_3$ nanospheres and nanorods, starting from gadolinium nitrate solution under weakly alkaline conditions, alkali metal carbonate/gadolinium nanomaterials with different morphologies were obtained after hydrothermal processing at different temperatures. After heat treatment, $Gd_2O_3$ nanospheres with good dispersion or nanorods of uniform size could be obtained. The average size of the $Gd_2O_3$ nanospheres was about 100 nm, and the nanorod diameter and length were about 50 nm and 200 nm, respectively. At temperatures below 70 °C, the hydrolysis of urea is very slow, but above 80 °C the urea begins to decompose and the reaction commences. As the temperature is increased to 160 °C, rod-like nanostructures are formed because the system reaches a supersaturation suitable for slow crystal growth with an isotropic orientation. If the hydrothermal reaction is initiated directly at 160 °C, the urea decomposition rate is too high, and crystals grow rapidly in all directions, forming a spherical nanostructure.

## 5.11 Composites of Dendrimers and Inorganic Nanoparticles

Dendrimers are a class of high polymers with multiple branches, regular structure, controllable molar mass, and a large number of functional groups on their surfaces. Their structures contain inner cavities, while their outer termini can be functionalized according to the need. On the basis of effective surface design, various functionalized and organic-modified nanoparticles can be synthesized by using dendrimer-modified inorganic nanoparticles, and inorganic particles as the grafting matrix have the advantage of low preparation cost.

Mo et al. (2006) used the Virtual Material (VM) software provided by the Institute of Chemistry of the Chinese Academy of Sciences to perform molecular dynamics simulations of lanthanum clusters under HO-PAMAM-2.5 G protection. They investigated the stability and interaction modes of HO-PAMAM-2.5/lanthanum composites modified with different proportions of hydroxyl groups in regular systems from the perspective of molecular structure and energy changes.

They proceeded to practical synthesis of HO-PAMAM-2.5/La nanocomposites, and used transmission electron microscopy (TEM) to characterize the obtained nanoparticles. Their results showed the experiment to be consistent with the molecular dynamics simulation. By reducing $La^{3+}$ to La with $NaBH_4$ solution, La clusters were formed in the inner cavities of HO-PAMAM-2.5. The use of HO-PAMAM-2.5 as a template and stabilizer circumvents the shortcoming of La cluster instability.

## 5.12 Fluorescence-Based Immunoassay of Rare Earth Nanoprobes

Fujian Institute of Research on the Structure, Chinese Academy of Sciences (FJIRSMCAS)' team (2014) collaborated with Huang's team of to develop a fluorescence immunoassay based on dissolution enhancement of rare earth nanocrystals to replace molecular probes with rare earth nanoprobes. The high concentration of rare earth cations in rare earth nanocrystals serves to increase the labeling level and greatly enhances the luminescence and detection sensitivity of the system.

The researchers used about 9 nm $NaEuF_4$ as the nano-fluorescent probe and β-naphthalene formyl trifluoroacetone (β-NTA) as the reinforcing agent. High-resolution fluorescence spectroscopy and elemental analysis revealed the mechanism of rare earth nanocrystal dissolution and luminescence enhancement. Highly sensitive detection of human broad-spectrum tumor markers was realized, with a detection limit of 0.1 pg/mL. The accuracy and reliability of the method were verified by measuring the coefficient of variation.

## 5.13 Rare Earth Nanoparticle Memory

According to information from the Max Planck Society, a team of experts at the Max Planck Institute of Quantum Optics has successfully located a single rare earth cation in a crystal and accurately measured its quantum mechanical energy state (May 28, 2014). This research shows that it is possible to store quantum information in ions, and should make a great contribution to the future development of quantum computers.

Many researchers are engaged in studying the modules that will build future quantum computers. Among them, the quantum system is the key element, and quantum dots or optical absorption defects (color centers) are the focus of many researchers. However, rare earth ions emit very little light and are difficult to detect. After more than six years of in-depth study, experts accurately positioned trivalent praseodymium ions ($Pr^{3+}$) in yttrium silicate crystals in several nanometers and measured their optical properties with unprecedented accuracy with the help of precision laser and microscope techniques.

Researchers have used lasers to excite single ions in crystals and then observed how they release the energy in the form of light over time. Because rare earth ions are not strongly affected by thermal or acoustic vibrations of the crystal, some of their energy states are abnormally stable, and take more than 1 min to return to the ground state, which is millions of times longer than for most quantum systems studied to date. When quantum information is stored in different energy states of atoms or ions, the ions can serve as the memory of quantum computers.

Research is aimed at using nano-antennas and micro-resonators to amplify the praseodymium signal hundreds or thousands of times. A single particle responding to less than 100 photons per second would allow the signal to be more easily seen.

## 5.14 Effects of Rare Earth Coating

Coating is a general term for any liquid or solid material coated on the surface of an object (solid-state coating) (Huang 2013).

Because of the special physical and chemical performance, rare earth elements can be widely used in coatings. The application of rare earth in coating industry mainly includes drying agent, curing agent, pigment, insulation coating, antibacterial and nano-ecological coatings and so on. For example, rare earth phosphate ($RPO_4$) and sulfate $(R_2(SO_4)_3)$ can be used as efficient curing agents in building coatings, such as in sodium silicate inorganic composite coatings.

Rare earth paints will play a role in reducing the radar-scattering cross-section, an aircraft does not reflect, or shows reduced reflection of radar waves, which reduces cross-section reach the stealth purpose. These future camouflage and stealth coatings will incorporate broadband, polymer, nano, and other materials in order to render them lightweight, multifunctional, bionic, and intelligent.

# References

Cardoso CED et al (2019) Recovery of rare earth elements by carbon-based nanomaterials—a review. Nanomaterials (Basel) 9(6):814

Dai Y, Hou YK, Long ZQ et al (2011) Characterization and properties of doped cerium oxide UV shielding materials. J Chin Soc Rare Earths 29(1):195

Escudero A et al (2017) Rare earth based nanostructured materials: synthesis, functionalization, properties and bioimaging and biosensing applications. De Gruyter, 29 June 2017

Fujian Institute of Material Structure (2014) http://tech.gmw.cn/2014-08/26/content_12795289.htm, 26 Aug 2014

Hong GY (2006) Preparation and assembly of rare earth nanomaterials. J Chin Soc Rare Earths 24(6):641

Huang SJ (2013) The big role of small coatings. China National Defense, 22 Oct 2013

Jia LP, Zhang DF, Pu XP (2008) Review of preparation methods for rare earth oxide nanomaterials. Chin Rare Earths 29(1):44

Liu XZ (2009) Rare earth fine chemical chemistry. Chemical Industry Press, Beijing, p 217

Liu KQ, Zhong MQ, Chen F et al (2001) Preparation of nano $CeO_2$ and their dispersion in PMMA. J Chin Soc Rare Earths 29(6):737

Liu LS, Xiong XB, Chen JL et al (2011) Research status of nano-powder surface modification. Chin Rare Earths 32(1):80

Mo ZL, Liu YZ, Chen H et al (2006) Molecular dynamics and preparation of HO-PAMAM-2.5G protected lanthanum cluster. J Chin Soc Rare Earths 24(1):12

Progress in the study of quantum memory by laser-localized crystals in Germany. http://www.laser-infrared.com/index.php?s=/Index/msg_detail/id/3922/cateid/25, 28 May 2014

Yin YD, Hong GY (2005) Morphology control of lanthanum hydroxide nanorods synthesized by hydrothermal microemulsion method. Chem J Chin Univ 26(10):1795–1797

Zhang Q, Hao PW, Li RX et al (2012) $Gd_2O_3$ morphology controllable synthesis of low-dimensional nanostructures. Chin Rare Earths 33(4):9

Zhou YH, Yang XF, Dong XT et al (2008) Advances in preparation of nano $CeO_2$ with different morphologies. J Chin Soc Rare Earths 26(3):257

# Chapter 6
# Solid Electrolytes Based on Rare Earth Oxides and Fluorides

## 6.1 Brief Description

Solid electrolytes are solid-state ionic conductors and belong to a branch of defect physics involving physics and chemistry. Morphologically, they include single crystal, polycrystalline, sintered body, and thin film electrolytes. Unlike fast ionic conductors, solid electrolytes are ordinary solid-state ionic conductors with low ionic conductivity.

In "Solid Electrolytes and Chemical Sensors" (Beijing: Metallurgical Industry Press, 2000), Wang CZ discussed the relevant properties and characteristics of solid electrolytes and chemical sensors. The book covered defect structures and the migration properties of solid electrolytes, solid-electrolyte materials, and research methods for determining solid-electrolyte structure, solid-electrolyte primary cells and the constraints affecting accurate measurement, the thermal mechanics of compounds, and the thermodynamics of alloy systems.

Solid-state ionic conductors, some with high ionic conductivities and low conductivity activation energies close to or even more favorable than those of molten salts, are often referred to as fast ionic conductors (FIC). They are formed when non-conductive ions in the crystal form a rigid skeleton, and there are more occupied positions within the lattice than the number of conductive ions. These positions are interconnected to form one-dimensional tunnel-type, two-dimensional plane-type, or three-dimensional conduction-type ion-diffusion channels, in which the conductive ions can move freely.

Solid electrolytes should have the following characteristics (Rapp and Shores 1970; Janke and Fischer 1977; Subbarao 1980; Hagenmuller 1984; Fischer and Janke 1991):

- $t_i > 0.99$, $t_e < 0.01$;
- the band-gap for electron transport should be greater than 3 eV;
- a low ion-migration activation energy;
- stable thermodynamic properties at operating temperatures;
- small phase change;

92                                    6  Solid Electrolytes Based on Rare Earth Oxides and Fluorides

- ions that are not susceptible to gain or lose electrons and change valence.

In terms of material composition and structure, solid electrolytes require high density and good mechanical strength.

In this chapter, we mainly discuss rare earth oxygen ion conductors, rare earth fluoride ion conductors, and other rare earth ion conductors, and their applications.

## 6.2  $ZrO_2(+Y_2O_3)$ Stable Solid Electrolyte (YSZ)

Most oxygen ion conductors are oxides of the group 4 metals or tetravalent rare metals, including rare earth metals. The research history of ion conduction in solids is very long. Yttrium-stabilized zirconium (YSZ) was the first solid oxide electrolyte to be studied. As early as 1897, Nernst pointed out that the current in this material might be due to the movement of oxygen anions ($O^{2-}$). Lattice defects of solid electrolytes were successively proposed by Frenkel in 1926 and Schottky in 1930.

Frenkel (1930) explained that some atoms in the crystal could be squeezed into the intersection sites in the lattice due to the deformation caused by thermal motion to form a gap atom, resulting in a vacancy. This is known as a Frenkel defect. Schottky (1935), Wagner and Schottky (1930) further pointed out that some atoms on the crystal surface could move to new positions and form vacancies if they have sufficient energy. A nearby atom could then occupy such a vacancy due to thermal motion, leaving a new vacancy, which is in turn transferred and gradually diffuses into the interior of the crystal to form an internal vacancy. This is known as a Schottky defect. Sometimes, both defects occur at the same time. Defects can also be introduced by doping or impurities.

Most of the above defects are ionic defects in crystals. Ionic crystals can also generate free electrons ($e'$) and electron vacancies (holes, $h^·$) under certain conditions. Defect formation is a thermodynamically reversible process with equilibrium states.

Using notation introduced by Kröger and Vink, one may write equations to illustrate the cause of crystal defects. The embedding mechanism for YSZ formation is thus:

$$Y_2O_3 + 2Zr_{Zr} + O_O = 2Y'_{Zr} + V_{\ddot{O}} + 2ZrO_2$$

In this equation, $Zr_{Zr}$ denotes in situ Zr in the lattice, $O_O$ denotes in situ O in the crystal, $Y'_{Zr}$ denotes Zr replaced by Y with a negative charge, and $V_{\ddot{O}}$ denotes an oxygen ion vacancy with two positive charges.

(1)  $ZrO_2-Y_2O_3$ system phase

The $ZrO_2-Y_2O_3$ system phase diagram and composition–temperature–conductivity diagram are shown in two figures, for example, the diagram of relationship between $ZrO_2-Y_2O_3$ system composition, temperature, and conductivity shows that the electrolyte conductivity within an 8–9% molar fraction of Y is the highest, and that the

6.2 $ZrO_2(+Y_2O_3)$ Stable Solid Electrolyte (YSZ)

composition of the system showing the highest conductivity appears at the low limit side, near the cubic phase region. The conductance activation energy of this composition is 0.7–0.8 eV at about 1000 °C. The oxygen ionic conductivity of this electrolyte is independent of the gas-phase oxygen partial pressure over a wide range thereof, but electronic conductivity is also found at very low oxygen partial pressures.

A deficiency of this electrolyte is its decrease of oxygen ionic conductivity, which is a problem that must be overcome if it is to be used as a fuel cell electrolyte.

Bodwal and Rajendran studied the aging problems of YSZ from the perspectives of particle size, phase combination, adhesion, raw material purity, and so on, and found that the grains and grain boundaries affected the electrical and mechanical properties. Successful operation requires micron size, high surface area, high activity, high density, no aggregation, and uniform grain distribution. The direct synthesis method and the co-precipitation method are commonly used to prepare powders. Generally, 8% $Y_2O_3$ (8-YSZ) is used as the solid electrolyte for fuel cells. This composition is near the C + t/C phase boundary in the two-phase region. A decrease in conductivity over time at 1000 °C can be observed by electron microscopy, which is accompanied by the formation of a very fine t-$ZrO_2$ precipitate. On adding 10% $Al_2O_3$ to 8% $Y_2O_3$–$ZrO_2$ and inspecting by scanning electron microscopy, it was found that $Al_2O_3$ did not participate in the reaction, but increased the material strength.

Experiments have shown that any $SiO_2$ impurity becomes distributed in the grain boundary region, which causes lattice disorder, phase separation, incomplete grain contact, porosity, lattice particle dislocation, misalignment, and the formation of a vitrified phase, which increases the resistance and decreases the conductivity of the material. Therefore, the purity of the raw materials should be high, so as not to form a grain boundary phase.

Suzuki Y (1996) deployed 8 mol% YSZ in a fuel cell for 114 days of continuous operation, and observed that the conductivity decreased over time. He concluded that this was due to a change of phase composition. Other researchers have studied relaxation phenomena by impedance spectroscopy, YSZ-Pt interface dynamics, the role of electron–ion conductors and YSZ, and the application of YSZ in various atmospheres. Wang CZ et al. studied the thermodynamic properties of the electronic-ionic conductor materials $La_{0.8}Sr_{0.2}MnO_3$ and $La_{0.9}Na_{0.1}MnO_3$ by a solid electrolyte electromotive force method (Hildrum et al. 1994). These compositions can be used as electrode materials for fuel cells.

(2) YSZ performance improvement

Mori et al. (2002) improved the properties of YSZ through heat treatment. They prepared two batches of powder by isostatic pressing. The first batch was directly heated to 1500 °C, held at this temperature for 4 h, and then cooled to room temperature. The second batch was heated to 1200 °C, held at this temperature for 40 h, then heated to 1500 °C, and held at this temperature for 4 h; the heating and cooling rates were both 200 °C/h. Particle size and ionic conductivity were determined. The samples were analyzed by scanning electron microscopy and high precision transmission electron microscopy. At liquid nitrogen temperature, appropriate thin samples

94 6 Solid Electrolytes Based on Rare Earth Oxides and Fluorides

were cut by means of an ion beam, and the bulk characteristics of each sample were determined by wide-scan XRD.

The results showed that the second batch had low impedance and capacitive reactance, and high conductivity. Highly pure, good quality samples are snow white. The improvement of YSZ quality was related to toughening through a heat-induced phase transition (Pastor et al. 1974; Porter and Heuer 1977; Lange 1982; Sheng and Xu 1985; Masaki 1986; Gao et al. 1988). The degree of sintering of the material determines its crystalline characteristics.

## 6.3 $Bi_2O_3$-$Y_2O_3$ Solid Electrolyte

$Bi_2O_3$ changes from a monoclinic $\alpha$-phase to a face-centered cubic phase at 730 °C, which is accompanied by an increase in its oxygen ion conductivity by three orders of magnitude. An $\sigma$ phase with high conductivity has a vacancy structure that is equivalent to 1/4 of the anion positions in the structure being vacancies, and these oxygen ion vacancies are regularly distributed. $O^{2-}$ can easily move between vacancies, as in $(Bi_2O_3)_{1-x}(Y_2O_3)_x$ doped with $Y_2O_3$ at $x = 0.25$, which has a conductivity of 1.3 $\times 10^{-2}$ S/cm at 500 °C (Takahashi et al. 1975, 1976; Wang and Xu 1984).

Meng et al. (1986) studied the impedance spectra and admittance spectra of $Bi_2O_3$-$Y_2O_3$, $Bi_2O_3$–$Nb_2O_3$, $Bi_2O_3$–$Nb_2O_3$–$Sm_2O_3$, $Bi_2O_3$–$Y_2O_3$–$Sn_2O_3$, and $Bi_2O_3$–$Y_2O_3$–$Pr_6O_{11}$ systems by impedance spectroscopy at 250–800 °C. They found that the form of the spectrum in the low temperature region is similar to that of the general solid electrolyte, but above about 500 °C, the inductive component appears in the circuit, located in the fourth quadrant of the spectrum, which reflects the specific behavior of the oxygen ion conductor. This means that oxygen ions in these materials produce a spiral motion. Watanabe and Kikuchi (1986) studied the relationship between polycrystalline transformation and lanthanide ion radius in $Bi_{0.775}Ln_{0.255}O_{1.5}$ systems by precise XRD and differential thermal analysis, and found that an $Ln_2O_3$ layered structure accounted for 20–30 mol% of each cubic crystalline solid solution. These systems undergo one or two reversible polycrystalline transformations with La, Pr, Nd, Sm, Eu, and Gd, whereas oxide systems containing Tb, Dy, Ho, Y, and Er show an irreversible phase transition.

As regards $Bi_2O_3$-$RE_2O_3$ systems, researchers are interested in their high oxygen anion conductivities, and hope to use them as solid electrolytes for fuel cells. Wang CZ et al. studied the relationship between $Bi_2O_3$-$Y_2O_3$ system composition, gas-phase $p_{O_2}$, and $Bi_2O_3$ activity in the temperature range 823–968 K (550–695 °C) by a solid electrolyte electromotive force method (Wang 2004). The $p_{O_2}$ value varied in the range $10^{-11}$–$10^{-8}$ Pa, indicating that if the solid electrolyte were to be used in a fuel cell at the reduction pole, it would gradually decompose (Huang et al. 1992). $Bi_2O_3$-based composites can be used as solid electrolytes for oxygen pumps. $Bi_2O_3$ decomposition is a gradual process, and the color gradually becomes lighter.

## 6.4 Rare Earth Fluoride-Ion Conductors

A fluoride ion conductor is an ion conductor with fluoride as the conduction ion. The applications of fluoride ion conductors include ions elective electrodes, gas detectors, solid-electrolyte battery materials, and so on (Sher et al. 1966; Sobolev et al. 1976; Fedorov et al. 1982).

The electronegativity of fluorine is 4.0, the highest of any element, and fluoride ion transport is very fast because of its small ionic radius. The fluorides of alkaline earth and rare earth metals form ionic crystals, electronic conductivity in which is negligible. $CaF_2$, $SrF_2$, YFS, and $LaF_3$, with fluorite-type structures and 8-, 9-, and 11-coordinated cations have large $F^-$ mobilities (Reau et al. 1976; Delcet et al. 1978). There are $MF_2$ and $LaF_3$ base solid solution in $MF_2$-$LaF_3$ (M = Ca, Sr, and Ba) system, both of these solid solution have the characteristics of large deviation from stoichiometry; have heterovalent eutectic with a large number of $F^-$ vacancies to obtain the practical $LaF_3$ solid solutions. Such solid solutions have more than tenfold higher $F^-$ conductivities than pure $LaF_3$, and the $MF_2$ content imparting the highest conductivity is about 5 mol.%.

Wang CZ team studied the relationships between synthesized poly crystal $F^-$ conductivity, composition, and temperature. In the range 423–773 K (K = °C + 273.15), the orders of magnitude of conductivity were $10^{-4}$–$10^{-2}$ S/cm, and the $F^-$ conductivity was highest with 5 mol% $CaF_2$, consistent with research results on single crystals. Single-crystal products of $LaF_3(CaF_2)$ and $LaF_3(EuF_2)$ are commercially available in the form of 1 mm thick sheets, which are mainly used for fluoride ions elective electrodes, whereas tubes need to be prepared in the laboratory from the polycrystalline materials.

## 6.5 Rare Earth Solid Electrolytes at High Temperature

Rare earth elements have unique electronic arrangements and physicochemical properties. Appropriate addition of trace rare earths to steel, cast iron, nonferrous metals, and their alloys can improve the material properties. There has been much research on the purification, metamorphism, and grain refinement of rare earth elements. According to the thermodynamic equilibrium law and kinetic factors, trace rare earth elements present in metals in atomic form can still play an alloying role (Xu et al. 1992; Xiao et al. 1994).

The determination methods for trace rare earths in metals involve long processes, are subject to many influencing factors. Wang CZ first studied rare earth sensors in 1992, and published the relevant research results with team in 1992–1997; the main work is as follows.

1. La-β-$Al_2O_3$ solid electrolytes

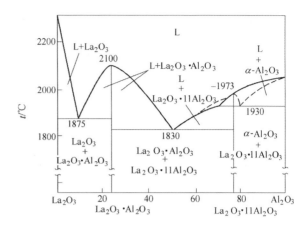

**Fig. 6.1** La$_2$O$_3$–Al$_2$O$_3$ system phase diagram

The β-Al$_2$O$_3$ is the non-chemical proportioning phase its composition is between Na$_2$O·5.3Al$_2$O$_3$ and Na$_2$O·8.5Al$_2$O$_3$. The metastable phase β″-Al$_2$O$_3$ is often generated during the preparation of β-Al$_2$O$_3$; β″-Al$_2$O$_3$ is irreversibly converted into β-Al$_2$O$_3$ at 155 °C. At high temperature, the delocalization of Na$^+$ is obvious. A new β-Al$_2$O$_3$-based electrolyte of monovalent ions and some divalent ions with small radius can be obtained by ion exchange reaction in a molten salt at 300–800 °C. However, due to the large radii of rare earth cations, it is difficult to prepare β-Al$_2$O$_3$ rare earth electrolytes by ion-exchange.

The La$_2$O$_3$-Al$_2$O$_3$ phase diagram is shown in Fig. 6.1. La$_2$O$_3$ can combine with Al$_2$O$_3$ to form La$_2$O$_3$·11Al$_2$O$_3$, that is, an La β Al$_2$O$_3$ phase.

The phases La$_2$O$_3$·11Al$_2$O$_3$ and Na$_2$O·11Al$_2$O$_3$ are similar, having cubic crystal structures and an open conducting phase. La$^{3+}$ is present in La-β-Al$_2$O$_3$, which can migrate and conduct electricity. La-β-Al$_2$O$_3$ has been prepared by direct solid-state synthesis, and XRD analysis confirmed its identity. Electrolyte tubes have been prepared by a hot-pressing casting method followed by firing at 1650–1700 °C. The ceramicized product was dense and strong, and was compressed by an isostatic pressing method. The La$^{3+}$ conductivity was determined by AC impedance spectroscopy. The conductivity increased from 10$^{-7}$ S/cm at 800 °C to 10$^{-4}$ S/cm at 1500 °C, which meets the requirements of a solid electrolyte used for the determination of La activity in steel and molten iron (Xiao et al. 1993; Zou et al. 1995).

2. La-β-Al$_2$O$_3$ solid-electrolyte rare earth sensor

Researchers have determined the La activities of different liquid slags by using an La-β-Al$_2$O$_3$ tube as the electrolyte, pure La as the reference electrode, and a molybdenum wire and molybdenum cermet rod as the reference electrode and tested electrode lead wire to form an La-sensing probe. At the same time, an oxygen-sensing probe was assembled to determine the activity of oxygen in iron solution, for which Cr (99.99%) and Cr$_2$O$_3$ (spectral purity) were used as the reference electrode, and molybdenum wire and cermet rods served as the reference and tested electrode lead wire.

### 6.5 Rare Earth Solid Electrolytes at High Temperature

For the standard free energy of $Cr_2O_3$, based on an analysis of some experimental data by researchers of the Japan Society for the Promotion of Science (JSPS), it was found that three studies yielded very similar figures, namely those of Mazandarany and Pehlke ($ThO_2$-$Y_2O_3$ as solid electrolyte), Jeannin and Richardson et al. ($H_2O$ mixture as reference electrode), and Jacob ($ThO_2$–$Y_2O_3$ as solid electrolyte). It was concluded that these experimental methods were scientifically credible, so the average value of these three data is recommended as (Japan Steel Sensors Commission 1989):

$$\Delta G^{\ominus}Cr_2O_3 = -(1,115,450 + 1,115,870 + 1,115,960)/3$$
$$+ (250.12 + 250.83 + 250.37)T/3$$
$$= -1,115,750 + 250.45T \pm 1255 \ (J/mol)$$

(temperature ranges are 877–1267 °C, 1027–1327 °C, 800–1199 °C, respectively).

According to the rules of using data, they can be used epitaxially as long as there is no phase transition in the high temperature section. Here, the $p_{O_2}$ value of the equilibrium with $[O]_{metal\ melt}$ indicates the chemical position of oxygen in melts. Free energy of dissolution data for oxygen in carbon-saturated iron solution has not been reported. The La activity in three different iron melts has been determined by La sensors, as briefly described below.

(1) Determination of La activity in carbon-saturated iron

Pure Fe (> 99.95%) is placed in a spectroscopically pure graphite crucible coated with such powder, and pretreated, melting in a high purity Ar of an $MoSi_2$ furnace with a sealed water jacket at 1450 °C. La is introduced into the melt in a quartz tube pre-filled with Ar. A high resistance (> $10^9$ $\Omega$) digital voltmeter interfaced to a computer is used to measure and record the EMF values of the two sensors.

Experiments have shown that the sensor response time is about 10 s. La activity and oxygen activity change regularly according to the La addition as the inner $p_{O_2}$ value in the furnace tube is very low. With increasing La content in the melt, $a_{La}$ increases and $p_{O_2}$ decreases.

(2) Determination of La activity in cast iron solution

Experiments are carried out in a 50 kg, 1000 kHz medium frequency induction furnace equipped with a magnesia crucible, using industrial raw iron at 1450 °C, with the furnace mouth open, that is, no protective atmosphere. The sensor reflects the La activity changes at the surface level after its portionwise addition; the activity increases as the La gradually melts. The adding rare earths by input or insertion method, the yield is still very low, and produce more rare earth oxides and oxygen sulfide, but after several applications, the probe remains intact.

(3) Determination of La activity in ductile liquid iron on a large scale

Field tests have been carried out in a large cast pipe plant with a weight of about 2.3 t per packet of molten iron and a temperature of 1300 ± 10 °C. Results have shown the rare earth to be mainly concentrated near the surface layer, making it easy to burn or oxidize, due to the influences of density, solubility, and kinetic diffusion rate. Previously reported low recovery rates may be ascribed to improper methods of adding the rare earths, the absence of a protective atmosphere, and process discontinuity. Ideally, rare earths should be added to molten metals in the form of "bullets", that is, wrapped in aluminum foil.

3. $LaF_3(CaF_2)$ lanthanum sensor

Researchers have prepared $La_{0.95}Ca_{0.05}F_{2.95}$ solid-electrolyte tubes by an isostatic pressing method and subjected them to burning under strict control of low $p_{O_2}$. $LaNi_5$-Ni alloy was used as a reference electrode. According to the La-Ni phase diagram, there are two phase regions in which Ni and $LaNi_5$ co-exist at temperatures below 1250 °C. Here, La chemical sites are only a function of temperature. That is to say, $LaNi_5$ alloy can provide a constant La position at a certain temperature (Xiao et al. 1994). Therefore, $LaNi_5$-Ni alloy is selected as one of the reference electrodes for a rare earth sensor, and Mo wire is used as the electrode lead wire to form the following battery to determine the La activity in molten aluminum.

Battery: $Mo|LaNi_5, Ni|La_{0.95}Ca_{0.05}F_{2.95}|[La]_{Al}|Mo$.
Reference pole: $LaNi_5 + 3F^- - 3e = [LaF_3]_{electrolyte} + 5Ni$.
Tested pole: $[LaF_3]_{electrolyte} + 3e = [La]_{Al} + 3F^-$
Battery reaction: $LaNi_5 = [La]_{Al} + 5Ni$.
The free energy of the battery reaction changes to:

$$\Delta G = \Delta G^{\ominus} + RT \ln \alpha_{La} = -3FE_{La}$$

Because the experimental temperature is higher than the melting point of La, pure liquid La is taken as the standard state, and $\Delta G^{\ominus}$ in the upper formula becomes:

$$\Delta G^{\ominus} = -\Delta G^{\ominus}_{(LaNi_5)}$$

From the above expressions, we obtain:

$$\alpha_{La} = \exp\left[\left(-3FE_{La} + \Delta G^{\ominus}_{(LaNi_5)}\right)/RT\right]$$

Xiao and Wang (1994) have measured:

$$\Delta G^{\ominus}_{(LaNi_5)} = -152,590 + 13.143T \pm 150 \text{ (J/mol)}$$

Therefore, provided that $E_{La}$ at a certain temperature can be measured, $\alpha_{La}$ can be calculated from the above formula, allowing $\gamma_{La}$ to be calculated from the relationship $\alpha_{La} = \gamma_{La} \cdot x_{La}$. Because there is an equilibrium relationship for deoxygenation by La, $2[La]_{Al} + 3[O]_{Al} = La_2O_{3(s)}$, and $\alpha_{La}$ and $\alpha_o$ in Al solution obey a deoxygenation constant $K = \alpha_{La}^2 \alpha_O^3$ relationship, oxygen in the Al solution is simultaneously determined by an oxygen-sensing probe.

## 6.5 Rare Earth Solid Electrolytes at High Temperature

La and Ni were melted in the water-cooled copper crucible of a magnetically controlled vacuum micro arc furnace according to $m_{La}:m_{Ni} = 7:93$, at which LaNi$_5$-Ni alloy is prepared according to the La-Ni phase diagram. After grinding the alloy under toluene protection in a small prototype, a sensor probe was assembled after XRD verification of the obtained phase.

The experiment was carried out at 750 °C in a corundum tube graphite crucible (spectroscopic purity) with a sealed water jacket under the protection of high-purity Ar. La was tightly wrapped with aluminum foil and added to the molten pool through the quartz tube at constant temperature. An La- and oxygen-sensing probe was then inserted into the molten pool to determine the respective sensing electromotive force values.

The two probes were automatically withdrawn under computer control and then lifted above the melt after a certain time; the experiment was stopped after repeated addition of La, for several times until the $E_{La}$ and $E_0$ changes became obvious. The response time of the lanthanum and oxygen sensor was 3–5 s, the measured value can be continuously stable for more than 2 min, and then the EMF value to decay. After inserting again, the original stable EMF value can be obtained. After the experiment, the La sensor probe was observed to be intact without cracks.

The La content in the dissolved state in the sample was separated by low-temperature water-free electrolysis, and plasma spectrum analysis was used to confirm the relationships o $E_{La}$-[La] and $E_0$-[La], as shown in Fig. 6.2, and the resulting relationship between ln $\alpha_{La}$, ln$\gamma_{La}$, and $x_{La}$.

The results for the determination of lanthanum and oxygen by the sensor are regular and the curve appears extreme value, which in accordance with the law of active metal deoxygenation. With the increase of La concentration in the melt, the interaction law of lanthanum-oxygen atom changes. Hence, the divalent interaction coefficient and divalent interpolation interaction coefficient should be considered in the calculation through application of the Taylor expansion.

According to the general thermodynamic calculation method, $\ln \gamma_{La}^0 = -14.35$ is obtained by extending the $\gamma_{La}$ and $x_{La}$ relationship curve to $x_{La} = 0$. When $\gamma_{La}^0 = 5.85 \times 10^{-7}$, corresponding to a 1% [La]$_{Al}$ solution as standard state, we obtain:

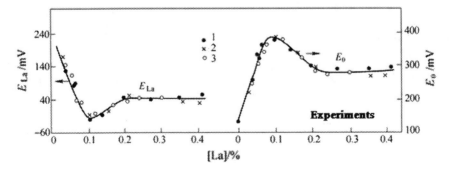

**Fig. 6.2** Relationship between $E_{La}$-[La] and $E_0$-[La]

100      6  Solid Electrolytes Based on Rare Earth Oxides and Fluorides

$$La_{(s)} = [La]_{1\%}$$

The standard free energy of dissolution of La in Al is:

$$-\Delta G_{La}^{\ominus} = RT \ln(\gamma_{La}^0) \cdot M_{Al}/100M_{La}) = -175.2 \ (kJ/mol)$$

Experimentally, $x_{La} < 8 \times 10^{-4}$, that is, the concentration of the dissolved state is very low, so the Al-La melt can be regarded as a regular solution. Consequently:

$$\varepsilon_{La}^{La} = -2 \ln \gamma_{La}^0 = 26.7$$

$$e_{La}^{La} = (M_{Al}/230M_{La}) \cdot \varepsilon_{La}^{La} = 0.024$$

These data are not reported in the literature, since La and Ce are similar lanthanides; the atomic number of Ce is only one larger than that of La, and their thermodynamic properties are very similar. Du et al. (1995) compared data obtained by the chemical equilibrium method for the Al-Ce system:

$$\gamma_{Ce}^0 = 7.64 \times 10^{-6} \ (700 \, ^\circ C)$$

Dewing et al. obtained:

$$\gamma_{Ce}^0 = 3.89 \times 10^{-8} (700 \, ^\circ C)$$

by a chemical equilibrium method.
  Wang CZ et al. obtained:

$$\gamma_{La}^0 = 5.85 \times 10^{-7} \ (750 \, ^\circ C)$$

Thus, these values are similar. Using different research methods and reagents of different purities, the results will inevitably be different. From the above, the lanthanum-sensing method used in this study may be deemed reliable.

4. Use of the lanthanum-sensing method to prove the existence of iron solution and solid-solution rare earths after solidification

In order to prove the existence of solid-solution rare earths, Wang and Wang (1996) studied the relationship between the activity ($\alpha_{La}$) and temperature for lanthanum during the solidification of molten iron. The lanthanum sensor was made of $LaF_3$-doped $CaF_2$ electrolyte as described above, simultaneously preparing an oxygen sensor. The battery form was:

$$Mo|La_{(l) \text{ or } (s)}|La_{0.95}Ca_{0.05}F_{2.95}|[La]_{\text{Carbon-saturated iron}}|Mo \text{ metal ceramic}$$

**Fig. 6.3** Relationships between $\alpha_{La}$-[La] and $p_{O_2}$-[La]

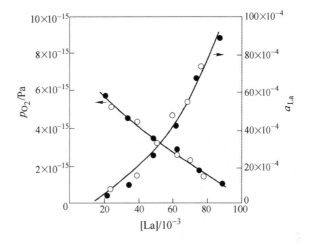

The equilibrium experiment was carried out at 1300 °C, and $E_{La}$ and $E_0$ were measured at different La concentrations. The experimental results are shown in Fig. 6.3, in which different symbols represent different groups of experimental data. The sensor probe response time was 2–3 s, allowing for continuous measurement until it was lifted. Since the oxygen partial pressure in the furnace was very low after introducing high purity Ar, the concentration of Law as calculated according to the amount added.

Experimental results concerning the solidification process of iron solution are drawn in figure, relationship between lg $\alpha_{La}$ solidification process and temperature. The solidification temperature of carbon-saturated iron solution is 1147 °C, and the lanthanum sensor remained responsive during and after the solidification until about 400 °C. After the experiment, inspection of the electrolyte tube showed that it was not corroded and the surface was smooth.

5. Lanthanum-sensing method to prove the presence of aluminum solution and solid-solution rare earth after solidification

Based on the above-mentioned battery form and experimental process, researchers studied the relationship between La activity, $p_{O_2}$, and temperature in the solidification process of liquid aluminum. The results are drawn in figures (omitted here).

6. YF$_3$(CaF$_2$) solid electrolyte yttrium sensor

Some provinces in southern China are rich in yttrium-based heavy rare earth ore. Research has shown that the addition of trace yttrium to aluminum can generate superior materials to those obtained by lanthanum addition. Based on a conductivity study of YF$_3$(CaF$_2$), YF$_3$ doped with 25 mol.% CaF$_2$ was selected as the solid electrolyte for a yttrium sensor. The electrolyte tube was fired at low oxygen partial pressure, and appeared bright and ceramicized.

The battery form of the yttrium sensor was:

$$Mo|Y_{(s)}|Y_{0.75}Ca_{0.25}F_{2.75}|[Y]|Mo$$

Reference pole: $Y + 3F^- - 3e = [YF_3]_{electrolyte}$
Pole to be tested: $[YF_3]_{electrolyte} + 3e = [Y] + 3F^-$
Battery reaction: $Y = [Y]$

$$\Delta G = \Delta G^\ominus + RT \ln \alpha_Y = -3FE_Y$$

If pure solid Y is taken as the standard state, then $\Delta G^\ominus = 0$. Substituting in the above formula, we obtain:

$$\alpha_Y = \exp\left(\frac{-3FE_Y}{RT}\right)$$

Similarly, yttrium was added to an aluminum solution to form a yttrium–aluminum alloy. The sensor response time was 3–5 s. During cumulative addition of yttrium in a yttrium–aluminum alloying process, the yttrium sensor could be used continuously. After use, it was seen to be intact. The results are shown in A study on the nonstoichiometry of tin oxides by coulometric titration.

Both the $\ln \alpha_Y - [Y]$ and $\lg p_{O_2} - [Y]$ diagrams (omitted here) show maxima and minima, indicating that yttrium addition to liquid aluminum follow the law of active metal deoxygenation. As yttrium is added, the electric field action of microscopic particles is modified, that is, the interaction relationships of yttrium atoms, oxygen atoms, and between yttrium and oxygen atoms are changed. Following the Taylor expansion relationship, it is necessary to consider the relationships between second and third-order interactions, as well as cross-interactions. It is the law of interatomic interactions in universal solutions.

## 6.6 Interaction Between Element Atoms in Metal Melts

The contents of the above sections all relate to an overall rule, that is, there are interactions between element atoms, for example in a molten metal. This may be illustrated through Taylor series expansion.

At a certain temperature, the activity of dissolved oxygen in a matrix metal is related to its concentration, but is also affected by the oxygen affinity and content of added alloying elements. Oxygen atoms and different alloying element atoms may interact, that is, attract or repel, and each element atom itself may engage in interactions, so the thermodynamic properties of the elements can be affected. A metal melt belongs to the broad category of solutions, and so the interaction between element atoms may be discussed by applying the general concepts of solutions.

The effect of the interaction between element atoms in solution on the thermodynamic properties of the element itself can be expressed in terms of a polynomial, where the variable is usually the mole fraction of solute. For example, the Taylor

## 6.6 Interaction Between Element Atoms in Metal Melts

series expansion of the partial molar excess free energy or activity coefficient of a solute in solution is expressed as:

$$\frac{G_2^E}{RT} = \ln \gamma_2 = \ln \gamma_2^0 + \left(\frac{\partial \ln \gamma_2}{\partial x_2}\right)_{x_2 \to 0} x_2 + \frac{1}{2}\left(\frac{\partial^2 \ln \gamma_2}{\partial x_2^2}\right)_{x_2 \to 0} x_2^2 + \cdots$$
$$+ \frac{1}{n!}\left(\frac{\partial^n \ln \gamma_2}{\partial x_2^n}\right)_{x_2 \to 0} x_2^n + \cdots \tag{6.1}$$

The zero valence term $\ln \gamma_2^0$ is the $\ln \gamma_2$ value in an infinitely dilute solution, and for a certain element it has a fixed value, which is a thermodynamic property. The first-order term is the first-order interaction coefficient, and using the symbols $\varepsilon_i^i$ and $\varepsilon_i^j$ it is defined as:

$$\varepsilon_i^i = \left(\frac{\partial \ln \gamma_i}{\partial x_i}\right)_{x_i \to 0} \tag{6.2}$$

$$\varepsilon_i^j = \left(\frac{\partial \ln \gamma_i^j}{\partial x_i}\right)_{x_i \to 0, x_j \to 0} \tag{6.3}$$

In Eqs. 6.2 and 6.3, a positive value of $\varepsilon$ indicates that the element atoms repel one another so that the activity coefficient increases, that is, the activity increases. Otherwise, a negative value of $\varepsilon$ indicates that the element atoms attract one another and the activity coefficient decreases.

Lupis CHP and Elliott JF proposed coefficients for second-order and third-order interactions, expressed by the symbols $\rho$ and $\tau$, respectively. For a multi-system solution, component 1 is set as the solvent, and components 2, 3, 4, ... are solutes. At a certain temperature, the Taylor expansion of activity coefficient of component 2 takes the following form:

$$\ln \gamma_2 = \ln \gamma_2^0 + \left(\frac{\partial \ln \gamma_2}{\partial x_2}\right)_{x_1 \to 1} x_2 + \left(\frac{\partial \ln \gamma_2}{\partial x_3}\right)_{x_1 \to 1} x_3 + \left(\frac{\partial \ln \gamma_2}{\partial x_4}\right)_{x_1 \to 1} x_4$$
$$+ \frac{1}{2}\left(\frac{\partial^2 \ln \gamma_2}{\partial x_2^2}\right)_{x_1 \to 1} x_2^2 + \frac{1}{2}\left(\frac{\partial^2 \ln \gamma_2}{\partial x_3^2}\right)_{x_1 \to 1} x_3^2 + \frac{1}{2}\left(\frac{\partial^2 \ln \gamma_2}{\partial x_4^2}\right)_{x_1 \to 1} x_4^2$$
$$+ \cdots + \left(\frac{\partial^2 \ln \gamma_2}{\partial x_2 \partial x_3}\right)_{x_1 \to 1} x_2 x_3 + \left(\frac{\partial^2 \ln \gamma_2}{\partial x_2 \partial x_4}\right)_{x_1 \to 1} x_2 x_4 + \cdots$$
$$+ \frac{1}{6}\left(\frac{\partial^3 \ln \gamma_2}{\partial x_2^3}\right)_{x_1 \to 1} x_2^3 + \frac{1}{6}\left(\frac{\partial^3 \ln \gamma_2}{\partial x_3^3}\right)_{x_1 \to 1} x_3^3 + \frac{1}{6}\left(\frac{\partial^3 \ln \gamma_2}{\partial x_4^3}\right)_{x_1 \to 1} x_4^3$$
$$+ \cdots \tag{6.4}$$

For each higher order term, the coefficients of $x$, $\gamma^0$, and the first-order interactions are defined as described above:

$$\frac{1}{2}\left(\frac{\partial^2 \ln \gamma_2}{\partial x_2^2}\right)_{x_1 \to 1} = \rho_2^2; \quad \frac{1}{2}\left(\frac{\partial^2 \ln \gamma_2}{\partial x_3^2}\right)_{x_1 \to 1} = \rho_2^3$$

$$\frac{1}{2}\left(\frac{\partial^2 \ln \gamma_2}{\partial x_4^2}\right)_{x_1 \to 1} = \rho_2^4; \quad \left(\frac{\partial^2 \ln \gamma_2}{\partial x_2^2 \partial x_3}\right)_{x_1 \to 1} = \rho_2^{(2,3)}$$

$$\left(\frac{\partial^2 \ln \gamma_2}{\partial x_2 \partial x_4}\right)_{x_1 \to 1} = \rho_2^{(2,4)}; \quad \frac{1}{6}\left(\frac{\partial^3 \ln \gamma_2}{\partial x_2^3}\right)_{x_1 \to 1} = \tau_2^2$$

$$\frac{1}{6}\left(\frac{\partial^3 \ln \gamma_2}{\partial x_3^3}\right)_{x_1 \to 1} = \tau_2^3; \quad \frac{1}{6}\left(\frac{\partial^3 \ln \gamma_2}{\partial x_4^3}\right)_{x_1 \to 1} = \tau_2^4$$

Thus, Eq. 6.4 can be written as follows:

$$\ln \gamma_2 = \ln \gamma_2^0 + \varepsilon_2^2 x_2 + \varepsilon_2^3 x_3 + \varepsilon_2^4 x_4 + \cdots + \rho_2^2 x_2^2 + \rho_2^3 x_3^2 + \rho_2^4 x_4^2$$
$$+ \cdots + \rho_2^{(2,3)} x_2 x_3 + \rho_2^{(2,4)} x_2 x_4 + \cdots + \tau_2^2 x_2^3 + \tau_2^3 x_3^3 + \tau_2^4 x_4^3 + \cdots \quad (6.5)$$

Here, $\rho_2^2$, $\rho_2^3$, and $\rho_2^4$, are the second-order interaction coefficients; $\rho_2^{(2,3)}$ and $\rho_2^{(2,4)}$ are the second-order cross-interaction coefficients; and $\tau_2^2$, $\tau_2^3$ and $\tau_2^4$, are the third-order interaction coefficients.

The above forms also apply to the activity coefficients of components 3 and 4. Generalizing the above, for the M-$i$-$j$-$k$...$m$ system of a multi-system solution, Eq. 6.5 can be written as:

$$\ln \gamma_i = \ln \gamma_i^0 + \sum_{j=2}^{m} \varepsilon_i^j x_j + \sum_{j=2}^{m} \rho_i^j x_j^2 + \sum_{j=2}^{m} \sum_{k=2}^{m} \rho_i^{j,k} x_j x_k + \sum_{j=2}^{m} \tau_i^j x_j^3 + \cdots$$

Mass fraction, as opposed to molar fraction, is commonly used to represent the compositions of real solutions; $f$ is used as the activity coefficient and the symbols $e$ and $r$ are used to represent the first- and second-order interaction coefficients, respectively. The interaction coefficient $e_i^i$ is defined as:

$$e_i^i = \left(\frac{\partial \lg f_i}{\partial [i]}\right)_{[i] \to 0} \quad (6.6)$$

Here, $i$ denotes the second element in the liquid medium besides the matrix metal. To obtain $e_i^i$, $\lg f_i$ is plotted against $[i]$, the curve is extended to $[i] = 0$ to draw a tangent extrapolating to an infinitely dilute solution, and then the slope of the tangent is obtained.

For an m-i-j ternary system:

$$f_i = f_i^i f_i^j$$
$$\lg f_i = e_i^i [i]$$

When the $j$ content is different, a different $f_i$ is obtained; $f_i^j$ is then determined, and ultimately $\lg f_i^j$:

$$e_i^j = \left(\frac{\partial \lg f_i^j}{\partial [j]}\right)_{[i]\to 0,[j]\to 0} \tag{6.7}$$

The same method as for a binary system can be applied, as, for example, for an $Fe - i - j - k \ldots m$ system. The expression is:

$$\lg f_i = \sum_{j=2}^{m} e_i^j [j] + \sum_{j=2}^{m} \gamma_i^j [j]^2 + \sum_{j=2}^{m}\sum_{k=2}^{m} \gamma_i^{j,k}[j][k] + \cdots \tag{6.8}$$

At a 1% mass fraction of the solute is used as the standard state, the $f$ used is the activity coefficient following Henry's law, and $f_i^0 = 1$, $\lg f_i^0 = 0$, so there is no $\lg f_i^0$ term in Eq. 6.8, all the above discussions are free-energy interaction coefficients.

The enthalpies and entropies of components in the solution can also be expanded by the Taylor series to determine first- and higher order interaction coefficients. The symbols and interrelationships of the various interaction coefficients already specified have been summarized into table, the items include order, free energy interaction coefficient $x/\%$, thermal enthalpy interaction coefficient $x/\%$, and entropy interaction coefficient $x/\%$. The relationship between the three interaction coefficients can be deduced from that between free energy, enthalpy, and entropy. For example:

$$e_i^j = \frac{h_i^j}{2.3RT} - \frac{S_i^j}{2.3R} \tag{6.9}$$

Researchers have also extensively studied oxygen activities in molten metals containing alloying elements.

# References

Delcet J, Heus RJ, Egan J (1978) Electronic conductivity in solid CaF$_2$ at high temperature. J Electrochem Soc 125(5):755

Du T, Han QY, Wang CZ (1995) Physical chemistry of rare earth alkali soil and its application in materials. Science Press, Beijing

Fedorov PP, Turkina TM, Sobolev BP (1982) Ionic conductivity in the single crystals of non-stoichiometric fluorite phases M$_{1-x}$R$_x$F$_{2+x}$ (M=Ca, Sr, Ba; R=Y, La-Lu). Solid State Ionics 6:331

Fischer WA, Janke D (1991) Metallurgical electrochemistry (trans: Wu XF). Northeast Institute of Technology Press, Shenyang

Frenkel J (1930) On the Electrical Resistance of Contacts between Solid Conductors. Phys Rev 36:1604

Gao L, Yan DS, Guo JK (1988) The effect of $ZrO_2$ particle size on phase change toughening in Y-TZP ceramics. Chin Sci 1A:95

Hagenmuller P (1984) Solid electrolytes, general principles, characteristics, materials and applications (trans: Chen LQ et al). Science Press, Beijing

Hildrum R, Brustad M, Wang CZ et al (1994) Thermodynamic properties of doped lanthanum manganites. Mater Res Bull 29(8):851

Huang K, Wang CZ, Xu XG (1992) Activity of $Bi_2O_3$ in $Bi_2O_3$-$Y_2O_3$ oxygen ion conductor. J Solid State Chem 98:206

Janke D, Fischer WA (1977) Physikalisch-chemischen eigenschaften oxid keramischer festektrolyte. Arch Eisenhüttenwes 48(5):255

Japan Steel Sensors Commission (1989) A new development of steel sensors—solid electronic sensors as the center. Japan Society for the Promotion Science, the 19th committee steel sensor subcommittee, Tokyo, pp 4151–4153

Lange FF (1982) Transformation toughening (Part 1–5). J Mater Sci 17:225

Masaki T (1986) Mechanical properties of $Y_2O_3$-stabilized tetragonal $ZrO_2$ polycrystals after ageing at high temperature. J Am Ceram Soc 69(7):519

Meng GY, Zhou M, Peng DK (1986) A new phenomenon—the inductive impedance in $Bi_2O_3$ based oxygen ionic conductors. Solid State Ionics 18–19:756

Mori T, Drennan J, Lee JH et al (2002) Improving the ionic conductivity of yttria-stabilised zirconia electrolyte materials. Solid State Ionics 154–155:529–533

Pastor RC, Pastor AC, Miller KT (1974) Congruently melting compounds of $CaF_2 \cdot r_c RF_3$: Part II. Mater Res Bull 9:1253–1259

Porter DL, Heuer AH (1977) Mechanisms of toughening partially stabilized zirconia (PSZ). J Am Ceram Soc 60(3–4):183

Rapp RA, Shores DA (1970) Solid electrolyte galvanic cells. In: Rapp RA (ed) Physicochemical measurements in metals research. Part II. Interscience Publishers, New York, London, Sydney, Toronto, pp 124–186

Reau JM, Lucat C, Campet G (1976) Application du tracé des diagrammes d'impédance complexe à la détermination de la conductivité ionique des solutions solides $Ca_{1-x}Y_xF_{2+x}$: Corrélations entre propriétés electriques et structurales. J Solid State Chem 17:123

Schottky W (1935) Z Physik Chem B29:353

Sheng XM, Xu J (1985) Toughening effect of fine dispersed $ZrO_2$ and its application in $Si_3N_4$. J Silicate 13(3):364

Sher A, Solomon R, Lee K (1966) Transport properties of $LaF_3$. Phys Rev 144(2):593

Sobolev BP, Fedorov PP, Seiranian KB et al (1976) On the problem of polymorphism and fusion of lanthanide trifluorides. II. Interaction of $LnF_3$ with $MF_2$ (M = Ca, Sr, Ba), change in structural type in the $LnF_3$ series, and thermal characteristics. J Solid State Chem 17:201

Subbarao EC (1980) Solid electrolytes and their applications. Plenum Press, New York

Takahashi T, Iwahara H, Arao T (1975) High oxide ion conduction in sintered oxides of the system $Bi_2O_3$-$Y_2O_3$. J Appl Electrochem 5:187

Takahashi T, Iwahara H, Arao T (1976) Electrical conduction in the sintered oxides of the system $Bi_2O_3$-BaO. J Solid State Chem 16:317

Wagner C, Schottky W (1930) Z phys Chem B11:163

Wang CZ (2004) Study on thermodynamics of chemical sensors and materials using solid electrolytes e. m. f. method. In: Proceedings of the 12th China conference on solid-state ionology. Rare Earth, vol 10, p 1

Wang CZ, Xu XG (1984) Ion conductance and electron conductance of $Bi_2O_3$-$Y_2O_3$ series high oxygen ion conductor. J Phys 33(2):221

Watanabe A, Kikuchi T (1986) Cubic-hexagonal transformation of yttria-stabilized $\Sigma$-bismuth sesquioxide, $Bi_{2-2x}Y_{2x}O_3$ (x = 0.215 − 0.235). Solid State Ionics 21:287

Wang P, Wang CZ (1996) Study on the activity of La sensor in the determination La liquid aluminum solidification. J Phys Chem 12(3):272–275

# References

Wang P, Wang CZ, Xu XG (1997) The application of a lanthanum sensor in the investigation of lanthanum activity in carbon saturated iron liquid solution and its solidification. Solid State Ionics 99:153

Xiao LS, Li GQ, Sui ZT, Wang CZ (1994) Study on impedance spectrum of solid electrolyte $La_{(1-x)}Ca_xF_{(3-x)}$. J Silicate 22(6):553–558

Xiao LS, Sui ZT, Wang CZ (1993) Study of dissolved state La activity in Al solution. Acta Metall Sin 29(8):49–54

Xiao LS, Yu HL, Wang CZ et al (1994) Study on $LaNi_5$ and $CeNi_5$ standard free energy of formation. J Chin Rare Earth Soc 12(1):15

Xu XG, Wang CZ, Li GG et al (1992) In: The sixth Japan-China symposium on science and technology of iron and steel, Chiba, Japan, Japan Iron Steel Association, pp 26–31

Suzuki Y (1996) Activation energy for electrical conduction of $Y_2O_3$-stabilized $ZrO_2$ containing 8 mol% $Y_2O_3$. Solid State Ionics 91:239

Zou KY, Wang CZ, Zhao NR (1995) Lanthanum sensor and activity of lanthanum in carbon saturated iron. Acta Metall Sin 31(17):195–199

# Chapter 7
# High-Temperature Proton Conductors Containing Rare Earths

## 7.1 Brief Introduction

High-temperature proton conductors were first reported in the 1980s, and have since been extensively researched. Iwahara et al. (1986, 1988) and Chandra (1988) focused on perovskite-type oxides. They studied the proton conductivity of $SrCe_{0.95}Yb_{0.05}O_{3-\alpha}$ in an atmosphere containing hydrogen or water vapor at 600–1000 °C. Such materials can serve as solid electrolytes to form primary batteries operable under both dry and humid air or oxygen (Shao et al. 2004). The battery electromotive force (EMF) generated can be exploited in chemical sensors. The use of such solid electrolytes to form fuel cells can generate electricity. Hydrogen can be precipitated from **an** electrolytic cell.

$SrCe_{1-x}M_xO_{3-\alpha}$ ($x = 0.05$–0.10) systems are prepared by doping $SrCeO_3$ with oxides of Y, Ga, Sc, Sm, Dy, or trivalent metals such as In. These materials are proton conductors in the presence of hydrogen or water vapor at about 600 °C. They become proton and $O^{2-}$ mixed conductors at above 1000 °C (Yajima et al. 1991). $BaCeO_3$-, $CaZrO_3$-, and $SrZrO_3$-based materials show similar properties (Iwahara 1995). Doped $SrCeO_3$ and $BaCeO_3$ matrices have high conductivities of the order of $10^{-3}$–$10^{-2}$ S/cm at 600–1000 °C.

Further studies revealed $SrCeO_3$- and $BaCeO_3$-based materials to be unstable in $CO_2$ atmosphere and below 800 °C, decomposing to $SrCO_3$ or $BaCO_3$ and $CeO_2$, and showing poor material strength. By adding 10 mol% $In_2O_3$ or $Sc_2O_3$ to $CaZrO_3$, $CaZr_{0.9}In_{0.1}O_{3-\alpha}$, and $CaZr_{0.9}Sc_{0.1}O_{3-\alpha}$ proton conductor were prepared. Although their proton conductivities are lower than those of the previously described materials, they have better stability and strength. Accordingly, the Japanese company TYK produces a $CaZr_{0.9}In_{0.1}O_{3-\alpha}$ proton conductor sensor for the on-line and rapid detection of hydrogen in liquid aluminum (Yajima et al. 1992). Chen et al. (1995) verified the applicability of such sensors for testing hydrogen in both static and flowing liquid aluminum.

Iwahara (1992) have focused on much research interest in relation to ion conductors and their applications. Researchers have studied high-temperature proton conductors from the perspectives of preparation, properties, proton conduction mechanism, and applications, as elaborated in the following sections.

© Science Press 2023
C. Wang, *Theory and Application of Rare Earth Materials*,
https://doi.org/10.1007/978-981-19-4178-8_7

## 7.2 Preparation of High-Temperature Proton Conductors

High-temperature proton conductors can be classed as functional ceramics, and so the preparation methods for ceramics can be applied, such as direct synthesis, co-precipitation, sol–gel methods, and so on. Direct synthesis by solid-state reaction is the most commonly used method. The process involves initial drying of analytically pure raw materials. These are then pulverized to micron size using agate or $ZrO_2$ (stabilized) tanks and balls in anhydrous ethanol, molded, and finally subjected to high-temperature synthesis.

The synthesized material is crushed to an average particle size of about 2 $\mu$m, formed into the desired shape according to the final use, and finally fired. Thermo-dynamically, the densification process of a sintered billet at high temperature is a process of reducing surface energy. The mass can be characterized by shrinkage, porosity, bulk density, or strength. Ca, Sr, and Ba oxides readily absorb water and $CO_2$, and so when materials containing these elements are prepared, carbonates are formed, the thermal decomposition of which produces oxides, which are reactive and readily combine with other raw materials (Yusa 2018).

The study of high-temperature proton conductors belongs to the field of defect physics. The research methods used are based on the effects of the action of elec-tromagnetic waves of different energies or high-speed electrons or particles. Lattice parameters, and the correlation, conductivity, and chemical shift of moving ions and skeleton ions are determined by X-ray absorption fine structure (EXAFS) analysis. The rotational characteristics of O–H are studied by infrared spectroscopy. The struc-tural characteristics of similar atoms are identified by neutron diffraction, and the migration motion mode of hydrogen ions is assessed by proton conduction.

The microstructure and morphology of ionic conductors are analyzed by scan-ning electron microscopy with an energy spectrum analyzer attachment. Secondary electron and backscatter electron imaging and section energy spectrum point anal-ysis are carried out to determine compactness, uniformity, and the segregation and loss of elements during the preparation. AC impedance spectroscopy is widely used to determine the relationship between ion conductivity and temperature in different hydrogen partial pressures and water vapor partial pressures. For the determination of positive holes or electronic conductivity in a proton conductor, the direct current (DC) polarization method can be used. According to the conductivity mediated by ions, electropositive holes, and electrons, the fraction of the total conductivity of each component can be obtained, that is, the transference number.

Differential heat and differential weight analysis can be used to determine the thermal regime of material synthesis, and to determine water vapor absorption and condensation. Iwahara and co-workers studied the process and quantity of water vapor condensed in a proton conductor during heating by a temperature-programmed desorption method in order to calculate the concentration of hydrogen in the sample.

$\Delta G^{\ominus}$ values of doped perovskite-type proton conductors have not been deter-mined.

## 7.3 Causes of Proton Conduction in Perovskite Materials

Compounds with the natural perovskite ($CaTiO_3$) structure and the chemical formula $ABO_3$ are commonly referred to as perovskites, where A represents a $+ 2$ valence cation, such as $Ca^{2+}$, $Sr^{2+}$, or $Ba^{2+}$, and B represents a $+ 4$ valence cation, such as $Ce^{4+}$ or $Zr^{4+}$.

An ideal perovskite-type structure has a cubic crystal system, but the lattices of many perovskite crystals are distorted. The tolerance factor formula is usually used to indicate the degree of distortion of a perovskite structure.

$$t = \frac{(r_A + r_O)}{\sqrt{2}(r_B + r_O)}$$

where $r_A$, $r_B$, and $r_O$ represent the ionic radii of A, B, and O, respectively. A high value of $t$ precludes synthesis of a material. The intrinsic lattice of a perovskite does not contain protons, but these are introduced from external water vapor or hydrogen. Taking $In^{3+}$-doped $CaZrO_3$ as an example, protonation of the oxide in the presence of water vapor or hydrogen is illustrated in the following. The $In^{3+}$ dopant occupies $Zr^{4+}$ positions and produces oxygen vacancies, $V_{\ddot{O}}$:

$$In^{3+} \rightarrow In'_{Zr} + \frac{1}{2}V_{\ddot{O}}$$

Electron Holes $h^{\cdot}$ and oxide ions are formed by the reaction of $V_{\ddot{O}}$ with dry $O_2$.

$$V_{\ddot{O}} + \frac{1}{2}O_2 \rightarrow 2O_O^x + 2h^*$$

The following reactions occur in the presence of water vapor:

$$H_2O + 2h^* = 2H^+ + \frac{1}{2}O_2$$

$$H_2O + V_{\ddot{O}} = 2H^+ + O_O^x$$

$$V_{\ddot{O}} + \frac{1}{2}O_2 = 2h^* + O_O^x$$

In $H_2$ atmosphere:

$$H_2 + 2h^* = 2H^+$$

All of the above reactions are reversible, and each has an equilibrium constant at a certain temperature and $p_{H_2O}$ or $p_{H_2}$, which determines the equilibrium constant of the overall reaction. When protons are liberated according to the above equation, their concentration is twice the water vapor concentration, that is:

$$[H^+] = 2[H_2O]$$

The proton concentration increases with increasing external $p_{H_2O}$ and decreasing temperature, as corroborated by secondary ion mass spectrometry. When $p_{H_2O}$ is reduced to 0.002 atm (about 200 Pa) at 600 °C, the proton content is about 0.02 mol/L.

In a proton conductor, the migration kinetics of protons is related to the lattice parameters of the material. Münch et al. (1977) studied the transport mechanism of protons in materials such as $BaCeO_3$, $SrTiO_3$, and $BaZrO_3$ by quantum molecular dynamics, and found that the migration of protons is related to the distance between $O^{2-}$ and the amplitude of ions in the lattice. They estimated the magnitude of the proton migration barrier and delineated possible migration pathways in the crystals. $H^+$ and $O^{2-}$ form weak O–H hydrogen bonds, allowing for rotational motion. With $O^{2-}$ regularly arranged throughout the lattice, $H^+$ can transfer from one rotating O–H bond to another. $H^+$ and $O^{2-}$ form a new O–H, re-rotation, and then form a new O–H bond again. Such a cycle spread, and the longest time change is $10^{-11}$ s. At a constant temperature, the diffusion of protons throughout the whole sample reaches dynamic equilibrium. When a sensor or fuel cell is formed, in the presence of an electric field, protons will move in a directional motion and conduct electricity.

Infrared spectroscopic analysis is considered to be the best method for determining the presence of O–H bonds. Researchers have carried out infrared spectroscopic analysis of materials such as $CaZr_{0.9}In_{0.1}O_{3-\alpha}$ and $BaCeO_3$ and $BaZrO_3$ with different levels of doping, and observed the characteristic rotational absorption bands of O–H bonds at 3500–3700 cm$^{-1}$, implying continuous rotational motion of the O–H bonds formed at $O^{2-}$. Kruth et al. (2007) studied proton migration in $Ba_{1-x}La_xCe_{0.9-x}Y_{0.1+x}O_{2.95}$ materials using a high-resolution electron diffractometer combined with an atomic model, and obtained similar results.

## 7.4 Discussion on Some Perovskite Proton Conductors

1. Stability and strength of $SrCeO_3$ and $BaCeO_3$ materials

In synthetic processes involving the sintering of $SrCO_3$ and $BaCO_3$ as raw materials, the materials will inevitably undergo a variety of phase transitions. Each phase transition will affect the free energy of material formation, and so will affect the structure and strength of the product. For example, Sr undergoes a phase transition at 589 °C, melts at 770 °C, and is vaporized at 1027 °C. SrO undergoes a phase transition at 327 °C. $SrCO_3$ has phase transition at 924 °C and decomposes at 1172 °C; Ba undergoes phase transition at 370 °C and 710 °C, BaO phase transition is at 997 °C, and $BaCO_3$ phase transition is at 806 °C, 968 °C, and 1127 °C, respectively. However, $CeO_2$ is a stabilized oxide at room temperature.

# 7.4 Discussion on Some Perovskite Proton Conductors

The light lanthanide Ce has an outer electron arrangement of $4f^1 5d^1 6s^2$, such that its 4f and 5d layers have several unfilled orbitals. A series of non-stoichiometric cerium oxides between $CeO_2$ and $Ce_2O_3$ is formed, depending on the gas-phase $p_{O_2}$ and temperature. When doped $SrCeO_3$ or $BaCeO_3$ is used as the electrolyte for a fuel cell, one pole is in an oxidizing atmosphere and the other is in a reductive atmosphere. The microscopic state of the material varies with the environment, and this affects the lattice strength. Therefore, although doped $SrCeO_3$ and $BaCeO_3$ materials have high proton conductivities, they cannot be used as electrolytes for fuel cells.

2. $CaZrO_3$ incorporating In or Sc

The conductivities of $CaZr_{0.9}In_{0.1}O_{3-\alpha}$ and $CaZr_{0.9}Sc_{0.1}O_{3-\alpha}$ are about two orders of magnitude lower than the proton conductivities of $BaCeO_3$-based materials at the same temperature. Their proton conductivities at 800 °C in hydrogen-containing atmosphere are about $8 \times 10^{-4}$ S/cm, making them suitable for use in chemical sensors. Chemical sensor application relies on the EMF of the battery $E$, requires no current to pass through, and does not require protons to have very high conductivity.

Wang CZ's team synthesized $CaZr_{1-x}In_xO_{3-\alpha}$ ($x = 0, 0.05, 0.10, 0.15$) powders by the direct synthesis solid-state reaction method. After molding and sintering, the samples were examined under Ar atmosphere containing water vapor at 600–850 °C (Li et al. 2012). The relationships between proton conductivity and conductance activation energy and temperature were delineated. At 800 °C, the undoped conductivity was determined as $4.64 \times 10^{-7}$ S/cm. At $x = 0.05, 0.10$, and 0.15, the proton conductivities were $3.06 \times 10^{-4}$, $3.89 \times 10^{-4}$, and $3.93 \times 10^{-4}$ S/cm, respectively. These experiments showed $x = 0.10$ to be optimal. At $x > 0.10$, the increase in conductivity was within the error range.

This study also highlighted a difficulty in obtaining $CaZr_{0.9}In_{0.1}O_{3-\alpha}$ in the desired ratio due to In losses (п.и.фЕДоров and Акчурин 2000). This is because, above 1200 °C, $In_2O_3$ begins to dissociate into $In_2O$ and $O_2$. The relationship between dissociation pressure and temperature can be expressed as:

$$lgp_{O_2} = 8.49 - 6314/T(1050 - 1300\,°C)$$

$In_2O_3$ can be reduced to In metal by $H_2$, CO, or C at 700–800 °C. As a result, in the preparation of $CaZr_{0.9}In_{0.1}O_{3-\alpha}$ material, it should take some measures such as covering and proper leaving pores. Iwahara et al. (1991) reported $CaZr_{0.9}Sc_{0.1}O_{3-\alpha}$ with Sc as the dopant instead of In. This material showed high stability and strength, as we have corroborated. However, $Sc_2O_3$ is expensive due to the low abundance of Sc.

3. Other high-temperature proton conductors containing rare earths

Researchers strive for proton conductors that combine the advantages of high ionic conductivity and good material density. The report edoped $SrZrO_3$, $BaZrO_3$, and $CaTiO_3$ and other mixed doped materials have high proton conductivities at 500–1000 °C. Cherry et al. 1995; Davies et al. 1999; Islam et al. 2004; Wu et al. 2005 calculated the energy required for proton entry into perovskite crystals by quantum simulations of $CaZrO_3$, $BaZrO_3$, $LaMO_3$ (M = Sc, Ga), and Ca-doped $LaYO_3$. They also assessed the possible orientation of O–H motion in the materials, the energy barrier for hydrogen-ion migration between two adjacent oxygen ions and charge re-interference and the lattice relaxation associated with hydrogen migration (Wu et al. 2005; Kendrick et al. 2007). Their results showed $Y^{3+}$ to be the best dopant, and $Yb^{3+}$ the second best, for $BaZrO_3$. High-resolution neutron diffraction studies of $Ln_{1-x}A_xMO_{4-(x/2)}$ (M = Ta, Nb, P) showed proton conductivity to be dominant in moist air at below 700 °C. A proton conductivity of $1 \times 10^{-4}$ S/cm was measured at 500 °C. Ahmed et al. studied Yb-doped $BaZrO_3$ materials, and found that the hydration process began at less than 200 °C. Infrared absorption spectroscopic analysis clearly indicated the presence of O–H in the materials.

Iwahara et al. (1991) studied $BaCeO_3$-based materials, incrementally substituting Ce by Zr ($0 \leq x \leq 0.9$). At high temperature and high $p_{O_2}$, mixed conduction by protons, oxide anions, and electron vacancies was evident. With increasing Zr content, the stability of the system to $CO_2$ gradually increased, but the proton conductivity gradually decreased. Upon partially replacing the Ba in $BaCe_{0.9}Nd_{0.1}O_{3-a}$ with Ca, the proton conductivity is approximately halved.

4. Defect chemical model study of $SrCe_{0.95}Y_{0.05}O_{3-\alpha}$

Both the thermodynamic and migration properties of proton conductors are affected by their defect structures, and the properties of the defect structures are affected by external conditions, such as the partial pressures of oxygen and water vapor in the gas phase. Strontium cerate-and barium cerate-based oxides readily react with acidic or even amphoteric gases, such as $SO_3$, $CO_2$, or $H_2O$, to produce sulfate, carbonate, and hydroxide, respectively.

Song et al. (2002) studied a defect chemical model of cerite with high-temperature proton conduction under funding from the U.S. Department of Energy to delineate quantitative relationships between defect concentration and gas-phase $p_{O_2}$ and $p_{H_2O}$. They studied previously characterized $SrCe_{0.95}Y_{0.05}O_{3-\alpha}$ materials (i.e., non-stoichiometric $ABO_3$-type compounds). In this work, the Poulsen FW viewpoint was adopted, the C language procedure was used, and comparisons were made with other calculations and experimental results.

The algorithm can be briefly described as follows:

(1) Description by mass balance, position balance, electrical neutrality, balance of internal ions and electrons, balance of sample and atmosphere.
(2) Fixing a defect concentration within a possible physical range.
(3) If a solid is in equilibrium with two gases, the partial pressure of one gas is given.

Defect chemistry and simulation:

The defect chemistry of a material needs to be described by ten particles for the non-stoichiometric phase of $SrCe_{0.95}Y_{0.05}O_{3-\alpha}$ expressed in Kröger–Vinknotation as follows: A position normal cation $Sr_{Sr}^x$ and cation vacancy $V_{Sr}''$; B position normal cation $Ce_{Ce}^x$, cation vacancy $V_{Ce}''''$, and replacing cation $Y_{Ce}'$; O position normal oxygen $O_O^x$, oxygen vacancy $V_{\ddot{O}}$; proton on oxygen position $OH_O$; electron $e'$; electron vacancy $h^*$.

## 7.4 Discussion on Some Perovskite Proton Conductors

Ten independent equations are required to solve these ten unknown values.

$$[sr_{Sr}^x] + [V_{Sr}''] = 1 \tag{7.1}$$

$$[Ce_{Ce}^x] + [Y_{Ce}'] + [V_{Ce}''''] = 1 \tag{7.2}$$

$$[V_{\ddot{O}}] + [O_o^x] + [OH_o^*] = 3 \tag{7.3}$$

The mass balance expressed by the amount of matter in the fraction is:

$$[Ce_{Ce}^x]/[Y_{Ce}'] = (1 - x)/x \tag{7.4}$$

$$[sr_{Sr}^x]/([Ce_{Ce}^x] + [Y_{Ce}']) = Z \tag{7.5}$$

Here, $x$ is the stoichiometric parameter for doping at the Ce position, and $Z$ is the A/B stoichiometric parameter.

Applying electrical neutrality:

$$2[V_{Sr}''] + [Y_{Ce}'] + 4[V_{Ce}''''] + n = 2[V_{\ddot{O}}] + [OH_o^*] + p \tag{7.6}$$

In accordance with the law of mass action:

$$nil \leftrightarrow 3V_{\ddot{O}} + V_{Sr}'' + [V_{Ce}''''] \tag{7.7a}$$

$$K = [V_{\ddot{O}}]^3[V_{Sr}''][V_{Ce}''''] \tag{7.7b}$$

As for internal electrical equilibrium, let $n$ and $p$ represent electron and electron vacancy concentrations:

$$K_1 = np \tag{7.8}$$

For external balance:

$$\frac{1}{2}O_{2(g)} + V_{\ddot{O}} = O_O^x + 2h^* \tag{7.9a}$$

$$K_{OX} = [O_O^x]p^2/\left(p_{O_2^{1/2}}[V_{\ddot{O}}]\right) \tag{7.9b}$$

$$H_2O_{(g)} + V_{\ddot{O}} + O_O^x = 2OH_o^* \tag{7.10a}$$

$$Kw = [OH_o^*]^2/(p_{H_2O}[V_{\ddot{O}}][O_O^x] \tag{7.10b}$$

116

Theoretically, the above ten equations can be solved since there are ten defect concentrations. Poulsen et al. (1985) applied mathematical methods to solve them. He selected the defect concentration as the independent variable, assumed to give $[V_{\ddot{O}}]$ and $p_{H_2O}$ with certain value, generating nine vacancy concentrations and ten independent equations. The step was:

$$[OH_o^*]^2 / (p_{H_2O}[V_{\ddot{O}}][K_W]) + [V_{\ddot{O}}] + [OH_o^*] - 3 = 0 \tag{7.11}$$

From the above quadratic equation, we can obtain $[OH_O]$ and $[O_O^x]$ from Eq. (7.3):

$$[O_O^x] = 3 - [V_{\ddot{O}}] - [OH_o^*] \tag{7.3'}$$

From Eqs. (7.1), (7.2), and (7.4):

$$[V_{Sr}''] = \left(1 - Z[Y_{Ce}']/x\right) \tag{7.12}$$

$$[V_{Ce}''''] = 1 - [Y_{Ce}']/x \tag{7.13}$$

From the following equation, we can derive $[Y_{Ce}']$:

$$\left(1 - Z[Y_{Ce}']/x\right)\left(1 - [Y_{Ce}']/x\right) = K_s/[V_{\ddot{O}}]^3 \tag{7.14}$$

The matrix ions of the material can be obtained from Eqs. (7.1) and (7.2):

$$[Ce_{Ce}^x] = 1 - [V_{Ce}''''] - [Y_{Ce}'] \tag{7.1'}$$

and

$$[Sr_{Sr}^x] = 1 - [V_{Sr}''] \tag{7.2'}$$

On the basis of electric neutrality, Eq. (7.6), $p$ can be obtained:

$$2[V_{Sr}''] + [Y_{Ce}'] + 4[V_{Ce}''''] + K_1/p = 2[V_{\ddot{O}}] + [OH_o^*] + p \tag{7.6'}$$

$n$ can then be calculated from the internal balance:

$$n = K_1/p \tag{7.8'}$$

The partial pressure of oxygen can be obtained from the above determined values:

$$P_{O_2} = [O_O^x]^2 p^4 / K_{Ox}^2 [V_{\ddot{O}}]^2 \tag{7.9}$$

The temperature applied in the simulation study was 700 °C, and the partial pressure of the gas phase was expressed in atm (1 atm = 101,325 Pa).

The experiment tests defect concentration of $SrCe_{1-x}Y_xO_{3-\alpha}$ at 700 °C, the relationship between proton and other defect concentrations and $p_{O_2}$, and the relationships between the defect concentrations of electrons, protons, electron vacancies, and oxygen vacancies with $p_{O_2}$ and $p_{H_2O}$.

## 7.5 Infrared Spectroscopy

1. Infrared spectral absorption method is highly characteristic

When a sample is irradiated with infrared radiation of different wavelengths or wavenumbers, an infrared absorption spectrum is formed because radiation of some specific wavelengths is weakened through absorption by the sample (Chen 1993).

Wavenumber refers to the number of waves per centimeter and is the reciprocal of wavelength. Different substances have different microstructures, absorb infrared radiation of different wavelengths, and produce a characteristic infrared absorption spectrum. The absorption peak intensity at a given wavelength depends on the concentration of a substance, allowing for quantitative analysis.

IR spectra can be divided into near-infrared (NIR, 13,300–4000 cm$^{-1}$), mid-infrared (MIR, 4000–400 cm$^{-1}$), and far-infrared (FIR, 400–10 cm$^{-1}$) wavenumber ranges. The FIR range mainly entails the rotational spectrum produced by transitions between the rotational energy levels of the molecule. In addition, it also includes lattice vibrational spectra produced by ionic crystals, atomic crystals, and molecular crystals, and the vibrational spectra of molecules of high relative molecular mass or having small bond force constants.

In an absorption spectrum, the wavenumber region 4000–1330 cm$^{-1}$ is known as the characteristic band region since it contains characteristic peaks such as those of hydroxyl and amino groups. The vibration of molecular groups is actually the vibration of chemical bonds, including stretching and deformation vibrations [bending vibration, angular vibration (AC)]. For different moieties with similar chemical bonds, the frequency of the stretching vibration decreases with increasing relative mass of the atoms. The characteristic O–H stretching peak appears in the range 3700–2800 cm$^{-1}$. The stretching vibration frequency appears in the high-wavenumber region, while the deformation vibration frequency appears in the low-wavenumber region. The various chemical bonds in a molecule interact with each other, so the basic vibrational frequency of a chemical bond or group is related to the bond strength, the relative atomic masses of the constituent atoms, and the vibrational mode. It can also be modified by electrical effects, spatial effects, and other properties of the adjacent groups.

2. Characteristics of infrared spectroscopy of inorganic compounds

Quantum chemistry points out that energy-level transition require certain conditions, that is, an energy-level transition can only occur if the change of molecular dipole moment is non-zero. Absorptions of inorganic matter in the MIR are mainly caused by lattice vibrations of the anions, so the absorption peak position is little affected by the cations. Generally speaking, increased mass of the cations only causes a slight shift in the absorption position to lower wave.

Water exists in the following forms in an inorganic lattice: (1) adsorbed water, (2) crystalline water, (3) structural water. These properties are reflected in the IR spectra. The relative peak intensities give an indication of which form of $H_2O$ is more prevalent. Additional **sharp** peaks between these two main peaks indicate high crystallinity.

Absorption spectra of a variety of proton conductors feature O–H peaks, and the rotating motion for proton transfer can be discerned. There is O–H in each moment, and the content changes are reflected in the infrared spectrum.

## 7.6 Application of Perovskite Proton Conductors

A $CaZr_{0.9}In_{0.1}O_{3-\alpha}$ sensor has been used in rapid on-line hydrogen measurement during the production of molten aluminum. Kurita et al. (1995) studied the relationship between proton conductivity at 350–1400 °C and gas-phase $p_{O_2}$ by an $H^+/O^+$ isotope tracer method. A dominant region map of $H^+$, $h^{\cdot}$, and $O^{2-}$ respective conductions was given, and the sensor was applied for hydrogen measurement in molten copper.

The behavior of hydrogen in the dehydrogenative heat treatment of steel and the dehydrogenation of liquid aluminum has both been studied by proton conductor sensing methods (Wang 2000).

1. Study on hydrogen behavior in dehydrogenative heat treatment of steel by proton conductor sensing

Hydrogen is a harmful element in steel. As early as 1940, the Chinese scholar Li X studied the self-fracture of a steel shaft from a crashed plane in England. He found that hydrogen had caused the steel to break. The transgranular fracture of stainless steel blades and the brittle fracture of railway frog steel are also hydrogen-induced fractures. In order to reduce the hydrogen content of steel, dehydrogenative heat treatment is often applied, and its progress has been monitored experimentally (Zhu and Li 2003). An in-house-constructed $CaZr_{0.9}In_{0.1}O_{3-\alpha}$ hydrogen sensor was used in the experiment; the atmosphere was composed of a high-purity Ar-$H_2$ mixture (standard ammonia) containing 1.16% $H_2$ as a reference electrode gas. Dehydrogenative heat treatment of two steel samples from a steel plant was studied. The composition of the steel samples is summarized up, for example, component units of steel samples (mass fraction, %) in Steel 1[#], C, Si, Mn, P, S, Al, N, Cr, V, and Ti is 0.021, 0.03, 0.26, 0.012, 0.008, 0.046, 0, 0, 0, and 0.13; in Steel 2[#], it is 0.058, 0.016, 0.23, 0.010, 0.013, 0.023, 0.002, 0.01, 0.001, and 0.001.

The EMF of the hydrogen sensor cell can be represented as:

## 7.6 Application of Perovskite Proton Conductors

$$E = \frac{RT}{2F} \ln \frac{p_{H_2}^{I}}{p_{H_2}^{II}}$$

In the formula:

$E$—battery EMF, V;

$R$—molar gas constant, 8.314 J/(mol K);

$F$—Faraday constant, 96,487 C/mol;

$T$—thermodynamic temperature;

$p_{H_2}^{I}$, $p_{H_2}^{II}$—partial hydrogen pressures of the two electrodes, one of which can be used as a reference electrode.

The experimental process and determination results were as follows:

The surface of the steel sample to be tested was cleaned. The sample was placed in the constant temperature zone of the furnace [(800 $\pm$ 1) °C], and the sensor probe was pressed against it. The furnace tube was purged with argon after threefold dehydration and purification. Data were collected by a computer, with an input impedance > $10^9$ $\Omega$ and a digital voltmeter of accuracy $\pm$ 0.01 $\mu$V. According to the above formula, $p_{H_2}$ could be calculated. A Solartron 1255 frequency response analyzer was used at constant temperatures of 600 and 800 °C, and the sensor was equipped with a 1286 electrochemical interface to measure the electrochemical impedance spectrum. The measured frequency range was $10^{-2}$–$10^6$ Hz, and calculations were performed with Zsimpwin software.

It was found that for the first steel sample the EMF of the hydrogen sensor suddenly increased from a negative value to 0.57–0.6 V. The corresponding hydrogen partial pressure was calculated at E 0.6 V, 710 °C, about 16,490 atm. After the experiment, the steel samples were taken out to observe their surfaces. The steel at the measured point was completely ruptured, which is consistent with the case that hydrogen-induced steel cracking. This phenomenon may have been due to defects in the steel interior near the contact point with the hydrogen sensor. In the process of heating, a large number of hydrogen has been focused in defects, and hydrogen molecules repeatedly impact the lattice. During cooling, hydrogen in steel is supersaturated, leading to a sharp, massive escape, which rapidly increases the hydrogen pressure. This is reflected by the hydrogen sensor, which shows a sharp increase in battery EMF, and the surface of the steel samples bursts.

The experimental results showed the reliability and feasibility of this test method with a hydrogen sensor. It can thus be used to detect the hidden danger of hydrogen in the production of key components.

2. Use of a proton conductor sensor for dehydrogenation of liquid aluminum

The solubility of hydrogen in aluminum increases with temperature. Liquid and solid aluminum coexist in equilibrium at its melting point of 660 °C. The solubility of hydrogen in liquid aluminum is 0.69 mL $H_2$/100 g Al, while that in solid aluminum is 0.034 mL $H_2$/100 g Al. Hydrogen solubility in solid aluminum has a ratio of 1:20.3 to that in liquid aluminum, so when liquid aluminum is solidified, supersaturated hydrogen will condense in large quantities. This leads to defects such as hair-line fractures and pores in the cast, which has a detrimental effect on the properties of aluminum and aluminum alloys.

Hydrogen in aluminum is mainly derived from environmental water vapor, and the partial pressure of water vapor in the air in more inland coastal areas is large, that is, the humidity is high, which has a great influence on the quality of aluminum alloys (Chen et al. 1995). In production lines for semi-continuous casting of aluminum and aluminum alloy, despite the application of an on-line degassing device, the degassing efficiency is limited. In a large plant, only 0.15 mL/100 g Al of hydrogen can be detected by a hydrogen sensor.

For further dehydrogenation to meet the needs of high-strength and high-toughness aluminum alloys, fluorine- or chlorine-containing air is used in all plants. However, this leads to environmental pollution and equipment corrosion. To this end, Wang CZ proposed a method of dehydrogenation using a proton conductor hydrogen pump in the flow trough. Its principle is as follows.

A direct current (DC) supply is connected to both sides of the proton conductor electrolyte tube to form an electrolyte-like device, in which the electrode and battery react as:

Anode:

$$H_{2(aluminum\ melt)} - 2e = 2H^+_{(proton\ conductor)}$$

Cathode:

$$2H^|_{(proton\ conductor)} + 2e = H_2 \uparrow$$

Battery reaction:

$$H_{2(aluminum\ melt)} = H_2 \uparrow$$

Cermets that do not react with liquid aluminum serve as electrode leads, the loop is connected to a mechanical pump, and the hydrogen produced in the proton conductor tube is continuously pumped into the atmosphere for the purpose of dehydrogenation of molten aluminum (Wang 2002). This work is described as follows.

Electrolytic dehydrogenation of molten aluminum was first conducted under laboratory conditions. In order to simulate the dehydrogenation of molten aluminum with different air humidities (equivalent to different water vapor partial pressures), high-purity Ar gas was used and saturated water vapor at different partial pressures was added (Wang 2000). The amount of aluminum was 450–500 g (accurately weighed), which was placed in a corundum crucible in a resistance furnace. The proton conductor tube above the aluminum melt was made of $BaCe_{0.9}Y_{0.1}O_{3-\alpha}$ and $BaCe_{0.9}Sm_{0.1}$ and was kept at 760–800 °C (temperature control accuracy $\pm$ 1 °C). The water vapor partial pressure was 4122.8–4754.7 Pa, and the external DC voltage was 0.4, 0.8, or 1.2 V. Utilization of a $CaZr_{0.9}In_{0.1}O_{3-\alpha}$ proton conductor sensor

## 7.6 Application of Perovskite Proton Conductors

allows the on-line detection of hydrogen in molten aluminum. The results showed an efficient dehydrogenation effect.

A survey of proton conductors used in the literature and those developed by our research group shows that the strength of $BaCeO_3$ matrix substrate conductors is poor. When such materials are used in proton conductor hydrogen pumps, they are susceptible to erosion by the flow of molten aluminum. For materials containing Zr instead of Ce, the conductivity is inferior to that of the $BaCeO_3$ base material, and the strength is not high. The strength of the composite perovskite material $BaCa_{1.18}Nb_{1.82}O_{9-\delta}$ is also poor. The incorporation of Ni, within a certain composition range, increases the material strength, but it is still not ideal.

The use of $CaZr_{0.9}In_{0.1}O_{3-\alpha}$ as a dehydrogenative proton conductor provides a suitably efficient dehydrogenation effect of molten aluminum of about 0.08 mL $H_2$/100 g Al (Li et al. 2012). Further research is under way. For materials with Zr instead of Ce, it has been reported in the literature that higher strength can be achieved, for example by sintering at pressures higher than 200 MPa at 1700 °C. Pulverization and sintering are also issues that need further study.

The method of solid-phase synthesis is accurate and simple, and is the most commonly used method for preparing high-temperature proton conductors and similar materials. Powder with an average particle size of several micrometers can be obtained by agate ball-grinding or $ZrO_2$ ball-grinding at a high rate (about 300 r/min).

The issues of pulverizing, synthesis, and sintering are discussed below.

### (1) Particle size of powders

Powders are aggregates of a large number of ions. The particle size, shape, and particle size distribution of a powder greatly influence its specific surface area and compressibility. They are predisposed to spontaneous agglomeration due to van der Waals forces between molecules, electrostatic attraction between particles, capillary force in the absorption of water, magnetic attraction between particles, mechanical entanglement force caused by rough particle surfaces, and so on. The particle size distribution characterizes the variation in particle size in a dispersed system, and is determined with a laser particle size analyzer.

### (2) Surface state and surface energy of powders

From a thermodynamic point of view, the free energy of a solid without an external force is the lowest stable state, and any comminuting deformation caused by an external force will make its free energy change. The energy consumed in crushing is mainly through two aspects: one is the surface energy needed to produce a new surface when cracks propagate; the other is the energy stored in the solid due to elastic deformation. When a material is crushed, its particle size is reduced and its physical and chemical properties are changed.

The surface of an ultrafine powder refers to one or more atomic layers. At the surface of the outer layer, the atoms are coordinatively unsaturated, can form dangling bonds, and thus show residual bonding ability and high reactivity.

Composites are characterized by XRD analysis, crushed to form the desired shape, and then fired to give the finished product. Sintering entails the densification of a raw billet at high temperature. If the optimum conditions are exceeded, the grain size grows and the strength decreases. Shrinkage rate, porosity, the ratio of bulk density to theoretical density, and other indicators can be considered to determine the best value.

# References

Chandra S (1988) Proton conductors and their applications in solid state electrochemical devices. In: Chowdari BVR, Radhakrishna S (eds) Proceedings of the international seminar solid state ionic devices. World Scientific, Singapore, pp 265–288

Chen YK (1993) Infrared absorption spectroscopy and its application. Shanghai Jiaotong University Press, Shanghai

Chen W, Wang CZ, Liu L (1995) Determination of hydrogen activity in molten aluminum alloy by sensing method. J Met 19:305

Cherry M, Islam MS, Gale JD et al (1995) Computational studies of proton migration in perovskite oxides. Solid State Ionics 77:207

Davies RA, Islam MS, Gale JD (1999) Dopant and proton incorporation in perovskite-type zirconates. Solid State Ionics 126:323

Islam MS, Slater PR, Tolchard JR et al (2004) This journal is The Royal Society of Chemistry. Dalton Trans 3001

Iwahara H (1992) High temperature type proton conducting ceramics and their applications. Ceramic 27(2):112

Iwahara H (1995) Technological challenges in the application of proton conducting ceramics. Solid State Ionics 77:289

Iwahara H, Chairman, Tanaka S (1986–1988) Chemo measuring by high performance chemical sensor systems. The Ministry of Education, Science and Culture, p 141

Iwahara H, Uchida H, Ogaki K, Nagato H (1991) Nernstian hydrogen sensor using $BaCeO_3$-based, proton-conducting ceramics operative at 200°–900°C. J Electrochem Soc 138(1):295

Kendrick E, Kendrick J, Knight KS et al (2007) Cooperative mechanisms of fast-ion conduction in gallium-based oxides with tetrahedral moieties. Nat Mater 6:871

Kruth A, Davies RA, Islam MS et al (2007) Combined neutron diffraction and atomistic modeling studies of structure, defects, and water incorporation in doped barium cerate perovskites. Chem Mater 19:1239

Kurita N, Fukatsu N, Ito K et al (1995) Protonic conduction domain of indium-doped calcium zirconate. J Electrochem Soc 142(5):1552

Li SL, Wang CZ et al (2012) $CaZr_{1-x}In_xO_{3-a}$ ($x$=0, 0.05, 0.10, 0.15) electrical properties of proton conductors. J Inorg Mater 87(4):427

Münch W, Seifert G, Kreuer KD et al (1977) A quantum molecular dynamics study of the cubic phase of $BaTiO_3$ and $BaZrO_3$. Solid State Ionics 97(1–4):39

п.и.фЕДоров, Акчурин (2000) Indium chemistry manual (trans: Zhang QY, Xu KM). Peking University Press, Beijing

Poulsen FW, Andersen NH, Clausen K et al (1985) Transport-Structure relations in fast ion and mixed conductors. Solid State Ionics 25:239

Shao W, Zhang HL et al (2004) Preparation of direct solid-state reaction and conductivity of $SrCe_{0.95}Yb_{0.05}O_{3-\alpha}$ protonic conduction. Chin J Nonferrous Met 14(1):42

Song SJ, Wachsman ED, Dorris SE et al (2002) Defect chemistry modeling of high-temperature proton-conducting cerates. Solid State Ionics 149:1

Wang CZ (2000) Solid electrolytes and chemical sensors. Metallurgical Industry Press, Beijing

Wang CZ (2002) On line detecting content and behavior of hydrogen and other elements in aluminum melt by solid electrolyte sensors. Nonferrous Met Process 31(2):24–26

Wu J, Davies RA, Islam MS et al (2005) Atomistic study of doped $BaCeO_3$: dopant site-selectivity and cation nonstoichiometry. Chem Meter 17:846

Yajima T, Iwahara H, Fukatsu N et al (1992) Measurement of hydrogen content in molten aluminum using proton conducting ceramic sensor. J Jpn Inst Light Met 42(5):263

Yajima T, Iwahara H, Koide K et al (1991) $CaZrO_3$-type hydrogen and steam sensors: trial fabrication and their characteristics. Sens Actuators B 5:145

Yusa H (2018) Structural relaxation of oxide compounds from the high-pressure phase. In: Tanaka I (ed) Nanoinformatics. Springer

Zhu JC, Li DC (2003) Dehydrogenation of alloy steel by heat treatment and its practical significance. Cast Technol 24(2):83

# Chapter 8
# Role and Application of Rare Earth Elements in Steel

## 8.1 Early Recognition

Studies in the early period of using rare earth elements in steel applications showed that the role of sulfur in steel is harmful except for easily chipped steel. For example, the hot embrittlement of steel during rolling is caused by FeS, and later, 99.9% of will be added manganese in production. The affinity of manganese for sulfur is higher than that of iron for sulfur. Limitations of using MnS were later found because the early majority of steel deoxidation was inadequate. Because of the presence of oxygen in steel, in hot rail steel, MnS inclusions are elongated, and the mechanical properties of steel are clearly reduced.

In the early days of the steelmaking era, the only solution to overcome manganese sulfide hazards was to replace it with elements with greater parathion than manganese. An increasing arrangement of negative free energy values generated according to the metal sulfide standard, titanium, zirconium, calcium and rare earth elements can be selected, but the nitrides and carbides of titanium and zirconium are very stable, affecting other properties. The oxides, oxygen sulfides and sulfides of rare earth elements are very stable. Since 1968, researchers at Jones & Laughlin Steel in the United States have added rare earths to many experimental furnaces, which proved to control sulfides. Vulcanization inclusions are eliminated and replaced by small, globular rare earth oxygen sulfides and sulfides. These results are directly applied to high strength and low alloy automotive steel with high cleanliness and low cost.

During 1970–1974, rare earths were widely used in pipeline steel. Germany, France and Japan also added rare earths to steel. The methods used varied according to the purpose and different steel species. Additives have a variety of shapes, such as block, plate, filamentous, etc. In Japan, for example, with extrusion molding products, each grain is 0.7–200 g; when choosing the addition method, the first consideration is to remove the oxygen in molten steel as much as possible, and the oxygen content can be less than 20 ppm when deoxidized with silicon and aluminum. Japan has used the projectile method to add such high speed addition of a rare earth, quickly melted and evenly distributed, immediately deoxidized and desulfurized. Rare earth oxides, oxygen sulfides and sulfides have small solubility products in molten steel, and are spherical in distribution, which can play the role of dispersion strengthening.

© Science Press 2023
C. Wang, *Theory and Application of Rare Earth Materials*,
https://doi.org/10.1007/978-981-19-4178-8_8

The continuous casting method has been widely used in recent years when adding rare earth elements, the use of encapsulated wire, the middle of a few mm of mixed rare earth core, coated with about 0.18 mm iron.

After 1975, Germany, Japan and others explicitly chose the low sulfur route. During injection, complete sulfide morphology control is usually lacking in the continuous casting billet. Afterward, researchers reconsidered the addition of rare earths, strictly controlling the amount of rare earth addition; or added a small amount of rare earth after spraying with calcium (Luyckx 1983; Ohmachi 1993; Chen 1987). In 1970–1976, the United States used a large amount of rare earth-added steel in high strength welded casing and drilling pipes in oil exploration wells; it was also used for submarine pipeline steel. However, it is necessary to control the amount of rare earth addition and the method of addition to obtain a good steel structure.

Researchers around the world have had more than half a century to understand the thermodynamics of steel and mechanisms of steel to overcome their problems. From the metallurgical physicochemical point of view, the roles of rare earths in steel include deoxidation; desulfurization; removing the harmful effect of Sn, Pb, As, Sb, Bi with low melting point; and the effects of solid solution and microalloying.

## 8.2 Rare Earth Oxides, Sulfide, Carbides and Nitrides

Steelmaking practice has proven that rare earths can do de oxygen in steel water, but also in de-steel bag and other refractory. The stability of CaO and MgO is lower than that of rare earth oxides at about 1600 °C. To compare with each other, it is usually based on 1 mol oxygen when determining the relation between the standard free energy of formation and temperature ($\Delta G^{\ominus}-T$) for various oxide formation reactions.

Taking a diagram as an example, it shows the relationship between the standard free energy of formation and temperature of some oxides, and approximate accuracy at 25 °C (Wang 2013). At the same temperature, the lower the position in the figure, the greater the negative value of its standard free energy of formation and the greater its stability. The special scale labeled on the periphery can directly read out the equilibrium oxygen partial pressure, the corresponding equilibrium ratio of $p_{H_2}/p_{H_2O}$ and $p_{CO}/p_{CO_2}$, to facilitate the study of the thermodynamic problems of the corresponding oxides by the $H_2$-$H_2O$ and $CO$-$CO_2$ equilibrium methods. Taking rare earth sulfides LaS and CeS as examples, they are the most stable class in various element sulfides.

On the relationship between the standard free energy of formation and temperature of sulfides, the stability of rare earth nitrides is poor, while that of rare earth carbides is worse. Rare earth carbides are not shown on the diagram, and such stability is similar to $CaC_2$, which reacts with water at room temperature to form acetylene.

The approximate accuracy at 25 °C is: ($A \pm 4$) kJ; ($B \pm 13$) kJ; ($C \pm 21$) kJ (it is ($C \pm 42$) kJ; ($D \pm 42$) kJ. Phase change point: melting point F; polycrystalline transition point T; boiling point B; square frame: phase transition point of the sulfide; and not framed one: phase change point of element. One cal $= 4.184$ J.

## 8.3 Deoxidation, Desulfurization Effect

There are more than 1000 parts containing rare earths from the former Soviet Union's tanks. Besides theoretical research, scholars in various countries have also carried out statistical and comparative work. For example, scholars in the former Soviet Union used the data of representative elements in reference books and periodicals to assess the interaction of the elements of the periodic table with oxygen, sulfur, phosphorus, nitrogen and carbon, including heat of formation $(- \Delta \cdot \overline{H})$ in the standard stateto compare the stability of each compound. The study lists the heat of formation data of Sc, Y, O, S, P, N, C.

The rare earth oxides and sulfides are stable, while phosphides, nitrides and carbides are gradually getting worse, which is in accordance with the diagram of the standard free energy of formation and temperature relationship of the aforementioned compounds, such as oxygen-, sulfur-, phosphorus-, nitrogen-, and carbon-elemental system.

## 8.3 Deoxidation, Desulfurization Effect

The addition of rare earths (RE) to molten steel requires deoxidation and desulfurization, for example:

$$2[RE] + 3[O] = RE_2O_{3(s)} \tag{8.1}$$

$$2[RE] + 2[O] + [S] = RE_2O_2S_{(s)} \tag{8.2}$$

$$2[RE] + 3[S] = RE_2S_{3(s)} \tag{8.3}$$

$$3[RE] + 4[S] = RE_3S_{4(s)} \tag{8.4}$$

$$[RE] + [S] = RES_{(s)} \tag{8.5}$$

In the case of Eq. (8.1), the equilibrium constant for the reaction at a given temperature is:

$$K = \frac{1}{\alpha_{[RE]}^2 \cdot \alpha_{[O]}^3}$$

Other reaction representations are similar. This representation is cumbersome and nonintuitive, so metallurgists reverse the reaction, for example, write $2[RE] + 3[O] = RE_2O_3(s)$ back to front as $RE_2O_3(s) = 2[RE] + 3[O]$, then $K = \alpha_{[RE]}^2 \cdot \alpha_{[O]}^3$. This constant is called the solubility product of the rare earth deoxidation reaction, indicating the solubility of the deoxidized product in the steel solution, also known as the deoxidation constant of the rare earth. Equations (8.2)–(8.5) are similarly also written back to front, called deoxidized sulfur constant, desulfurization constant, etc.

The smaller these constant values the smaller the solubility of the product, the easier it is to form precipitates or inclusions. The free energy change of deoxidation and desulfurization is reciprocal to that of the solubility product reaction.

Wei (1978) gave a conference report entitled "Physical and chemical problems of rare earth steel smelting," which analyzed the behavior and function of rare earths in steel, the order of formation of rare earth oxides and sulfides, and took Ce as an example. The data of solubility product of different rare earth compounds are shown in a table.

(1) Comparison of $CeO_2$ and $Ce_2O_3$ generation possibilities

If the free energy, $\Delta G$, of the reaction is negative, then the solubility product reaction $\Delta G$ is positive,

$$2[Ce] + 4[O] = 2CeO_{2(s)}, \tag{8.6}$$

the solubility product reaction to

$$\Delta G_1^{\ominus} = RT\ln K_1$$

$$2[Ce] + 3[O] = Ce_2O_{3(s)}, \tag{8.7}$$

the solubility product reaction to

$$\Delta G_2^{\ominus} = RT\ln K_2$$

Form (8.6) and (8.7), then

$$Ce_2O_{3(s)} + [O] = 2CeO_{2(s)} \tag{8.8}$$

$$\Delta G_3 = \Delta G_3^{\ominus} + RT\ln\frac{1}{\alpha_{[O]}}$$

$$= RT\ln\frac{K_1}{K_2 \cdot \alpha_{[O]}}$$

For (8.8) to proceed spontaneously, $\Delta G_3$ must be negative, that is

$$\frac{K_1}{K_2 \cdot \alpha_{[O]}} < 1$$

By Table 8.1, $K_1 = 4 \times 10^{-11}$, $K_2 = 3 \times 10^{-21}$, substituting into the formula, $a_{[O]}$ must be greater than 0.53%. Suppose that $a_{[O]} \approx [O]$, i.e., need $[O] > 0.53\%$ to generate $CeO_2$.

At 1600 °C, the saturated oxygen content in steel is 0.23%, so in practice, adding Ce to molten steel can only generate $Ce_2O_3$, but not $CeO_2$.

## 8.3 Deoxidation, Desulfurization Effect

**Table 8.1** Chemical composition of iron bombardment

| Element | C | Si | Mn | S | Ni | Cu | Pb | As |
|---|---|---|---|---|---|---|---|---|
| Mass fraction (%) | 0.001 | 0.005 | < 0.001 | 0.004 | < 0.05 | 0.001 | < 0.001 | < 0.004 |
| Element | Zn | Mg | Ca | Al | Cd | Bi | Sb | Sn |
| Mass fraction (%) | < 0.0005 | < 0.0005 | < 0.0001 | < 0.003 | < 0.0001 | < 0.0001 | < 0.001 | < 0.0005 |
| Element | Ag | Ga | In | N | P | O | | |
| Mass fraction (%) | < 0.0001 | < 0.0001 | < 0.0001 | < 0.006 | < 0.005 | 170 ppm | | |

*Notes* FeS analytical purity; Ce 99.99%

### (2) Comparison of $Ce_2O_3$ and $Ce_2O_3S$ generation possibilities

The molecular formula for rare earth oxysulfide is $RE_2O_2S$.

Consider

$$2[Ce] + 2[O] + [S] = Ce_2O_2S$$

$$\Delta G_4^\ominus = RT \ln K_4 \tag{8.9}$$

Form (8.9)–(8.7),

$$Ce_2O_{3(S)} + [S] = Ce_2O_2S + [O]$$

$$\Delta G_5^\ominus = RT \ln \frac{K_4}{K_2} \tag{8.10}$$

$$\Delta G_5 = \Delta G_5^\ominus + RT \ln \frac{\alpha_{[O]}}{\alpha_{[S]}}$$

$$= RT \ln \frac{K_4}{K_2} + RT \ln \frac{\alpha_{[O]}}{\alpha_{[S]}}$$

$$= RT \ln \frac{K_4 \cdot \alpha_{[O]}}{K_2 \cdot \alpha_{[S]}} \tag{8.11}$$

If (8.10) proceeds to the right, the $\Delta G_5$ must be negative, that is:

$$\frac{K_4 \cdot \alpha_{[O]}}{K_2 \cdot \alpha_{[S]}} < 1$$

or

$$\frac{\alpha_{[S]}}{\alpha_{[O]}} > \frac{K_4}{K_2} = \frac{1.3 \times 10^{-20}}{3 \times 10^{-21}} = 4.33$$

If the activity coefficients of [S], [O] are 1, then

$$[O] < 0.23[S]$$

When [S] = 0.03% in molten steel, if [O] > 0.0069%, $Ce_2O_2S$ is generated.

The possibility of CeS, $Ce_2S_4$, $Ce_3S_4$ and $Ce_2S_3$ generation can be calculated in the same way:

When

$\alpha_{[S]} < 0.006\%$,

$$\frac{1}{3}Ce_3S_4 = CeS + \frac{1}{3}[S], \quad \Delta G \text{ equals negative.}$$

When.

$\alpha_{[S]} < 0.13\%$,

$$\frac{1}{2}Ce_2S_3 = \frac{1}{3}Ce_3S_4 + \frac{1}{6}[S], \quad \Delta G \text{ equals negative.}$$

The above calculations show that when the oxygen content of molten steel is > 60 ppm, adding Ce can generate $Ce_2O_3$, if the molten steel is deoxidized, [O] < 60 ppm, generates $Ce_2O_2S$. The generation of this compound can reduce [O] to very small values.

After removing [S] with [Ce] to generate $Ce_3S_4$, when reducing to 0.006%, continue to remove [S], generating CeS. So for the deoxidized liquid steel, after the addition of rare earth, the order of the formation of inclusions is generally:

$$Ce_2O_2S \rightarrow Ce_3S_4 \rightarrow CeS.$$

A schematic diagram of the obtained rare earth compounds in a practical smelting operation indicates the generation order of rare earth sulfur oxides and sulfides in production, that is, first to be generated is $Ce_2O_2S$, and then the rare earth sulfides, which indicates that the thermodynamic calculation and the actual situation are consistent. $CeO_2$ has been found in Fe–Ce–O equilibrium experiments (Vahed and Kay 1976).

The study suggested that the data given should be used, for example, on basic thermodynamic data of Ce compounds in molten iron (1600 °C), when $CeO_{2(s)} = [Ce] + 2[O]$, equilibrium constant (solubility product) $K$ is $7.94 \times 10^{-10}$, and Lg $K$ is $- 9.10$; when $CeO_{3(s)} = 2[Ce] + 3[O]$, equilibrium constant (solubility product) $K$ is $4.90 \times 10^{-18}$, and Lg $K$ is $- 17.31$. Wei SK team's method used the thermodynamic data to determine the conversion relationship of deoxidized desulfurization products of various rare earths.

# 8.5 Thermodynamic Experiments in Liquid Iron                               131

## 8.4 Vahed and Kay's Work

In March 1974, Wilson WG, Kay DAR, and Vahed A, cooperated in research to publish "Predicting the behavior of rare earths in steel with thermodynamics and phase equilibrium" in the Journal of Metals, which has attracted great interest from researchers in rare earth steel throughout the world.

The solubility product of rare earth oxides, oxygen sulfides and sulfides in steel was calculated using the standard of various rare earth compounds to form free energy combined with the dissolved free energy of oxygen, sulfur and rare earths in pure iron solution. The free energy of dissolution of the rare earths used in the iron solution was not reported in the literature, and the authors calculated it according to the phase diagram.

They calculated a diagram of the standard free energy of formation and temperature of more rare earth oxides, and the relationship between the standard free energy of formation and temperature of three more accurate rare earth oxygen sulfides.

Comparing the solubility product data of each rare earth compound, it is concluded that the order of formation in iron and steel solution is: oxide, oxygen sulfide, $RE_xS_y$, RE (As, Sb, Al, etc.), nitride, carbide. Increasing RE amount can generate $\Delta G$ corrected compounds.

The standard free energy of formation (J/mol) and temperature (1100–2000 K) relationship data for common La, Ce sulfides are as follows: LaS, $123,250 + 25.3$ T; CeS, $132,480 + 24.8$ T; $Ce_3S_4$, $161,060 + 33.0$ T; $Ce_2S_3$, $175,600 + 38.0$ T. The experiment obtains the solubility product data at 1,900 K of rare earth oxides and sulfides. Vahed and Kay supplemented the data of 1900, 1950, 2000 K, obtaining the data of the relationship between solubility product and temperature (in molten steel, using Henry's law, 1% as the standard state).

For example, $CeO_2$, dissolution product is $h_{Ce} \times h_O^2$, the solubility product in 1900 K, 1950 K, and 2000 K is $4.0 \times 10^{-11}$, $1.6 \times 10^{-10}$, and $6.0 \times 10^{-10}$, respectively; to $Ce_2O_3$, they are $h_{Ce}^2 \times h_O^3$, $3.0 \times 10^{-21}$, $3.0 \times 10^{-20}$, and $2.8 \times 10^{-19}$, respectively; to $La_2O_2S$, they are $h_{La}^2 \times h_O^2 \times h_S$, $7.3 \times 10^{-22}$, $6.7 \times 10^{-21}$, $5.3 \times 10^{-20}$, respectively.

The results were published in 1976.

From the total data, they were less than 1 to several magnitude orders, larger than the experimental values. The experiment summarized Ce, O, S diagram of Henry's activity relationship, and relationship between the formation of cerium oxide, cerium oxysulfide and cerium sulfide in Fe-Ce-O-S system.

In 1974, Mclean and Lu (1974) also analyzed the thermodynamics of rare earths in steel, but the results are rarely cited.

## 8.5 Thermodynamic Experiments in Liquid Iron

In the early stage of rare earth steel production, the solubility product was measured mainly from field sampling, including inclusions, the figures are generally large. Langenberg and Chipman (1958) determined the solubility product of CeS at 1600 °C, obtaining $[Ce] \cdot [S] = 1.5 \times 10^{-5}$, the error is $\pm 0.5 \times 10^{-3}$. The experimental furnace has a sealing device and sample sampling port, the furnace tube is a quartz tube, with corundum (pure $Al_2O_3$), 30 kW spark discharge device, supply furnace power, temperature measurement was with an optical pyrometer. The temperature measurement error is $\pm$ 15 °C in the range 1550–1625 °C, using CeS and the MgO crucible, the raw material is electrolytic iron, pre-deoxygenated to 0.1%. The vacuum valve is closed, and the Ar gas for welding after 600 °C of calcium chip deoxidation is input. When the furnace tube is slightly positive, the Ar gas entry port is closed so that the experiment is carried out in an Ar atmosphere.

Sample and sampling with quartz tubes. Ce and S were analyzed using chemical methods. Taking the sulfur content as the horizontal coordinate and the cerium content as the vertical coordinate, a linear relationship was obtained, but the data points are scattered. Further, they obtained $[Ce]\cdot[S] = 1.5 \times 10^{-3}$ (the error is $\pm 0.5 \times 10^{-3}$), it is too big. Langenberg and Chipman first evacuated the furnace for 1 h and then heated it to 1600 °C, closed the vacuum valve, introduced the purification Ar gas; the sulfur sample used was generally analytical pure.

According to the published years of the article, another characteristic study was the study by Janke and Fischer (1978) on the deoxidation equilibrium of cerium, lanthanum and europium in molten iron. They believe that deoxidation and desulfurization constants measured by previous researchers were the results obtained by chemical analysis, including the data of inclusions. For example, the solubility product at 1600 °C of $Ce_2O_3$ is $1.0 \times 10^{-14}$–$1.8 \times 10^{-9}$, and the $La_2O_3$ solubility product is $2.5 \times 10^{-13}$–$6 \times 10^{-9}$, values are even higher than the Al deoxidation constant $\alpha_{Al}^2 \cdot \alpha_O^3 = 2.42 \times 10^{-14.37}$. Apparently, the Ce, La deoxygenation constants are wrong. Therefore, it is important to select suitable crucible materials for determination, it is appropriate to use CaO or $ThO_2$ crucibles.

The above review shows that a high sensitivity method is necessary for the determination of particularly small solubility product data, while the solid-electrolyte EMF method is most suitable. If the choices of the solid-electrolyte and reference electrodes are reasonable, much more accurate data can be obtained than by the chemical equilibrium method.

A $La_2O_3$, $Ce_2O_3$ crucible was used for the determination of the La, Ce deoxidation constant, and the isostatic pressing method was used to press it by itself and burn it at high temperature. For the Th deoxidation constant determination, a market-purchased $ThO_2(+Y_2O_3)$ crucible was used. The experimental work is as follows. Regarding the solid-electrolyte oxygen sensor, a CaO stable $ZrO_2$ tube with a closed end is applied with a layer of $ThO_2(+Y_2O_3)$ solid electrolyte to form a double-walled cell to reduce the electronic conductivity of the solid electrolyte. The battery formed is

$$(+) \text{ Rh}|\text{Air (Reference)}|ZrO_2(+CaO)|ThO_2(+Y_2O_3)|\text{Iron Melten}|Fe_{(S)}(-)$$

The experiment was carried out in an electric furnace with a sealed controllable atmosphere (1600 °C). After each addition of deoxidized elements, the sensor (preheated in the upper part of the crucible) was inserted immediately, and the response time was 10–20 s; a melt sample was obtained from a fine quartz tube

## 8.5 Thermodynamic Experiments in Liquid Iron

and quenched in water. The rare earth content and Si content were analyzed by chemical analysis. Si is less than 0.02%, and so the amount of $SiO_2$ reduced by the rare earth is negligible.

$$\alpha_{[O]} = \exp\left(\frac{\Delta G_O^\ominus}{RT}\right) \cdot \left[\left(p_{e'}^{1/4} + p_{\text{Reference}}^{1/4}\right) \cdot \exp\left(-\frac{EF}{RT}\right) - p_{e'}^{1/4}\right]^2$$

In the formula,

$R$—Molar gas constant, 8.3143 J/(K mol);

$F$—Faraday's constant, 96,487 C/mol;

$T$—Absolute temperature, K;

$\Delta G_{[O]}^\ominus$—Change of dissolved free energy value of oxygen in pure Fe;

$p_{e'}$—Characteristic oxygen partial pressure for electronic conduction of solid electrolytes;

$p_{\text{reference}}$—Reference electrode oxygen partial pressure (0.21 atm).

The experimental results are summarized into diagrams, such as deoxidized equilibrium of, La, Ce and Al in pure Fe melt, and relationship between deoxidized element activity and oxygen activity in pure Fe melt, at 1600 °C.

Then, at 1600 °C,

$$K_{La_2O_3} = [La]^2 \cdot \alpha_{[O]}^3 = 4.074 \times 10^{-19} \ (0.010 - 0.60\% \ [La])$$

$$K_{Ce_2O_3} = [Ce]^2 \cdot \alpha_{[O]}^3 = 9.376 \times 10^{-18} (0.026 - 0.47\%[Ce])$$

$$K_{HfO_2} = [Hf]^2 \cdot \alpha_{[O]}^2 = 8.710 \times 10^{-12} (0.006 - 1.20\% \ [Hf])$$

The values of $K_{La_2O_3}$ and $K_{Ce_2O_3}$ are smaller than those measured by previous researchers, but still much larger than those calculated by Wilson, Vahed and Kay. The authors of this book thought that in the Fe extremely dilute solution, [RE] and [O] were very far away from a microscopic perspective, it is not easy to grow deoxidized products by diffusion and collision nucleation, so the experimental value cannot reach the theoretical calculation value, with the simple structure of the compound, relatively easy nucleation.

Han (1987) used the radioisotope tracer atomic method to study the thermodynamics of the formation of rare earth compounds, which is also an innovative work. In nature, most elements exist in a mixed form of two or more atoms with the same chemical properties and the same number of protons, but different atomic mass numbers, and occupy the same position in the periodic table, they are called isotopes. With stable and unstable isotopes, unstable isotopes spontaneously undergo nuclear decay and can emit various rays, so they are also called radioisotopes.

Common nuclear decays include $\alpha$, $\beta$, $\beta^+$, $\gamma$, etc., which emit $\alpha$, $\beta$, $\beta^+$, $\gamma$ and other rays, respectively. Half-life is usually used to characterize radioisotope attenuation. The half-life is $T_{1/2}$, the time required for the radioisotope to decay to half the original radioactivity. Its decay formula is:

$$I = I_0 \exp\left(\frac{-0.693t}{T_{1/2}}\right)$$

Provided the half-life of a radioisotope is known, the formula can be used to calculate the radioactivity at different decay times.

A nuclear reaction can occur by bombarding nuclei with neutrons, protons, $\alpha$ particles, etc. Most of the elements produced after the nuclear reaction are radioactive. Currently, radioisotopes of all elements in the periodic table can be produced by nuclear reaction. The radiation emitted by each radioisotope has a characteristic energy. Thus, the radiometric intensity or content of the isotope can be known by measuring the ray energy.

Researchers first discovered radioactive material from the fact that radiation can fog sensitive photographic films. Radioisotope self-radiography is a kind of radiography that uses the radiation produced by the radioactive sample itself to show the distribution and content of radioactive material in the sample. If a sample containing radioactive material is attached to a photographic film in a dark room or directly coated with nuclear latex on the sample, the part of the film (or latex film) close to the radioactive atom is sensitive to radiation emitted by the radioactive atom over time. When the film (or latex film) is developed, the distribution of radioactive material can be seen from the light-sensitive blackening, and the degree of concentration in each part of the sample can be determined by measuring the blackness of each part with a micrometer.

Han et al. (1985) used a radioisotope self-ray-tracer atomic method to carry out many chemical studies of metallurgical materials. The research on the rare earth desulfurization reaction in pure iron is as follows: The reaction equilibrium of Ce, La and Nd formation of sulfides in molten iron was studied at 1600 °C, a MgO crucible was used to avoid the reaction of the rare earth with the crucible and the balance time was increased to ensure balance.

Electrolytic iron with an oxygen content of about 170 ppm was deoxidized using calcium in a carbon tube furnace Ar atmosphere. The melt oxygen content was determined by $ZrO_2(+MgO)$ solid-electrolyte oxygen sensor. The calcium deoxidized to about 10 ppm. The iron sample was condensed in an Ar atmosphere, the surface oxide skin was removed, then the iron was melted in an Ar atmosphere, and a certain amount of chemical impurity was added to make $a_{[S]}/a_{[O]} > 100$, so as not to produce rare earth oxygen sulfides after adding the rare earth. The content of rare earth in the dissolved state, the content of rare earth in inclusions and the total content of rare earth were determined.

The result was: in pure Fe solution at 1600 °C, $K_{CeS} = 2.80 \times 10^{-6}$, $K_{LaS} = 7.41 \times 10^{-7}$, $K_{NdS} = 2.57 \times 10^{-6}$. These are similar to the theoretical calculations of Wilson, Vahed and Kay. The study also calculated the mutual coefficients between rare earths and sulfur:

$$e_S^{Ce} = -1.88, \quad e_S^{La} = -1.51, \quad e_S^{Nd} = -1.51$$

Han QY also studied the deoxidation constant of La in molten iron. The radioisotope tracer atomic method is not easy to be applied because of the need for a safe radioisotope laboratory. Han studied in the former Soviet Union when the area of study in China expanded to metallurgical melt research in many aspects.

Du T et al. (1995), in addition to studying the deoxidation and desulfurization reactions of rare earths in Fe solution, extended the method to a nickel-based solution and so on, and determined a series for Fe of deoxidation, deoxidized sulfur and desulfurization constants.

The experiments were carried out in a molybdenum wire furnace, Ar gas protection after purification, MgO crucible. The melt oxygen activity was determined using a $ZrO_2(+MgO)$ oxygen sensor and sampled with a quartz tube. The equilibrium products were separated by electrolysis in a low-temperature anhydrous electrolyte. The electrolysis temperature was $\leq 10\ ^{\circ}C$, current density $\leq 50\ mA/cm^2$, organic electrolyte was

$$1\%\ LiCl + 2\%\ Glycerol + 2\%\ Citric\ acid + 95\%\ Methanol$$

The rare earths in rare earth sulfides were dissolved in dilute aqua regia and determined using inductively coupled high-frequency plasma torch spectroscopy. The total rare earth content minus rare earth sulfide was used to obtain the dissolved rare earth content. Then, the electrolyte and washing solution were combined and MgO evaporated to methanol-free in a water bath. HCl was added to make the solution acidity 5%. Finally, the solution content of rare earths was determined using plasma spectrometry.

According to the data obtained, this method has good accuracy, and the amount of destruction of rare earth sulfides was < 0.03%. The dissolved sulfur content was obtained by chemical analysis of the total sulfur content minus the sulfur content in rare earth compounds. Wang (2000) studied rare earth sensors to prove the existence of solid-solution rare earths until room temperature, see Chap. 6.

## 8.6 Thermodynamic Studies on the Nucleation of Rare Earth Compounds

The nucleation of inclusions in iron melts and the natural process of growing are seldom known. Elliott and other researchers (1970) have reported studies related to nucleation reactions for Fe–O–Al and Fe–Si–O melt $Al_2O_3$ and $SiO_2$. Several theories can be used to deal with nucleation phenomena.

According to the classical theory, the work of forming spherical nuclei in a metal melt is:

$$w = 4\pi\gamma^2\sigma + 4/3\pi\gamma^3(\Delta G/V)$$

136          8 Role and Application of Rare Earth Elements in Steel

In the formula,

$\sigma$—interfacial tension between the matrix and the new phase;

$r$—particle radius;

$\Delta G$—the difference in free energy between the matrix and the new phase (calculated by bulk material);

$V$—molar volume of the new phase.

The morphology of the formed nuclei was observed with a microscope. Wang et al. (1987) experimentally studied the thermodynamics of nucleation reactions in CeS and $Ce_2O_2S$ in a pure Fe solution at 1600 °C according to the methods of Turpin and Elliott. Raw material: secondary electron bombardment of iron (purchase from the Central Iron & Steel Research Institute), the chemical composition of which is shown in Table 8.1.

The deoxidized iron sample was melted and a certain amount of FeS was added to make the Fe–S parent alloy. Studying the CeS nucleation reaction, $a_{[S]}/a_{[O]} > 100$ is maintained to avoid the formation of rare earth oxygen sulfides and for the $Ce_2O_2S$ nucleation reaction, the $a_{[S]}/a_{[O]}$ was adjusted to another value.

The experiment was carried out in a molybdenum wire furnace with sealed water jacket and sample casting and sampling device, and the temperature control accuracy was $(1600 \pm 2)$ °C, atmosphere in the furnace, $p_{O_2}$ is about $10^{-18}$ Pa. Quartz tube sampling took place a certain time after each addition of rare earth. After sample preparation, the amounts of rare earth and sulfur were analyzed, and the phase analysis was carried out with an electron probe and scanning electron microscope. The nucleation samples were obtained experimentally (Kippenhan and Gschneidner 1970). This research work also calculated the relationship between the interfacial energy of other rare earth compounds and the matrix and supersaturation, as shown in Fig. 8.1, which indicates that the supersaturation of the generated CeS, LaS simple compounds is small, while supersaturation of the generated $Ce_3S_4$ is large because of the complex lattice arrangement of complex compounds.

Li and Suito (1997) used a $3Al_2O_3 \cdot 2SiO_2$ tubular solid-electrolyte and $ZrO_2(+9\%$ MgO mole fraction) plug-in solid-electrolyte battery EMF method at 1600 °C to study Fe–Al–M (M = C, Te, Mn, Gr, Si, Ti, Zr and Ce) supersaturated oxygen activity in the melt (M mass fraction in Fe is 0.0017–0.41%), the iron raw material was high purity electrolytic iron.

The experiment was carried out in a tubular $LaCrO_3$ furnace with a sealing sleeve with a corundum crucible. After electromotive force measurement, the crucible was removed and quenched under high He gas flow to analyze the contents of relevant elements.

## 8.7 Effects of Rare Earths and Sn, Pb, As, Tb, Bi, etc.

Some iron ores, like Guangxi iron ore, contain fusible metals such as Sn, Pb, As, Bi which easily make steel brittle, so it is generally believed that rare earths form high melting point compounds with these elements, which reduces the vulnerability.

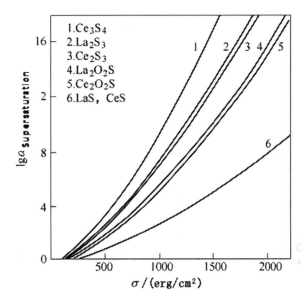

**Fig. 8.1** Relationship between interfacial energy and supersaturation of Fe–rare earth compounds

For example, in the La-Bi series, the La$_4$Bi$_3$ melting point is 1670 °C, while that of LaBi$_2$ is 932 °C (Department of Metals, Sun Yat-sen University 1978).

Some reaction phase diagrams are summarized up, for example, Ce-Te partial phase diagram of binary system, and La-Te, Ce-Te, Ce-Pb, Nd-Te, and Er-Se binary system phase diagram.

Wilson et al. (1974) gave the relationship between the standard free energy of formation and temperature of rare earth and low melting point metal compounds, rare earth oxides, oxygen sulfides, and sulfides.

The stability of rare earths and low melting point metal compounds is much lower than that of rare earths and oxygen and sulfur compounds. Only when rare earths react with oxygen and sulfur in molten steel can they form compounds with low-melting-point metals, and only high melting point compounds can float away. Therefore, it is necessary to analyze the phase diagram in detail, master the oxygen and sulfur conditions in the molten steel and strictly control the metallurgical conditions.

To eliminate the harmful effects of lead and tin, adding an appropriate amount of yttrium is better than adding the same amount of cerium, but if yttrium is insufficient, the effect is not good, which is due to the formation of low-melting-point compounds.

## 8.8 Solid-Soluble Rare Earths in Iron and Steel and Alloying Action

1. Solid-soluble rare earths

The influence of rare earths on the properties of steel is related to the trace solid solution rare earth and alloying. Han QY and others determined the amount of rare earth in the solution using an organic solution electrolytic inclusion separation and radioisotope tracer atomic method; Du T et al., using an organic solution to separate the inclusion and plasma spectrum analysis, obtained similar results; this book's author Wang CZ thought that the amount of rare earth in steel in atomic form can be inferred from the accurate determination of rare earth deoxidation constant, deoxidized sulfur constant or desulfurization constant. The amount of oxygen can be measured using an oxygen sensor and sulfur is analyzed using a chemical method, for example,

$$2[RE]_{Fe} + 2[O]_{Fe} + [S]_{Fe} = RE_2O_2S_{(s)}$$

$$K_{RE_2O_2S} = \alpha^2_{[RE]} \cdot \alpha^2_{[O]} \cdot \alpha_{[S]}$$

Rare earths in atomic form exist in Fe solution, which is not a solid solution, the meaning of "solid" is solid Fe or steel. When the molten steel solidifies, two phenomena will occur: the first is that the solubility of rare earth compounds decreases with temperature, and the solubility product decreases with temperature. The content of rare earth and other elements before solidification can be obtained from the relationship between solubility product and temperature. The second phenomenon is the direction, distribution and alloying of rare earths at the beginning of solidification or later.

An Fe atom has a radius of 1.27 Å or 0.127 nm (1 nm = 10 Å) (Xiao 1985), while the atomic radii (nm) of rare earth elements are in order,

$\alpha$-Sc, 0.16406; $\alpha$-Y, 0.18012; $\alpha$-La, 0.18791; $\alpha$-Ce, 0.172; $\beta$-Ce, 0.18321; $\gamma$-Ce, 0.18247; $\alpha$-Pr, 0.18279; $\alpha$-Nd, 0.18214; $\alpha$-Pm, 0.1811; $\alpha$-Sm, 0.18041; Eu, 0.20418; $\alpha$-Gd, 0.18013; $\alpha$'-Tb, 0.1784; $\alpha$-Tb, 0.17833; $\alpha$'-Dy, 0.1774; $\alpha$-Dy, 0.17740; Ho, 0.17661; Er, 0.17566; Tm, 0.17462; $\alpha$-Yb, 0.19451; $\beta$-Yb, 0.19392; Lu, 0.17349.

High temperature metal radius (nm) of rare earth metals:

$\beta$-Sc, 1337 °C, 0.166; $\beta$-Y, 1478 °C, 0.183; $\beta$-La, 325 °C, 0.1875; $\gamma$-La, 887 °C, 0.190; $\delta$-Ce, 757 °C, 0.184; $\beta$-Pr, 821 °C, 0.184; $\beta$-Nd, 883 °C, 0.184; $\beta$-Pm, 890 °C, 0.183; $\beta$-Sm, 450 °C, 0.18176; $\gamma$-Sm, 922 °C, 0.182; $\beta$-Gd, 1265 °C, 0.181; $\beta$-Tb, 1289 °C, 0.181; $\beta$-Dy, 0.1381, 0.180; $\gamma$-Yb, 763 °C, 0.198.

The atomic radii of the above rare earth elements, whether at room temperature or high temperature, are much larger than the Fe atomic radius. Atoms are generally regarded as spherical. When unequal diameter heteroatoms make up an alloy, it will produce elastic strain or distortion energy that will cause lattice distortion and increase the internal energy; the rare earth atoms will make the iron atoms produce elastic strain so that the distance between atoms changes, the internal energy increases.

Based on the characteristics of rare earth elements and related experimental results, Qiu (1987) comprehensively reviewed the grain boundary existence form of rare earths and its influence on grain boundary state. Mclean and Northcott deduced

## 8.8 Solid-Soluble Rare Earths in Iron and Steel and Alloying Action

the formula of equilibrium grain boundary segregation concentration of solute atoms using statistical mechanics.

$$C_{\text{segregation}} = \frac{Cm A \exp(Q/RT)}{1 + Cm A \exp(Q/RT)}$$

In the formula.

$C_m$—Solubility of solute atoms in crystals, $C_m \ll 1$;

$A$—Vibration entropy factor in the grain boundary region;

$Q$—Difference of distortion energy between the distribution of solute atoms in the crystal and grain boundary;

$R$—Molar gas constant;

$T$—Thermodynamic temperature.

From the formula, we can see that the rare earth atoms are polarized in grain boundaries and nearby regions, which makes the system energy decrease.

A Gibbs adsorption model can also be used to predict the range and quantity of grain boundary segregation. For binary alloy systems, Cahn and Hilliard derived the maximum equilibrium segregation formula of solute atoms in the grain boundary region according to the Gibbs equation.

$$\gamma_2^0 = \frac{\sigma_1}{KT} \left(1 + \ln\frac{X_e}{X_o}\right)^{-1}$$

In the formula, $\sigma_1$—Surface tension of component 1(solvent);

$X_e$—Solubility limit of component 2 (solute) in component 1;

$X_o$—Mean concentration of solute in the matrix.

Estimating La segregation amount in grain boundaries by the formula, the solubility limit of La in Fe is $X_o = 0.001$ at 785 °C (1058 K), and the surface energy of Fe is $\sigma_1 = 850$ erg/cm$^2$. Taking La average concentration in Fe of $X_o = 0.0001$ (atomic fraction) and substituting these values into the formula, we get,

$$\Gamma_{\text{La}}^0(\text{Fe}) \approx 2 \times 10^{15} \text{ atom / cm}^2$$

As a result, for La the concentration effect of segregation at grain boundaries is strong. Other rare earth atoms can be estimated by similar methods.

2. Grain boundary enrichment of rare earths and their forms

Some researchers have tried to use electron probes to analyze solutions of rare earths in the grain boundary of rare earth steel, but did not obtain satisfactory results, which is mainly due to the low analytical sensitivity of an electron probe; ion probe analysis sensitivity is high; in the following years, the accuracy of phase analysis equipment has increased.

In Study Method of Metallurgical Physics and Chemistry (4th Edition), Lin Q proposed some detection equipment sensitivity; this book's author demonstrated the presence of solid-soluble rare earths using La, Y transducers (see Chap. 6).

## 8.9 Role of Electroslag Containing Rare Earth Slag

Some experiments show that rare earths have the effect of hindering the precipitation of some carbides and nitrides along the grain boundary, or breaking the carbides and nitrides, making sheets become "island" shaped, which hinders the formation and expansion of grain boundary cracks.

The 52 Institute used $Al_2O_3$-$CaF_2$ as-cast fracture defect appeared on the longitudinal fracture specimen after normal tempering treatment of $PCrNi_3MoV$ steel with remelted old slag. They later used strips of rare earth oxide, such as $CeO_2$–CaO–$CaF_2$ (30:20:50), $RE_2O_3$–CaO–$CaF_2$ (30:20:50) slag system electroslag remelting, which improved the as-cast fracture surface. Later, the Wang CZ team participated in their ESR test.

Wang and Yang (1979) studied the $La_2O_3$–$Al_2O_3$–$CaF_2$, $Ce_2O_3$–CaO–$CaF_2$, $CeO_2$–CaO–$CaF_2$ isoreactivity line of a slag system using Zou YX's guidance method (Wang and Zou 1965, 1980), and results are shown in Figs. 8.2 and 8.3, 8.4 and 8.5 (Wang et al. 1984, 1986).

The isoreactivity lines of the $La_2O_3$–$Al_2O_3$–$CaF_2$ slag system are shown in Figs. 8.2 and 8.3; those of $Ce_2O_3$–CaO–$CaF_2$ are summarized in Fig. 8.4 ($Ce_2O_3$ isoreactivity line in mole fraction at 1600 °C); those of $CeO_2$–CaO–$CaF_2$ are summarized in Fig. 8.5 ($CeO_2$ isoreactivity lines expressed as mass fractions at 1600 °C).

According to the thermodynamic law that the activity of a rare earth oxide increases with temperature, the amount entering steel increases, the distribution of resistance spin or fragmentation of AlN, VC, VN at a grain boundary, the toughness increased. Thus, the role of rare earth atoms is to break the harmful compound distribution (Gou and Huang 1962).

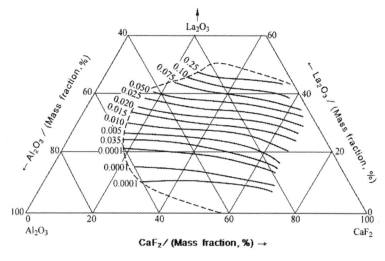

**Fig. 8.2** $La_2O_3$ isoreactivity lines expressed as mass fractions at 1600 °C

8.9 Role of Electroslag Containing Rare Earth Slag 141

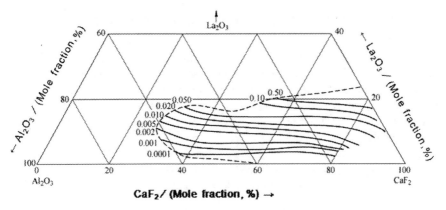

**Fig. 8.3** La$_2$O$_3$ isoreactivity line in mole fraction at 1600 °C

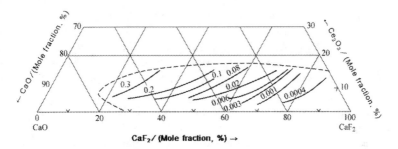

**Fig. 8.4** Ce$_2$O$_3$ isoreactivity line in mole fraction at 1600 °C

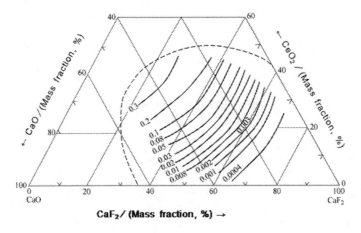

**Fig. 8.5** CeO$_2$ isoreactivity lines expressed as mass fractions at 1600 °C

## 8.10 Permeability of Slag for Electroslag Remelting

The $La_2O_3$ isoreactivity line depending on the mass fraction at 1600 °C indicates that the slag has some permeability. Zhou and Wang (1995) used a platinum–rhodium crucible and iron crucible slag in a molybdenum wire furnace, the relationship between the $p_{O_2}$ values of the rare earth oxide slag system and $Al_2O_3$-$CaF_2$ slag system and the partial pressure of gas phase oxygen was determined using a solid-electrolyte oxygen sensor.

The oxygen sensor battery design is as follows:

Pt or Mo|Cr, $Cr_2O_3$|$ZrO_2$(+MgO)| gaseous phase and slag oxygen|Pt or Mo

To avoid the reaction of solid electrolyte and slag, some tubes were treated with platinum on the outer wall. The experiment used a platinum–rhodium crucible when the atmosphere was air and changing the $p_{O_2}$ of the Ar–$O_2$ mixture gas. Two oxygen-sensing probes were used to determine the oxygen content of the gas phase and slag. Experiments showed that when the $p_{O_2}$ of the gas phase changes, the slag $p_{O_2}$ traces the change rapidly, and then gradually tends to balance. Under air pressure, the $p_{O_2}$ in the slag gets to vary from $10^{-4}$ atm at 1393 °C to $10^{-3}$ atm at 1530 °C. With purified Ar, the gaseous phase $p_{O_2}$ was reduced to $10^{-19}$ atm from 0.208 atm. Finally, $p_{O_2}$ in the slag sank to $10^{-12}$–$10^{-10}$ atm depending on the different slag strips. In the slag strip containing $CeO_2$, there are many nonstoichiometric phases between $CeO_2$–$Ce_2O_3$, and each has a corresponding equilibrium $p_{O_2}$ value; the change is complicated.

The experiment was done under static conditions, but in actual production, the effect of the arc makes slag more susceptible to oxygen in the air. Therefore, the easily oxidized elements are burned, so electroslag remelting should be carried out under the protection of Ar gas. This also avoids the oxidation of rare earths in steel.

## 8.11 Rare Earth Applications in Some Steel Types

Rare earths show strong deoxidation and desulfurization effects; the oxide, oxygen sulfide and sulfide inclusions formed in molten steel are spherical, and the dispersion is refined and evenly distributed in the steel. The rare earth atoms in the dissolved state are much larger than the matrix iron atoms because they are not easy to dissolve each other and are squeezed to the grain boundary during solidification. In this process, the long-range nature of polarization can affect the distribution of hundreds of atoms, driving away the carbide, nitride or carbon nitride of some elements in the grain boundary aggregation or fragmentation. Therefore, if adding rare earths to steel, the addition method, form and appropriate quantity can improve the properties of all steel, and has a characteristic effect on some kinds of steel. Several classes of steel are summarized below.

# 8.11 Rare Earth Applications in Some Steel Types

(1) Low temperature oil and gas pipeline steel: Oil pipeline steel used in a cold zone is very strict for low-temperature brittleness. This kind of well pipe must withstand the test of deep high pressure, hydrogen sulfide, $CO_2$ and other harsh conditions. The addition of rare earths can solve these problems, but the amount of rare earth added must be controlled.

(2) Low alloy steel: The impact strength of low alloy steel (0.14–0.2% C, 0.1–0.2% Mo, Cr, V) by adding a rare earth is several times that of steel without the rare earth.

(3) High alloy steel: High-alloy steel includes stainless steel, heat-resistant steel, high-strength stainless steel, high-speed steel and high-manganese steel. Zhao XC and Song WS reviewed the effect of adding rare earths to high-alloy steel.

(4) Structural steel: The addition of a rare earth to low carbon structural steel can improve the toughness and tensile strength at room temperature.

(5) Rail: Adding a rare earth to rail steel can increase wear resistance and hardness by 25–30% or more.

(6) Silicon steel plate: The addition of rare earths to silicon steel plate is very good. The addition of a rare earth can prevent or reduce loss, and produce nondirectional silicon steel plate with a high magnetic field and good characteristics.

(7) High-alloy steel.

High-alloy steel includes stainless steel, heat resistant steel, high-strength stainless steel, high-speed steel and high-manganese steel. Zhao XC et al. review the effect of adding rare earth to high-alloy steel, for example:

Adding rare earths to W18Cr4VMn high speed steels can significantly improve thermoplasticity; 9Cr18 and Cr12Ni25Mo3Si2Nb steels with rare earths can also significantly improve their thermoplasticity; for ZGMn13, OCr13Ni12Mn9N improves properties, fluidity and grain refinement.

The work-hardening rate of OOCr12Ni9Cu2TiNb martensite aged stainless steel with rare earth is low, the cold-working property is good, the age-hardening effect of OCr17Ni7Al semi-austenite precipitated hardened stainless steel is improved, and various performance indexes are improved. Creep fracture strength and plasticity of Cr18Ni18Si2 stainless steel were clearly improved. Rare earths have the effect of hindering the formation and development of grain boundary wedge cracks, can hinder the precipitation of nitrides along grain boundaries, improve the lasting strength of steel such as Cr18Si2, Cr18Ni8N, Cr25Ni20 and Cr24Ni7N at different temperatures, strengthen grain boundaries and disperse impurity elements in grain boundaries.

Rare earths can improve the chemical stability of high-alloy steel and reduce the oxidation of OOCr11Ni10Mo2Ti steel at 800–900 °C in air to 1/20. Major oxides, from poorly protective needle-like iron oxides to good protective $Cr_2O_3$, this is because the rare earth oxide is solidly dissolved in the $Cr_2O_3$ of the main oxide, refining the grain of the oxide, improving the plasticity of the oxide skin, making it difficult to crack and fall off.

Wang and Ye (1985) studied the relationship between the standard free energy of formation and temperature of $Y_2O_3 \cdot Cr_2O_3(2YCr_2O_3)$ to understand the thermodynamic stability of $Y_2O_3 \cdot Cr_2O_3$. They used $ZrO_2(+MgO)$ as a solid electrolyte to form the following cell:

$$Mo|Cr, \; Cr_2O_3|ZrO_2(+MgO)|Y_2O_3 \cdot Cr_2O_3, Y_2O_3, Cr|Mo \; 909 - 1113\,°C$$

$$\text{Battery reaction is } Y_2O_{3(s)} + Cr_2O_{3(s)} = Y_2O_3 \cdot Cr_2O_{3(s)}$$

$Y_2O_3$ and $Cr_2O_3$ can generate $Y_2O_3$ and $Cr_2O_3$ protective layer beneficial to steel, at about 847 °C. The reactions of other rare earth oxides and $Cr_2O_3$ can also be explained by the actual many kinds of steel experiments to produce a $RE_2O_3 \cdot Cr_2O_3$ protective layer, which increases the corrosion resistance of steel at medium and high temperatures. The application of rare earths in cast steel is similar to that in rail steel, and the reason is similar.

## 8.12 Surface Infiltration of Rare Earths in Steel

Some equipment has special properties on the surface of some steel parts, and some special-shaped parts need certain properties, so the surface infiltration of a rare earth is needed for this purpose. Because the radius difference between solute and solvent atom is more than 25%, the solubilizing theory was not available until 1968. Since 1979, the Harbin University of Technology, Shenyang University of Technology, Wuhan Institute of Material Protection and Shanghai University of Technology have carried out research and practical application of rare earth surface infiltration with different methods.

Experimental studies have shown that rare earth atoms can diffuse into the matrix. For example, to solve the nitrogen-induced embrittlement problem of chromium alloy in modern air engine blade materials, Stephens has studied the surface of yttrium, lanthanum and europium single and composite layers.

The mechanism of rare earth infiltration to improve the surface properties of steel, Lv and Lv (1987) believes that in addition to purification and metamorphism, there is a microalloying effect, similar to molten steel plus rare earth. Polarization sphere theory should also be fully considered. In this book, the author, Wang CZ, thinks that surface physics can help to understand the essential problem of rare earth surface infiltration.

## 8.13 Production of Rare Earth-Containing Steel

There have been many successful applications of rare earth-containing steel. For example, some of Baotou Iron and Steel Co.'s products have excellent properties such as strip steel, rail steel and wind power plate; the technology has been applied to new steel, such as automobile, armored plate and rare earth rail, the effect is good. Since 2005, this company has developed high strength steel and exported the rail to Brazil, the United States and other countries. The addition of rare earths in steel makes the rail high temperature plastic with low crack sensitivity, so the rolled rail has high strength, good toughness and easy welding. In addition, the technology of multifurnace continuous casting of rare earth steel also makes steel products tougher and more resilient.

The high strength rare earth rail of Baotou Iron and Steel Co. is laid on the Beijing-Baotou railway line. With the rapid development of railway transportation, the characteristics of heavy train axle load, fast driving speed, and high traffic density cause the rail life to be shortened and affect the driving safety. So need to develop higher strength of heat treatment rail. Baotou Iron and Steel Company developed high-strength rail with rail strength 1080 MPa, quenching strength 1280 MPa. The addition of rare earth in steel makes rail have good high temperature plasticity and low crack sensitivity, which is suitable for continuous casting production. The rolled rail has high strength, good toughness and easy welding.

Other institutions such as Tiangong Group, Southeast University and the General Institute of Steel Research cooperated to manufacture high quality rare earth high speed steel by horizontal continuous casting, developed a new adaptive PID control remelting technology and completed production automatic control, etc.

## References

Baotou Steel rare earth. The research project of rare earth steel of Baotou Steel Group has made new progress. http://www.cnree.com/news/xitujike/2/9454.html. 27 Nov 2013

Chen XY (1987) Application of rare earths in steel abroad. In: Yu ZS (ed) Application of rare earths in steel. Metallurgical Industry Press, Beijing, pp 56–64

Department of Metals, Sun Yat-sen University (1978) Rare earth physicochemical constants. Metallurgical Industry Press, Beijing

Du T, Han QY, Wang CZ (1995) Physical chemistry of rare earth alkali soil and its application in materials. Science Press, Beijing

Forward G, Elliott JF, Kuwabara T (1970) Formation of silica and silicates during the solidification of Fe-Si-O Alloys. Metall Mater Trans B 1(10):2889–2898

Gou QQ, Huang SX (1962) Analytical wave functions of atoms. J Phys 18(2):63

Gschneidner KA, Kippenham N, MacMaster OD. Thermochemistry of the rare earths. Report IS-RIC-6, Rare Earth Information Center, Ames, Iowa

Gschneidner KA, Kippenham N. Thermochemistry of the rare earth carbides, nitrides and sulfides for steelmaking. Rare Earth Information Center, Ames, Iowa 50010, Report IS-RIC-5

Han QY, Dong YC, Feng X et al (1985) Equilibria between rare earth elements and sulfur in molten iron. Metall Trans B 16B:785–792

Han QY(1987) Physical chemistry of rare earths in iron and steel smelting. In: Yu Z (ed) Application of rare earths in steel. Metallurgical Industry Press, Beijing, pp 34–50

Janke D, Fischer WA (1978) Deoxidation equilibria of cenum, lanthanum and hafnium in liquid iron. Arch Eisenhüttenwes 49(9):425–430

Kippenhan N, Gschneidner KA (1970) Rare-earth metals in steels. Molybdenum Corporation of America, IS-RIC-4

Langenberg FC, Chipman J (1958) Equilibrium between cerium and sulfur in liquid iron. Trans Metall Soc AIME 5:290–293

Li GQ, Suito H (1997) Galvanic cell measurements on supersaturated activities of oxygen in Fe-Al-M (M=C, Te, Mn, Cr, Si, Ti, Zr, and Ce) melts. Metall Mater Trans B 2BB 4:251–258

Lv ZJ, Lv QG (1987) Rare earth surface treatment of steel. In: Yu ZS (ed) Application of rare earths in steel. Metallurgical Industry Press, Beijing

Luyckx LA (1983) Chinese rare earths (foreign rare earth application album), Suppl 17–33

Mclean A, Lu SK (1974) The Thermodynamic behaviour of rare earth elements in molten steel. Met Mater 10:452–457

Ohmachi RJ (1993) Application of rare earth elements in metals. In: Shiokawa JR, Zhai YS, Yu ZH (eds) The latest application technology of rare earths. Chemical Industry Press, Beijing, pp 229–239

Qiu JF(1987) The existence form of rare earth in grain boundary and its influence on grain boundary state. In: Yu ZS (ed) Application of rare earths in steel. Metallurgical Industry Press, Beijing, pp 251–262

Vahed A, Kay DAR (1976) Thermodynamics of rare earths in steelmaking. Metall Trans B 7B:375–383

Wang CZ (2000) Solid electrolytes and chemical sensors. Metallurgical Industry Press, Beijing, pp 518–541

Wang CZ (2013) Metallurgical physical and chemical research methods. Version 4. Metallurgical Industry Press, Beijing, pp 106–128

Wang CZ, Yang DY (1979) Determination of rare earth oxide activity in slag. Acta Metall Sin 15:433–437

Wang CZ, Gustafsson S, Li YZ, et al. (1987) Nucleation thermodynamics of rare-earth inclusions in liquid iron. Chin J Met Sci Technol 3: 102–106

Wang CZ, Ye SQ, Hu YN, Du QS (1986) Activity study of $Ce_2O_3$ in $Ce_2O_3$-$CaO$-$CaF_2$ ternary slag system. Acta Metall Sin 22:77–79

Wang CZ, Ye SQ (1985) In: Proceeding of the international conference of rare earth development and applications, Beijing, p 1248

Wang CZ, Ye SQ, Yu DY, Guo WQ (1984) $La_2O_3$ activity in $La_2O_3$-$Al_2O_3$-$CaF_2$ ternary slag system. Acta Metall Sin 05

Wang CZ, Zou YX (1965) $La_2O_3$ activity in liquid $La_2O_3$-$CaF_2$ and $La_2O_3$-$CaF_2$-$CaO$-$SiO_2$ slag. J Northeastern Univ (Nat Sci) 2(5):1980; Acta Metall Sin 16:190–194

Wang CZ, Zou YX (1980) $La_2O_3$ activity in liquid $La_2O_3$-$CaF_2$ and $La_2O_3$-$CaF_2$-$CaO$-$SiO_2$ slag. Acta Metall Sin 16:190–194

Wang CZ, Ye SQ, ZX (1985) Thermodynamic properties of $Y_2O_3 \cdot Cr_2O_3$ composite oxides. J Phys 34(8):1017–1025

Wei SK (1978) Physical and chemical problems of rare earth steel smelting. In: Meeting on smelting process and addition methods of rare earth steel (Liuzhou), Sept

Wilson WG, Kay DAR, Vahed A (1974) The use of thermodynamics and phase equilibria to predict the behavior of the rare earth elements in steel. J Met 14–23

Xiao JM (1985) Alloy energy. Shanghai Science and Technology Press, Shanghai, pp 229–302

Zhou CX, Wang CZ (1995), Permeability of rare earth oxide electro slag. In: Du T, Han QY, Wang CZ (eds) Physical chemistry of rare earth alkali soil and its application in materials. Science Press, Beijing

# Chapter 9
# Application of Rare Earths in Nodular and Vermicular Cast Iron

## 9.1 On Nodular and Vermicular Cast Iron

In 1752, Swedish chemist discovered a new kind of ore; scientists later identified it as ceria and the element cerium. Several uses of cerium, such as light-emitting enhancers, flint stones, electromagnetic carbon rods, were discovered. In 1947, Morrogh H in the U.K. found that if cerium (Ce) and other rare earth elements were added to hypereutectic gray cast iron and incubated, they formed alloys with Si, Mn, and Zr, and that graphite was rendered spherical if the content of Ce in molten iron exceeded 0.02% (Zeng and Wu 1987). At almost the same time, Gangnebin AP in the U.S. pointed out that with the addition of magnesium to molten iron and subsequent inoculation with ferrosilicon, when the content of manganese in molten iron was more than 0.04%, spherical graphite could be obtained.

Since then, large-scale industrial production and research on ductile iron have continued apace. It has been found it is necessary to add rare earths, usually in conjunction with magnesium to completely spheroidize graphite. At first, the obtained worm-shaped products were considered inferior. After 1977, however, opinions changed about this cast iron, as it was found to have a high strength coefficient and good casting properties, allowing product walls to be made thinner and the fabrication of lightweight vehicles. Compared with spherical cast iron, vermicular cast iron does not need heat insulating cover and the cost is low (Shiokawa 1993).

In China, researchers successfully developed rare earth/magnesium-doped ductile iron between 1959 and 1964. The research focused on the manufacture of ductile iron from high-sulfur pig iron and the identification of new spheroidizing agents suitable for productive casting. Since 1970, research has mainly been directed towards improving the quality of ductile iron in terms of its overall mechanical properties and stability, as well as its mass production (Xie 1981; Xue and Cao 1981; Wang 1981; Sun and Wang 1984; Li 1983; Chen et al. 1985; Liu et al. 1994). At present, the focus of ductile iron research is increasingly directed towards endowing it with high strength, toughness, and hardness (Wuxi Diesel Engine Plant, Xi'an Jiaotong University, Shanghai Internal Combustion Engine Plant 1981; Niu and Zhang 1983; Liu

© Science Press 2023
C. Wang, *Theory and Application of Rare Earth Materials*,
https://doi.org/10.1007/978-981-19-4178-8_9

et al. 1994). Thus, the roles of rare earths in ductile iron and vermicular graphite cast iron are widely studied in China (Linebarger et al. 1983; Zhang 1983; Zhu et al. 1984; Huang et al. 1987; Qian et al. 1987; Xu et al. 1987; Zhu and Zhang 1987; Jiang et al. 1994; Shen et al. 1994; Xiong et al. 1994). Sheng et al. (1987) studied the effects of trace and small amounts of rare earth alloys on the microstructure and properties of different carbon contents and sulfur-containing hot metal.

Mixed rare earth (RE > 99%) and rare earth ferrosilicon alloys (of which there are two types: RE about 30%; RE about 21%) have been studied experimentally. The result showed that using hypoeutectic, eutectic, and super eutectic iron melt with a sulfur content < 0.08% got good results, the compressive strength increasing by 19.6–49 MPa. With increasing addition of rare earth alloys, the change of mechanical properties has obvious regularity, showing a double peak effect; the tensile strength of cast iron gradually increases. Flake graphite is transformed into worm-like graphite, along with variable amounts of spherical graphite. However, when excess alloy is added, it has a negative impact on performance due to the formation of rare earth carbides or other compounds.

For example, at $1 \text{ kgf/nm}^2 = 9.8 \text{ MPa}$, the relationship between the addition of rare earth ferrosilicon alloy (e.g. REFeSi) and mechanical properties ($\sigma_b$) is shown in a diagram.

Basic tests were carried out in laboratory. Samples were melted in an intermediate-frequency induction furnace, monitoring the temperature with a fast platinum–rhodium thermocouple. The graphite morphologies and matrix structures of the samples were observed with a full-phase microscope and a scanning electron microscope. The fluidity was also studied. The processing temperature was 1420–1450 °C, the alloy size was 3–8 mm, and the processing capacity was about 8 kg each time.

MnS in initial iron melt was polygon, but when adding rare earth, formed approximately spherical complexes containing rare earth, manganese, and iron, which served to improve the properties of cast iron.

Sheng et al. (1994) found that hot metal shows a double-peak effect with increasing rare earth addition, which is a valuable finding in relation to cast iron. This double-peak phenomenon is found in both iron and steel metallurgy and non-ferrous metallurgy. For example, Yang YJ and the book author Wang CZ in 2004 have studied the relationship between the activity and concentration of yttrium in Y-Al alloys (Wang 2000; Yang and Wang 2004); for a homemade $YF_3$ (+$CaF_2$) solid-electrolyte yttrium sensor, the two-peak phenomenon was seen, as shown in Figs. 9.1, 9.2 and 9.3.

$E_Y$–[Y] is first valley then peak; while the $\ln \alpha_Y$–[Y] is first peak then valley; the $E_0$–[Y] is also first peak then valley. The peak and valley order is related to the representation of coordinates. The actual interaction relationship remains unchanged. A theoretical calculation of the relationship between O content and Al content in the Fe-Al-O system at 1600 °C is summarized in Fig. 9.4 (relationship between oxygen and Al content of Fe-Al-O system (1600 °C)). The lowest point of the oxygen content is at $[Al]_{Fe} \approx 0.081\%$, consistent with that in the industrial deoxygenation of aluminum.

9.1 On Nodular and Vermicular Cast Iron

**Fig. 9.1** $E_Y$–[Y] curve

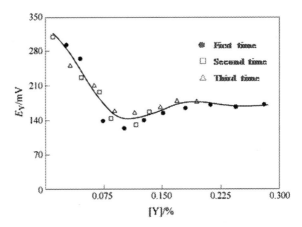

**Fig. 9.2** $\ln \alpha_Y$–[Y] curve

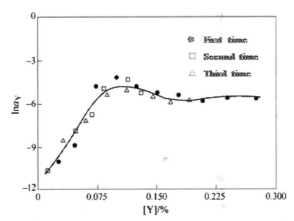

**Fig. 9.3** $E_0$–[Y] curve

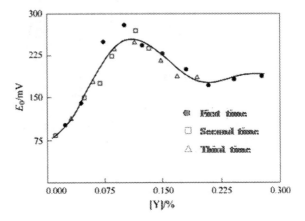

**Fig. 9.4** Relationship between oxygen and Al content of Fe-Al-O system (1600 °C)

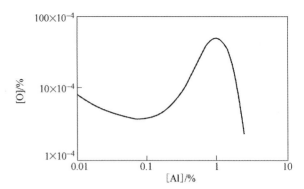

In nonferrous metals, the oxygen activity coefficient $f_O$ also shows a double peak with increasing alloy element concentration. Janke and Fischer obtained the Cu-P-O systems curves, the relationship between $\lg f_O$ and [P] at 1150 and 1600 °C by EMF experiments (Janke and Fischer 1973), as shown in Fig. 9.5.

The $\lg f_O$ and [P] relationship curves obey the equations:

$$\lg f_O = -5.91[P] + 10.34[P]^2 - 6.80[P]^3 + 1.48[P]^4 \ (1150\,°C)$$

$$\lg f_O = -2.48[P] + 3.71[P]^2 - 2.01[P]^3 + 0.36[P]^4 \ (1600\,°C)$$

**Fig. 9.5** Relationship between oxygen activity coefficient and P content in copper melt

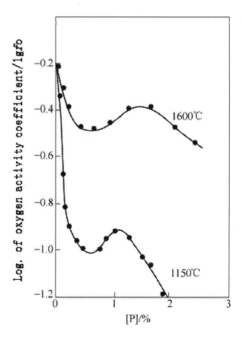

## 9.2 Double-Peak Effect of the Taylor Expansion Equation

There are many similar examples, but according to the actual production needs, the alloy element content is often low, so only a single peak is seen. The second peak would be seen at higher concentrations.

## 9.2 Double-Peak Effect of the Taylor Expansion Equation

The bimodal effect on physical and chemical behavior of elements in a melt and the physical properties after solidification with the change of the composition shows that the field strength of the microscopic particles also changes with the content of the alloying elements in the melt (or after solidification). This results in interactions changes, such as between atoms themselves and between different element atoms; and the properties also change as composition change.

Taylor expressed this law of change by a mathematical formula involving primary, secondary, and tertiary differentials, etc (Sigworth and Elliott 1974a, b; Pierre 1977; Lupis 1983). This formula is a mathematical description of various changes. The Taylor expression was used by Wagner C to describe the relationships between the activity coefficients of metallurgical melt components and the composition, and has become known as the Wagner formula.

For example, let component 1 be a solvent, and 2, 3, and 4 be solutes. If pure matter is the standard state, then for component 2 the relationship between activity coefficient and composition can be expressed as:

$$\ln\gamma_2 = \ln\gamma_2^0 + \left(\frac{\partial \ln \gamma_2}{\partial x_2}\right)_{x_1 \to 1} x_2 + \left(\frac{\partial \ln \gamma_2}{\partial x_3}\right)_{x_1 \to 1} x_3 + \left(\frac{\partial \ln \gamma_2}{\partial x_4}\right)_{x_1 \to 1} x_4$$

$$+ \frac{1}{2}\left(\frac{\partial^2 \ln \gamma_2}{\partial x_2^2}\right)_{x_1 \to 1} x_2^2 + \frac{1}{2}\left(\frac{\partial^2 \ln \gamma_2}{\partial x_3}\right)_{x_1 \to 1} x_3^2 + \frac{1}{2}\left(\frac{\partial^2 \ln \gamma_2}{\partial x_4^2}\right)_{x_1 \to 1} x_4^2$$

$$+ \cdots + \left(\frac{\partial^2 \ln \gamma_2}{\partial x_2 \partial x_3}\right)_{x_1 \to 1} x_2 x_3 + \left(\frac{\partial^2 \ln \gamma_2}{\partial x_2 \partial x_4}\right)_{x_1 \to 1} x_2 x_4$$

$$+ \cdots + \frac{1}{6}\left(\frac{\partial^2 \ln \gamma_2}{\partial x_2^3}\right)_{x_1 \to 1} x_2^3 + \frac{1}{6}\left(\frac{\partial^3 \ln \gamma_2}{\partial x_4^3}\right)_{x_1 \to 1} x_4^3$$

In the formula, each higher order is:

$$\frac{1}{2}\left(\frac{\partial^2 \ln \gamma_2}{\partial x_2^2}\right)_{x_1 \to 1} = \rho_2^2; \quad \frac{1}{2}\left(\frac{\partial^2 \ln \gamma_2}{\partial x_3^2}\right)_{x_1 \to 1} = \rho_2^3; \quad \frac{1}{2}\left(\frac{\partial^2 \ln \gamma_2}{\partial x_4^2}\right)_{x_1 \to 1} = \rho_2^4;$$

$$\left(\frac{\partial^2 \ln \gamma_2}{\partial x_2^2 \partial x_3}\right)_{x_1 \to 1} = \rho_2^{(2,3)}; \quad \left(\frac{\partial^2 \ln \gamma_2}{\partial x_2 \partial x_4}\right)_{x_1 \to 1} = \rho_2^{(2,4)}; \quad \frac{1}{6}\left(\frac{\partial^3 \ln \gamma_2}{\partial x_2^3}\right)_{x_1 \to 1} = \tau_2^2;$$

$$\frac{1}{6}\left(\frac{\partial^3 \ln \gamma_2}{\partial x_3^3}\right)_{x_1 \to 1} = \tau_2^3; \quad \frac{1}{6}\left(\frac{\partial^3 \ln \gamma_2}{\partial x_4^3}\right)_{x_1 \to 1} = \tau_2^4$$

This can also be written as:

$$\ln \gamma_2 = \ln \gamma_2^0 + \varepsilon_2^2 x_2 + e_2^3 x_3 + e_2^4 x_4 + \cdots + \rho_2^2 x_2^2 + \rho_2^3 x_3^2 + \rho_2^4 x_4^2$$
$$+ \cdots + \rho_2^{(2,3)} x_2 x_3 + \rho_2^{(2,4)} x_2 x_4 + \cdots + \tau_2^2 x_2^3 + \tau_2^3 x_3^3 + \tau_2^4 x_4^3 + \cdots$$

Here, $\rho_2^2$, $\rho_2^3$, and $\rho_2^4$ are the second-order interaction coefficients, $\rho_2^{2,3}$ and $\rho_2^{2,4}$ are the second-order cross-phase interaction coefficients, and $\tau_2^2$, $\tau_2^3$, and $\tau_2^4$ are the third-order interaction coefficients. The above forms also apply to the activity coefficients of components 3 and 4. See Sect. 6.5 in Chap. 6.

## 9.3 Process Conditions for Stable Production

Qualified vermicular graphite cast iron is only obtained when the residual amounts of rare earths, magnesium, and other elements are in very narrow ranges. Worm-like graphite is formed on going from flake graphite to spherical graphite, and so the properties of vermicular graphite cast iron to some extent resemble those of ductile iron and gray cast iron.

Sheng and Ying (1987) studied the technological characteristics of stable production from cupola hot metal of rare earth vermicular graphite cast iron.

## 9.4 Metallurgical Factors Affecting Production

The requisite amount of vermicularizer must be added to the hot metal so that the residual amounts of rare earth, magnesium, and other elements remain within a predetermined range. If their contents are too low, graphite will remain in flake form. If their contents are too high, the graphite will become spherical. The production of ductile iron from cupola iron melt, due to high sulfur content therein, further increases the difficulty of stable production (Sanbongi et al. 1972).

The results of research on the production stability of vermicular graphite cast iron are given in an article by Zeng DB et al. A certain correspondence between graphite morphology and residual RE and Mg was evident. A worm-like graphite content of more than 50% can only be obtained if the residual amounts of RE and Mg are in the region between the two slashes in a diagram. The upper-right region of the diagram is dominated by spherical graphite, whereas the lower-left region of the diagram is dominated by flake graphite.

This shows that vermicularizing agents with a high RE/Mg ratio have a weaker ability to form spherical graphite than those with a low RE/Mg ratio, although the range for obtaining vermicular graphite cast iron is much wider. When using RE–Mg alloy with high RE/Mg content to produce vermicular iron, it is easy to exceed

## 9.4 Metallurgical Factors Affecting Production

the permissible content range for obtaining vermicular iron and thereby affect the production stability.

1. Thermodynamic and kinetic factors affecting graphite morphology

The nucleation length of metal and graphite in cast iron mainly depends on the driving force, and the solid/liquid interface of the metal is uneven, that is, "rough surface". The energy of metal atoms in the liquid is easily adsorbed on this rough surface, and growth can easily occur in various crystallographic directions. Nonmetallic carbon is very different from a metal; its solid/liquid interface is smoother at the atomic level, adventitious carbon is not easily adsorbed at such an interface, and growth can only occur in several limited crystallographic directions; the growth process needs a greater driving force.

The way in which graphite nucleation proceed is related to the distance between atoms in different crystal planes and their binding force. Graphite is a densely packed hexagonal crystal with a nearest distance between two adjacent atoms in the basal plane of 1.42 Å (1 Å $= 10^{-10}$ m), while the furthest distance between atoms is 2.84 Å, but in the prism, the nearest distance 1.42 Å and the furthest distance is 3.64 Å. This indicates that graphite is compact in the basal plane and loose in the prism, and the binding energy between atoms in the basal plane is much greater than that between atoms in the prism.

From a thermodynamic point of view, carbon atoms are more easily adsorbed on the basal plane, i.e., they tend to form spherical graphite and vermicular graphite. This has been confirmed by the growth of spherical graphite when almost sulfur-free, oxygen-free cast iron solidifies in vacuo. If iron contains sulfur and oxygen, these are easily adsorbed on the prism, because their atomic radii (oxygen 0.66 Å, sulfur 1.04 Å) are much smaller than the distance between the atoms. Consequently, the interfacial energy is greatly reduced, and becomes lower than that of the basal plane. Graphite grows preferentially in the presence of rare earth elements with large atomic radii, but their density is different from that of iron, and they do not readily diffuse in molten iron. Moreover, the oxygen and sulfur compounds produced by rare earth elements have high melting points, and tend to encrust the outer surface of the alloy. This further hinders the dissolution and diffusion of rare earth elements in molten iron, inhibiting their good effects.

In order to effectively utilize rare earths for spheroidization and vermicularization, their reaction kinetics in hot metal must be improved.

2. Measures to improve the kinetics of rare earth vermicular treatment

A vermicularizing agent may be prepared by the gasification of a low-boiling-point metal in hot metal. The rare earth metal and low-boiling-point metal are prepared together, and the characteristics of gasification of the low-boiling-point metal in the hot metal are exploited to break through the shell of oxygen and sulfur compounds coated on the surface of the alloy. The hot metal is stirred during the upward bubbling process of the low-boiling-point metal, which accelerates the diffusion rate of the rare earth metal in the hot metal and regulates its treatment effect. Two kinds of alloys are commonly used for better results.

(1) The main components of the alloy are:

$$[RE]6-8\%, [Mg]9-10\%, [Si] \leq 44\%, [Ca] \leq 5\%;$$

$$\text{Density } 4-4.1 \, g/cm^3; \quad \text{melting point } 980-1180 \, °C$$

Some producers have adjusted the proportion of each alloy element. It has been demonstrated that the higher the content of rare earth in the alloy, the wider the creep range. High-magnesium and low-rare earth alloys react intensely and the effect is unstable; the low-magnesium and high-rare earth reaction is gentler, the decrease of magnesium content makes graphite flake easily and causes graphite to decline. Studies have suggested that middle amount of magnesium and rare earths, that is, [Mg] 9–10%, [RE] 4–6%, are ideal vermicularizers.

(2) Rare earth zinc-magnesium alloys are a new type of creep agent designed in China. There are two types: one is rare earth zinc-magnesium–aluminum alloy, [RE] 20.9%, [Mg] 3.31%, [Zn] 2.85%, [Al] < 1.0%, [Si] 38.22%, residual iron, density 4.29 g/cm$^3$; the other is rare earth zinc-magnesium alloy: [RE] 20.8%, [Mg] 3.63%, [Zn] 2.58%, [Si] 38.98%, residual iron, density 4.36 g/cm$^3$. This mixture has a low boiling point of 907 °C; it has the same gasification effect as magnesium, but zinc does not have the ability to spheroidize graphite. Therefore, the tendency to form globular graphite during creep treatment is reduced, and the likelihood of obtaining vermicular graphite cast iron is increased. Indeed, vermicular graphite cast iron has been successfully produced in factories.

3. Bottom detonation

Rare earth ferrosilicon and rare earth ferrosilicon alloy are layered at the bottom of the package, and the dynamic reaction conditions are improved by the stirring of the hot metal induced by the gasification and explosion potential of rare earth magnesium ferrosilicon alloy.

4. Tapping trough rendering and double-slot vortexing rendering

The result of tapping trough rendering is not ideal; the double-slot vortexing method has simple structure, convenient operation and stable vermicularization.

## 9.5 Thermodynamic Study of the Effects of Arsenic, Antimony, and Bismuth

As discussed earlier, rare earth can inhibit the harm of arsenic, antimony and bismuth to steel. Is this the same for cast iron? Wang CZ has participated in the Liuzhou steel plant to cast molten iron with rare earth removal of white metal hazards. The test results confirmed that the strength of the cast iron was increased, and the graphite was spherical or spheroidal (Zhu and Wang 1981).

## 9.5 Thermodynamic Study of the Effects of Arsenic, Antimony, and Bismuth

Iron ore in some areas of southern China contains trace arsenic. Without special milling of ores or roasting treatment, the arsenic in the ore will be transferred to pig iron or steel during smelting, rendering them brittle when hot. Many procedures have been explored for the removal of arsenic from molten iron and steel, such as high-speed blowing in oxygen, chlorination, and vacuum smelting.

1. Adding yttrium experiment

In recent years, hot metal has been pretreated in order to improve metallurgical processing. Wang and Shi have discussed the possibility of arsenic removal while dephosphorization and desulfurization with hot metal pretreatment. At the same time, through the Fe-$C_{saturation}$-Y and Fe-$C_{saturation}$-Y-As systems experiment, explored the inhibitory effect of rare earths on the arsenic hazard. A study of arsenic removal with several ratios of two slag systems, $Na_2O$-CaO-$SiO_2$ and $Na_2CO_3$-CaO-$CaF_2$, showed that the graphite in cast iron is agglomerated. This indicated that arsenic is not a spheroidizing element of graphite, but is in fact anti-spheroidizing, so it renders cast iron brittle. Addition of a small amount of yttrium to the cast iron melt and rare earths are found to promote graphite spheroidization.

2. Thermodynamic study of Fe-$C_{saturation}$-Ce, Fe-$C_{saturation}$-Ce-Sb, and Fe-$C_{saturation}$-Ce-Bi systems

The Ce content in mixed rare earths is the highest, and it is a good graphitic spheroidizing element. Wang et al. (1987) and Wang and Wang (1988) studied the thermodynamics of Fe-$C_{saturation}$-Ce, Fe-$C_{saturation}$-Ce-Sb, and Fe-$C_{saturation}$-Ce-Bi systems.

The hot metal in a blast furnace is saturated with carbon. Therefore, to study the physicochemical properties of hot metal in a laboratory, it is necessary to prepare carbon-saturated iron in a graphite crucible.

The raw materials and treatments were as follows:

- Graphite crucible: made of spectroscopically pure graphite rod and graphite block; vacuum degassed at about 200 °C prior to use.
- Pure iron: secondary electron bombardment of iron (> 99.9%).
- Cerium: > 99.99%.
- Antimony: > 99.97%.
- Bismuth: > 99.95%.
- Carbon-saturated iron: in order to obtain a low-oxygen melt, and make carbon melting reaction get equilibrium in the experiment, low-temperature (1150 °C) carbon-saturated iron was first prepared and stirred with a spectroscopically pure graphite rod. After equilibration, samples were withdrawn through a quartz tube under Ar and set aside.

The experimental set-up and methods were as follows:

The experiment was carried out in an $MoSi_2$ furnace with a sealed water jacket. The sample was placed in a corundum tube under a purified Ar atmosphere. The constant-temperature band was 5 cm, in a range $\pm$ 1 °C, the Ar > 99.99%, through silica gel and $P_2O_5$ double stage dehydration, and magnesium furnace deoxygen

(560 °C). In order to avoid the influence of residual air in the furnace, titanium sponge was placed in the 10 cm length zone below the constant temperature area of the furnace. Titanium oxygen reaction is at 1300 °C, and according to the Ti-O equilibrium oxygen partial pressure, $\rho_{O_2}$ is $3.6 \times 10^{-26}$ atm (1 atm = 101,325 Pa).

To verify the feasibility of the experimental set-up and method, the solubility of C in Fe–C binary alloys was first determined. The experimental temperatures were 1315, 1415, and 1515 °C. The equilibration time was determined by experiments at 1315 °C. The C content was analyzed by a combustion energy method combined with infrared spectroscopy. The equilibration time was 4 h. In order to ensure that equilibrium had been reached, the Fe-C$_{saturation}$ binary experiments at all three temperatures were conducted for 6 h.

Experimental results were consistent with the data of Chipman and Floridis (1955). They determined the solubilities of carbon in Fe-C$_{saturation}$ binary mixtures at 1315, 1415, and 1515 °C as 4.68%, 4.93%, and 5.19%, respectively, compared with 4.69%, 4.93%, and 5.19% in this work, which can be described as fully consistent. To Fe-C$_{saturation}$-Ce ternary alloy, in carbon pre-saturated condition, dissolution equilibrium of cerium is reached within 30 min.

In experiments with a cerium content of less than 0.2% (mass fraction), a large graphite crucible was used. In experiments with high cerium content, we used a graphite holder capable of holding eight small crucibles. For high cerium content, ternary alloy containing cerium was used as the raw material. In order to prevent direct contact between cerium and the graphite crucible to produce cerium carbide prematurely, the cerium was first loaded into a carbon-saturated hollow iron tube, which was in turn placed in the middle of a small crucible. Before heating, the working furnace was filled with Ar. After an equilibration time of 4 h, the whole graphite bracket was quenched and carbon and cerium were sampled and analyzed. For Fe-C$_{saturation}$-Ce-Sb and Fe-C$_{saturation}$-Ce-Bi quaternary system experiment, adopt large crucible method. When Fe-C$_{saturation}$-Sb or Fe-C$_{saturation}$-Bi equilibration at experimental temperature and 5 h, then add cerium, and take samples every hour to determine the balance time.

The experimental results for the Fe-C$_{saturation}$-Ce ternary system are summarized in Fig. 9.5 (cerium content, (a) expressed in mass fraction; (b) expressed in mole fraction), omitted here. The slope of each line in the diagram is > 0, which indicates that cerium increases the solubility of carbon, and $N_C$ and $N_{Ce}$ were evaluated by linear regression.

$$1300\,°C, \ N_C = 0.1850 + 2.81 N_{Ce}$$

$$1400\,°C, \ \ N_C = 0.1937 + 2.74 N_{Ce}$$

$$1515\,°C, \ \ N_C = 0.12032 + 2.56 N_{Ce}$$

At $N_{Ce} = 0$, the Nc in the above three formulas is the extrapolation value of carbon solubility in Fe-C binary solution, which is basically consistent with the Fe-C binary system.

## 9.5 Thermodynamic Study of the Effects of Arsenic, Antimony, and Bismuth

The relationship between the solubility of carbon in Fe-C-Ce ternary solution and the content of cerium can also be written as follows:

$$N_C = N_C^C + mN_{Ce}$$

Here, $N_C^C$ is the intercept of a straight line, and $m$ is the slope.

From $N_C^C$ and the $m$ value at each temperature, the relationship with temperature is as follows:

$$N_C = 0.052 + 0.846 \times 10^{-5}/T + (0.582 + 3550/T)N_{Ce} \quad 1300-1515\,°C$$

Use the lattice ratio as the concentration coordinate:

In the processing of experimental data, a relatively simple relationship of thermo-dynamic quantities is sought, such as the lattice ratio concentration relationship:

$$Z_i = \frac{N_i}{N_1 + \sum N_j \gamma_j}$$

In the formula, $N_1$ is the molar coefficient of the solvent; $N_i$ are $N_j$ are the molar fractions of the solutes; and $\gamma_j$ is a parameter. For the displacement solute $\gamma_j = +1$; for the gap solute $\gamma_j = \frac{1}{b}$, where $b$ is the number of gap positions corresponding to atoms in each lattice.

Chipman J discussed the thermodynamics of several iron-based solutions, assuming $b = 1$, and applied the ratio of lattice to concentration to the Fe-C-Ce ternary solution, with carbon as the interstitial solute and cerium as the displacement solute, $\gamma_C = -1, \gamma_{Ce} = 1$; then:

$$Z_C = \frac{N_C}{1 - 2N_C}; \quad Z_{Ce} = \frac{N_{Ce}}{1 - 2N_C}$$

Under carbon-saturation conditions, $\alpha_C = 1$, that is, $\psi_C Z_C = 1$, where $\psi_C$ is the activity coefficient of carbon corresponding to the specific lattice concentration, and $-\ln \psi_C = \ln Z_C$. The linear relationship is obtained by plotting $\ln \psi_C$ versus $Z_{Ce}$ at each concentration, and the equations are:

$$1300\,°C, \ \ln \psi_C = 1.225 - 15.52Z_{Ce}$$
$$1400\,°C, \ \ln \psi_C = 1.151 - 13.86Z_{Ce}$$
$$1515\,°C, \ \ln \psi_C = 1.072 - 12.12Z_{Ce}$$

The corresponding activity interaction coefficient with the ratio of lattice to concentration $Z_i$ is:

$$\theta_i^j = \left( \frac{\partial \ln \psi_i}{\partial Z_j} \right)_{Z_j \to 0}$$

Under the carbon potential, the interaction activity coefficient is defined as:

$$\theta_C^{*Ce} = \left(\frac{\partial ln\psi_C}{\partial Z_{Ce}}\right)_{\substack{\alpha_C \\ Z_{Ce} \to 0}}$$

$ln\,\psi_C$ can be expressed as:

$$ln\,\psi_C = ln\,\psi_C^b + \theta_C^{*Ce}Z_{Ce}$$

Here, $ln\,\psi_C^b$ is the $ln\,\psi_C$ value when $Z_{Ce} = 0$.

Plots of $ln\,\psi_c^b$ and $\theta_c^{*Ce}$ versus $1/T$ at the above three temperatures are shown in a diagram.

$$ln\,\psi_C^b = -0.147 + 2170/\,T$$
$$\theta_C^{*Ce} = 15.09 - 48410/T$$

Combining the two:

$$ln\,\psi_C = -0.147 + 2170/\,T + (15.09 - 48,410/\,T)Z_{Ce}$$

From this formula, the carbon solubility $Z_C$ at the experimental temperature in the concentration range can be calculated:

The interaction activity coefficient directly reflects interaction properties and scale. According to Wagner definition, the first-order interaction activity coefficient between components i and j is expressed as:

$$\varepsilon_i^j = \left(\frac{\partial \ln \gamma_i}{\partial N_j}\right)_{\substack{N_1 \to 1 \\ N_j \to 0}}$$

By this definition, the activity coefficient of component i can be expressed as:

$$ln\,\gamma_i = ln\,\gamma_i^0 + \sum_{j=2}^{m} \varepsilon_i^j N_j + 0(x^2)$$

$$\varepsilon_i^j = \left(\frac{\partial \ln \gamma_i}{\partial N_j}\right)_{\substack{N_1 \to 1 \\ N_j \to 0}}$$

where $\gamma_i^0$ is the activity coefficient of component $i$ in an infinitely dilute solution, and $0(x^2)$ denotes higher order terms. The second-order and cross-activity interaction coefficients are defined as:

## 9.5 Thermodynamic Study of the Effects of Arsenic, Antimony, and Bismuth

$$\rho_i^j = \frac{1}{2}\left(\frac{\partial^2 \ln \gamma_i}{\partial N_j^2}\right)_{N_1 \to 1}$$

$$\rho_i^{i,j} = \frac{1}{2}\left(\frac{\partial^2 \ln \gamma_i}{\partial N_i \partial N_j}\right)_{N_1 \to 1}$$

Due to the high carbon concentration in a carbon-saturated solution, the higher order interaction coefficients need to be considered using the Wagner expression. Hence, Lupis and Elliott (1967) defined interaction coefficients suitable for determining the component activity in a carbon-saturated solution as:

$$\varepsilon_C^{*i} = \left(\frac{\partial \ln \gamma_C}{\partial N_i}\right)_{\substack{\alpha_C = 1 \\ N_i \to 0}} = -\left(\frac{\partial \ln N_C}{\partial N_i}\right)_{\substack{\alpha_C = 1 \\ N_i \to 0}}$$

$$\rho_C^{*i} = \frac{1}{2}\left(\frac{\partial^2 \ln \gamma_C}{\partial N_i}\right)_{\substack{\alpha_C = 1 \\ N_i \to 0}} = -\left(\frac{\partial^2 \ln N_C}{\partial N_i}\right)_{\substack{\alpha_C = 1 \\ N_i \to 0}}$$

$$\rho_C^{*i,C} = \frac{1}{2}\left(\frac{\partial^2 \ln \gamma_C}{\partial N_i \partial N_C}\right)_{\substack{\alpha_C = 1 \\ N_i \to 0}} = -\left(\frac{\partial^2 \ln N_C}{\partial N_i \partial N_C}\right)_{\substack{\alpha_C = 1 \\ N_i \to 0}}$$

The relationship between interaction activity coefficients of components in carbon-saturated solution and that in dilute solution can thus be derived:

$$\varepsilon_C^{*i} = \left(\varepsilon_C^i + \rho_C^{i,C} N_C^*\right)/\left(1 + \varepsilon_C^C N_C^* + 2\rho_C^C N_C^{*2}\right)$$

If this formula is simplified by not considering the exchange term $\rho_C^{i,C}$, we obtain:

$$\varepsilon_C^i = \varepsilon_C^{*i}\left(1 + \varepsilon_C^C N_C^* + 2\rho_C^C N_C^{*2}\right)$$

According to the above formula, interaction activity coefficient $\varepsilon_C^i$ can be obtained from the experimental $\varepsilon_C^{*i}$ data for a carbon-saturated solution. Experimentally, $\varepsilon_C^{Ce}$, $e_C^{Ce}$, and $e_{Ce}^C$ in an infinitely dilute solution were calculated from $\varepsilon_C^{*Ce}$ as follows:

When 1823 K (1550 °C), $\varepsilon_C^{Ce} = -46.44$, $e_C^{Ce} = -0.078$, $e_{Ce}^C = -0.91$

When 1848 K (1575 °C), $\varepsilon_C^{Ce} = -45.63$, $e_C^{Ce} = -0.076$, $e_{Ce}^C = -0.089$

When 1873 K (1600 °C), $\varepsilon_C^{Ce} = -44.82$, $e_C^{Ce} = -0.075$, $e_{Ce}^C = -0.88$

Their relationships with temperature at 1823–1873 K (1550–1600 °C) are as follows:

$$\varepsilon_C^{Ce} = -110{,}800/T + 14.33$$

$$e_C^{Ce} = -206/T + 0.035$$
$$e_{Ce}^C = -205/T + 0.22$$

Sanbongi K and Han QY also obtained $e_{Ce}^C$ experimentally. At 1873 K (1600 °C), Sanbongi obtained $e_{Ce}^C = -0.077$, while Han obtained $e_{Ce}^C = -0.43$. The author Wang CZ obtained $e_{Ce}^C = -0.88$. It is difficult to comment on which result more accurate due to the different methods used is.

Regarding $CeC_2$ formation during the experiment, at cerium contents > 1.2%, a brass-colored substance was formed, which agglomerated within the melt. It rapidly deliquesced in the air, changing from a hard bulk to a powder, and acetylene ($C_2H_2$) was given off when it was placed in water. When the substance was identified by X-ray diffraction analysis, it was not successful due to reaction with collodion.

According to the theoretical analysis, it can be seen from the Ce-C binary phase diagram that two compounds between $CeC_2$ and $Ce_2C_3$ are stable. Thermodynamic calculations were carried out based on literature data:

$$Ce_{(1)} + 2C_{(s)} = CeC_{2(s)}, \quad \Delta G^{\ominus} = -20{,}370 - 6.45\,T \text{ (cal)} \tag{9.1}$$

$$2Ce_{(1)} + 3C_{(s)} = Ce_2C_{3(s)}, \quad \Delta G^{\ominus} = -45{,}000 - 3.5\,T \text{ (cal)} \tag{9.2}$$

Combining the above two forms, we get:

$$2CeC_{2(s)} + C_{(s)} = Ce_2C_{3(s)}, \Delta G^{\ominus} = -4260 + 9.4\,T \text{ (cal)} \tag{9.3}$$

According to Eq. (9.3), at T > 453 K (180 °C), $\Delta G^{\ominus} > 0$, this implies that $CeC_2$ is more stable than $Ce_2C_3$ in the presence of carbon at above 180 °C. Separating out $CeC_2$ from Fe-C-Ce ternary solution, the solubility of carbon and cerium is shown in a table, for example, when temperature/K is 1573 K (1300 °C), [C]/% is 4.86, NC is 0.1927, [Ce]/% is 0.76, and NCe is 0.00257; when 1673 K (1400 °C), the data is 5.14, 0.2022, 0.92, and 0.00308. Generally, the Ce content is relatively low.

Experimental results for Fe-C-Ce-Sb and Fe-C-Ce-Bi (at 1350 °C) are summarized in two diagrams. The respective linear equations are:

$$N_C - 0.1888 + 1.10N_{Ce}, \quad N_{Ce} < 7.0 \times 10^{-4}$$
$$N_C - 0.1887 + 1.26N_{Ce}, \quad N_{Ce} < 7.0 \times 10^{-4}$$

When Sb or Bi is added to molten iron, it will interact with any Ce present. Indeed, Ce can reduce the segregation of Sb or Bi at grain boundaries, and also affect the nucleation and growth of Sb or Bi on C.

To observe the effect of Ce in pure iron solution on graphite spheroidization, Ce was added by using carbon-saturated iron after deoxygenation and desulfurization as raw material; some are to add Ce, Sb or Bi, heating to 1350 °C and quenching. A batch of full-phase sample was prepared, each of mass 8 g.

## 9.5 Thermodynamic Study of the Effects of Arsenic, Antimony, and Bismuth

Each sample was ground with dry abrasive paper, polished with W5 diamond grinding paste, and photographed with a full-phase microscope. At $100\times$ magnification, the graphite spheroidization effect could be observed, the order number is from 1 to 5, and $B_1$ to $B_4$, the items are $[Ce]/\%$, $[C]/\%$, and as-cast all-phase structure (spheroidization and size levels) (JB1802-76). For example, when order number is 1, $[Ce]/\%$ is 0.73, $[C]/\%$ is 4.84, and as-cast all-phase structure (spheroidization and size levels) (JB1802-76) is spheroidization level 2, size level 3. To $B_1$–$B_4$, the items also include 0.2%Bi—0.2%Sb.

Taking $B_3$ and $B_4$ as examples, the full-phase photos are shown in a diagram.

The commonly used spheroidizing agents in cast iron production incorporate Ce, other RE, Y, and Mg. Based on previous experience, the spheroidizing abilities of these agents can be judged as follows:

- The shape of graphite: the more round the graphite, the stronger the spheroidization ability of the agent.
- The lower the consumption (mass fraction) of a spheroidizing agent, the stronger its spheroidizing ability.
- According to the foregoing discussion, worm-like cast iron has the most comprehensive properties, and spheroidizing agents are mainly mixed light rare earths. This should also be considered from an economic point of view, and varies from place to place. In China, for example, southern ore is rich in heavy rare earths and yttrium, and factories, especially small ones, almost all use yttrium as a spheroidizing and vermicularizing agent.

Studies have shown that the crystal structure is affected by the order of addition of Sb or Bi. Before adding Sb or Bi, the sample has smooth surfaces, fine dendrites, strong densification, and toughness. After their addition, the sample has rough surfaces, many pores, and a poor texture. Adding rare earths can inhibit the harmful effects of Sb and Bi on the properties of cast iron. Sb is a spheroidizing interfering element in cast iron, but also has a stabilizing pearlite function. Du et al. (1995) investigated the use of rare earths and magnesium to remove or weaken the adverse effects of Sb, while retaining its favorable factors.

A sample of corundum crucible was melted in a $MoSi_2$ furnace, and Sb and rare earths were added together to the carbon-saturated samples. Sb contents were < 0.002%, 0.029%, 0.032%, 0.080%, and 0.082%, respectively. The cooling methods studied were wet pouring and water quenching. These samples were followed by grinding and polishing, full-phase microscope structure observation, and scanning electron microscopy and electron probe analyze the distribution of Ce, Sb, and particle analysis to ductile iron.

Taking rare earth ferrosilicon-treated ductile iron as a benchmark, when the Sb content is 0.029%, no flake graphite is present in ductile iron; when the Sb content is 0.032%, very little flake graphite is formed and having other non-uniformity shape of graphite; when it is 0.08%, a large amount of flake graphite is formed, and increases with increasing Sb content. The experiments of Morrogh H showed that Sb content is harmful to Mg-treated cast iron at 0.004%; when the graphite ball deteriorates, Ce addition can completely restore the ductile shape. Electron probe analysis indicated

Ce- and Sb-enriched particles, and Ce can reduce Sb activity, and CeSb compounds can be formed to reduce the amount of free Sb, thereby inhibiting its harmful effects on ductile iron.

# References

Chen XC, Yi SS, Wang ZL (1985) Metallurgical behavior of rare earth ferrosilicon, magnesium, aluminum, calcium polyspheroidizing agent in hot metal. J Chin Soc Rare Earths 3(4):19–25

Chipman J, Floridis TP (1955) Activity of aluminum in liquid Ag-Al, Fe-Al, Fe-Al-C, and Fe-Al-C-Si alloys. Acta Metall 3(5):456–459

Du T, Han QY, Wang CZ (1995) Physical chemistry of rare earth alkali soil and its application in materials. Science Press, Beijing, pp 330–334

Huang HS, Qiu HQ, Sheng D (1987) Research, production and application of rare earth vermicular graphite cast iron in china. In: Application of rare earth in steel. Metallurgical Industry Press,Beijing, pp 94–106

Janke D, Fischer WA (1973) The activity of oxygen in phosphorous copper and nickel melts. Arch Eisenhüttenwes 44(1):15–18

Jiang BH Su GQ et al (1994) Research on rare earth emulsifiers. In: Proceedings of the third annual academic conference of rare earth science in China, 3 parts, pp 185–188

Li XH (1983) Study on rare earth ductile iron. Chin Rare Earths 1:43–47

Linebarger HF et al (1983) Study on the influence of rare earth on the formation of fragment graphite in large section ductile iron. Chinese Rare Earths supplement (foreign rare earth application album) (trans: Cai), pp 8–17

Liu BC, Yang XH et al (1994) Effects of antimony and rare earth elements on graphite morphology in super large ductile iron castings. In: Proceedings of the third annual academic conference of China Rare Earth Society, 3 sub books, pp 171–174

Liu JH, Lou EX, Fang G (1994) Effect of rare earth and magnesium on graphitization of ductile iron. J Chin Soc Rare Earths 12(1):46–49

Lupis CHP, Elliott JF (1967) Part I: Acta Metall 1966(14):529–538; Part II: Acta Metall (15):265–276

Lupis CHP (1983) Chemical thermodynamics of materials. North-Holland, New York, Amsterdam, Oxford

Niu YY, Zhang Z (1983) Study on the influence of rare earth on the formation of fragment graphite in large section ductile iron. Chin Rare Earths 3:28–40

Pierre GRS (1977) The solubility of oxides in molten alloys. Metall Trans 8B:215–217

Qian HC, Koo ZX et al (1987) Review on the research of ductile iron with rare earth system. In: Application of rare earth in steel. Metallurgical Industry Press, Beijing, pp 173–182

Sanbongi K et al (1972) Iron and steel smelting technology. Asakura Publishing Co., Ltd., Tokyo

Shen ZJ, Zhang QH et al (1994) Application of rare earth in austenitic cast iron. In: Proceedings of the third annual academic conference of China Rare Earth Society, 3 sub-books, pp 166–170

Sheng D, Ying ZT (1987) Process characteristics of cupola hot metal stable production of rare earth vermicular graphite cast iron. In: Application of rare earth in steel. Metallurgical Industry Press, Beijing, pp 117–126

Sheng D, Ying ZT (1986) Production of rare earth vermicular graphite cast iron by cupola hot metal in China. Chin Rare Earths 5:52–56

Sheng D, Huang HS et al (1987) Micro rare earth gray cast iron microstructure and properties. In: Application of rare earth in steel. Metallurgical Industry Press,Beijing, pp 81–93

Sheng D, Zheng YP, Yang RD (1994) Study on factors affecting elongation of cast rare earth ductile iron pipes. In: Proceedings of the third annual academic conference of China Rare Earth Society, 3 sub-books, pp 192–197

# References

Shiokawa J (1993) The latest application technology of rare earths (trans: Zhai YS, Yu ZH). Chemical Industry Press, Beijing, pp 240–243

Sigworth GK, Elliott JF (1974a) The thermodynamics of dilute liquid copper alloys. Can Metall Q 15(2):455–461

Sigworth GK, Elliott JF (1974b) The thermodynamics of liquid dilute iron alloys. Met Sci 8:298–310

Sun XJ, Wang DB (1981) The relationship between tin ink growth and under cooling. Ball Iron 32–35

Wang HZ (1981) Inner inoculation of ductile iron. Ductile iron, 46 and 24 (in two places)

Wang CZ, Wang SL, Guo W (1987) Thermodynamics of carbon dissolved in Fe-Ce, Fe-Y, Fe-Ce-Sb and Fe-Ce-Bi melts. Inorg Chim Acta 140:183–186

Wang SL, Wang CZ (1988) Thermodynamics of carbon dissolution in Fe-Ce, Fe-Ce-Sb, Fe-Ce-Bi melt. J Met 24(5b):32–38

Wuxi Diesel Engine Plant, Xi'an Jiaotong University, Shanghai Internal Combustion Engine Plant (1981) Microstructure and fracture toughness of rare earth magnesium ductile iron. Ball Iron 2:8–15

Xie ZC (1981) Study on the mechanism of black block in internal spheroidization. Microstructure and fracture toughness of rare earth magnesium ductile iron. Ball Iron 25–31

Xue DS, Cao YJ (1981) Study on the mechanism of chromium in ductile iron. Ball Iron 2:1–7

Xu JC et al (1987) Study on spheroidization ability of mixed rare earth metals. In: The application of rare earths in steel. Metallurgical Industry Press, Beijing, pp 144–151

Xiong ZM, Zhang SM, Shi DC et al (1994) Effect of rare earth and antimony on gray cast iron. J Chin Soc Rare Earths 12(2):150–153

Yang YJ, Wang CZ (2004) Impedance spectroscopy of $Y_{1-x}Ca_xF_{3-x}$-YOF solid electrolytes. J Chin Rare Earth Soc Spec Issue 22:308–311

Zeng YC, Wu DH (1987) Application of rare earths in ductile iron. Application of rare earths in steel. Metallurgical Industry Press, Beijing, pp 135–143

Zhang WS (1983) Economic benefit analysis of rare earth vermicular graphite cast iron ingot mould. Chin Rare Earths 2:50–52

Zhu FL, Wang CZ (1981) Study on the inhibition of arsenic hazard by using rare earth elements in iron and steel. Guangxi Metall 1:35–40

Zhu YK, Peng YQ, Yu XB et al (1984) Study on new production methods of vermicular graphite cast iron. J Chin Soc Rare Earths 2(1):64–69

Wang CZ (2000) Solid electrolytes and chemical sensors. Metallurgical Industry Press, Beijing, pp 361–382

Zhu ZH, Zhang WD (1987) Study on primary crystallization process of rare earth and magnesium creep iron application of rare earth in steel. Metallurgical Industry Press, Beijing, pp 107–116

# Chapter 10
# Role of Rare Earths in Non-ferrous Metals and Alloys

## 10.1 Role of Rare Earths in Ni, Cu, and Al Solutions

The rare earth elements have deoxygenation and desulfurization functions in liquid Fe and in the similar liquid Ni. The question then arises as to what effects they may have in Cu and Al fluids. Du T. and co-workers have studied the thermodynamics of rare earths in Ni, Cu, and Al fluids (Du et al. 1995).

(1) Thermodynamics of rare earths in Ni-based solution

To study the thermodynamics of Y, Ce, and La in Ni-based solutions, the purity of the Ni raw material was 99.99%, and the purities of Y, Ce, and La were > 99.95%. The experimental set-up and research methods were similar to those in Fe solution. The reaction with sulfur is illustrated below.

The experiment was carried out in a molybdenum wire furnace equipped with a sealing device and an electric melting magnesium oxide crucible. The protective atmosphere was highly pure Ar free from water and oxygen. Oxygen activity in the melt was measured with a $ZrO_2$ (+ MgO) solid-electrolyte oxygen-sensing probe. Samples were withdrawn through a quartz tube and quenched. The products in solid equilibrium samples were separated by electrolysis with a low temperature anhydrous electrolyte. The electrolytic temperature was $\leq 10\ °C$, the current density was $\leq 50\ mA/cm^2$, and the organic electrolyte was composed of 1% LiCl + 2% glycerol + 2% citrate + 95% methanol. Y, Ce, and La in rare earth sulfides were dissolved in diluted aqua regia and determined by inductively coupled high-frequency plasma-torch spectroscopy. The content of rare earth minus the content of rare earth sulfides is obtained. The dissolved sulfur content is obtained from the total amount of chemical analysis minus the sulfur content in rare earth inclusions.

Methanol was evaporated from the combined electrolyte and washings by heating over a water bath. The residue was acidified with hydrochloric acid, and the solid-solution rare earth content was determined by plasma spectroscopy. The sulfur

© Science Press 2023

C. Wang, *Theory and Application of Rare Earth Materials*,

https://doi.org/10.1007/978-981-19-4178-8_10

content in the dissolved state was calculated by subtracting that in rare earth inclusions from the total amount determined by chemical analysis. Rare earth inclusions were identified as YS, CeS, and LaS by X-ray diffraction analysis.

The relationships between desulfurization constant and temperature for Y, Ce, and La in Ni solution were as follows.

For Ni-Y-S solution systems, it is:

$$YS_{(s)} = [Y]_{Ni} + [S]_{Ni} \quad K'_{YS} = [\%\,Y][\%\,S]$$
$$K_{YS} = \alpha_Y \cdot \alpha_S = f_Y[\%\,Y] \cdot f_s\,[\%\,S]$$
$$\lg K_{YS} = \lg K'_{YS} + e_Y^Y[\%\,Y] + e_S^Y[\%\,Y] + e_S^S[\%\,S] + e_Y^S[\%\,S]$$

Applying the measured $e_Y^Y = 0.004$ and the derived $e_S^S$, we get:

$$e_Y^Y[\%\,Y] \leq 4.8 \times 10^{-5} - 7.6 \times 10^{-5}, \quad e_S^S[\%\,S] \leq 6.0 \times 10^{-4} - 7.6 \times 10^{-4}$$

In the low concentration range, these two terms may be omitted and the equation may be reorganized as:

$$\lg K'_{YS} = \lg K_{YS} - e_S^Y\{[\%\,Y] + 2.778[\%\,S]\} \tag{10.1}$$

For Ni-Ce-S solution systems, it is:

$$CeS_{(s)} = [Ce]_{Ni} + [S]_{Ni} \quad K'_{CeS} = [\%\,Ce][\%\,S]$$
$$K_{CeS} = \alpha_{Ce} \cdot \alpha_S = f_{Ce}\,[\%\,Ce] \cdot f_s[\%\,S]$$
$$\lg K_{CeS} = \lg K'_{CeS} + e_{Ce}^{Ce}[\%\,Ce] + e_S^{Ce}[\%\,Ce] + e_S^S[\%\,S] + e_{Ce}^S[\%\,S]$$

Again, the equation can be reorganized as:

$$\lg K'_{CeS} = \lg K_{CeS} - e_S^{Ce}\{[\%\,Ce] + 4.375[\%\,S]\} \tag{10.2}$$

For Ni-La-S solution systems, the same treatment gives:

$$\lg K'_{LaS} = \lg K_{LaS} - e_S^{La}\{[\%\,La] + 4.341[\%\,S]\} \tag{10.3}$$

By plotting:

$$\lg K'_{RES} - \{[\%\,RE] + x[\%\,S]\}$$

for the systems at 1500 °C, 1550 °C, and 1600 °C according to Eqs. (10.1)–(10.3), the results shown in Figs. 10.1, 10.2 and 10.3, respectively, were obtained.

Tangents of the composition lines in the low concentration range $\{[\%\,RE] \times [\%\,S]\} \rightarrow 0$ gave intercepts that corresponding to the logarithmic values of the desulfurization constants of Y, Ce, and La in liquid Ni. Regression of their relationships

## 10.1 Role of Rare Earths in Ni, Cu, and Al Solutions

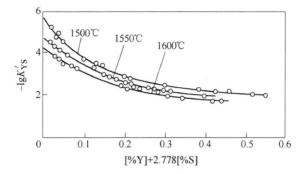

**Fig. 10.1** Relationship between $-\lg K'_{YS}$ and Y, S concentration in Ni liquid

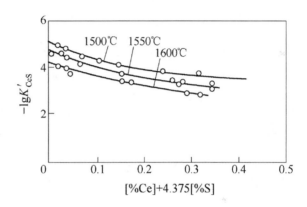

**Fig. 10.2** Relationship between $-\lg K'_{CeS}$ and Ce, S concentration in Ni liquid

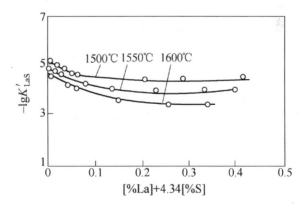

**Fig. 10.3** Relationship between $-\lg K'_{LaS}$ and La, S concentration in Ni liquid

with temperature yielded:

$$\lg K_{YS} = -47{,}000/T + 20.86 \ (r = 0.99) \qquad (10.4)$$

$$\lg K_{CeS} = -29{,}792/T + 11.50 (r = 0.98) \qquad (10.5)$$

$$\lg K_{LaS} = -9996/T + 0.25 \ (r = 0.99) \tag{10.6}$$

The relationships between the standard free energies of generation and temperature for the Y, Ce, and La desulfurization equilibrium products YS, CeS, and LaS in the liquid were:

$$[Y] + [S] = YS_{(s)} \quad \Delta G^{\ominus} = -899{,}560 + 399.68T \ \text{(J/mol)} \tag{10.7}$$

$$[Ce] + [S] = CeS_{(s)} \quad \Delta G^{\ominus} = -570{,}280 + 220.10T \ \text{(J/mol)} \tag{10.8}$$

$$[La] + [S] = LaS_{(s)} \quad \Delta G^{\ominus} = -191{,}330 + 4.60T \ \text{(J/mol)} \tag{10.9}$$

Other thermodynamic quantities could also be obtained.

(2) Thermodynamics of rare earths in Cu-based solution

Similar to the thermodynamic study of rare earths in Ni-based solution, equilibrium data for the desulfurization reaction could be obtained. Here, we considered thermodynamic results for the Cu-Ce-O system as an example. The result of room temperature identification of the equilibrium product of the Cu-Ce-O system was $CeO_2$, because $Ce_2O_3$ was unstable at room temperature, being oxidized to $CeO_2$. $Ce_2O_3$ was stable at 1200 °C. The equilibrium reaction in this system was:

$$Ce_2O_{3(S)} = 2[Ce] + 3[O]$$

With pure matter as the standard state, and $Ce_2O_3$ at 1200 °C as the solid phase:

$$\alpha_{Ce_2O_3} = 1$$

the deoxygenation equilibrium constant was:

$$K = \alpha_{Ce}^2 \cdot \alpha_O^3$$

so that

$$K' = [\% \, Ce]^2 \cdot \alpha_O^3$$

and

$$K = K' f_{Ce}^2$$

It follows that:

$$\lg K = \lg K' + 2 \lg f_{ce}$$

## 10.1 Role of Rare Earths in Ni, Cu, and Al Solutions

**Fig. 10.4** Relationship between $-\lg K'_{Ce_2O_3}$ and [% Ce] in Cu solution (1200 °C)

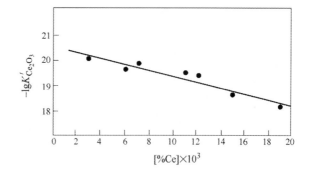

$$-\lg K' = -\lg K + 2\left(e^{Ce}_{Ce}[\% Ce]\right) + e^{O}_{Ce}[\% O] \tag{10.10}$$

Because $\alpha_O$ was very low, [% O] ≈ $\alpha_O$. In a plot of [% Ce] + [% O] versus $-\lg K'_{Ce_2O_3}$, as [% Ce] + [% O] → 0, the intercept on the *x*-axis represented lg*K*. This was because [% O] was very small, such that [% Ce] + [% O] ≈ [% Ce]. The linear relationship between

$$-\lg K'_{Ce_2O_3}$$

and [% Ce] in Cu solution is summarized in Fig. 10.4, that is, relationship between $-\lg K'_{Ce_2O_3}$ and [% Ce] in Cu solution (1200 °C).

At 1200 °C, the Ce deoxygenation constant in the Cu solution was:

$$\lg K = -20.62, \quad K = 2.40 \times 10^{-21}$$

The oxygen activity was greatly reduced, indicating that the deoxygenation ability in Cu solution was very strong.

(3) Thermodynamics of rare earths in Al-based solution

In alloys with Al, rare earths can play a purifying role by removing contaminants. Sun and Du (1994) have studied the thermodynamics of Cu, Si, and Ni in Ce and Al solutions.

For the Al-Ce-X (X = Cu, Si, and Ni) systems, the direct equilibrium method of metal solution-reactive product was used. Mixtures were heated in a corundum crucible in a molybdenum wire furnace under a protective Ar atmosphere. After attaining constant temperature, the requisite amount of solute element was added. After equilibration, a sample was withdrawn from the middle part of the crucible through a quartz tube and quenched in water. Cu and Si were analyzed by inductively coupled plasma spectroscopy. Ni was analyzed by flame atomic absorption spectrometry. The equilibrium products were identified by X-ray diffraction analysis and transmission electron microscopy. Because the equilibrium product from the experimental system was an intermetallic compound of small particle size, it was difficult to separate it from the Al matrix, and X-ray diffraction analysis could not be performed. Based on the Ce-Cu binary phase diagram, $CeCu_6$ was the only stable compound at 850 °C. The results showed that $CeCu_6$ was indeed present in the Al solution.

The equilibrium products of Al-Ce-Si and Al-Ce-Ni were identified as CeSi and $Ce_7Ni_3$, respectively.

The three systems involved the following reactions:

$$[Ce] + 6[Cu]$$
$$CeSi_{(l)} = [Ce] + [Si]$$
$$Ce_7Ni_{3(l)} = 7[Ce] + 3[Ni]$$

The product was a solid or liquid phase, the activity $a$ was 1, so the above equilibrium constant:

$$K_{Ce_iM_j} = \alpha_{Ce}^i \cdot \alpha_M^j$$

was:

$$K'_{Ce_iM_j} = [\% \, Ce]^i [\% \, M]^j$$

It followed that:

$$K_{Ce_iM_j} = (f_{Ce} \cdot [\% \, Ce])^i \cdot (f_M [\% \, M])^j$$
$$\lg K_{Ce_iM_j} = \lg K'_{Ce_iM_j} + i\left(e_{Ce}^{Ce}[\% \, Ce] + e_{Ce}^M[\% \, M]\right) + j\left(e_M^M[\% \, M] + e_M^{Ce}[\% \, Ce]\right)$$

Because the interaction between elements was generally small, $e_{Ce}^{Ce}[\% \, Ce]$ and $e_M^M[\%M]$ in the above expression could be omitted, and substituting

$$e_M^{Ce} = (M_M/M_{Ce})e_{Ce}^M$$

we obtained:

$$-\lg K'_{Ce_iM_j} = -\lg K_{Ce_iM_j} + e_{Ce}^M(i[\% \, M] + jA[\% \, Ce]) \qquad (10.11)$$

i, j, A of the three systems were:

$$\text{Al-Ce-Cu system,} \quad i = 1, \quad j = 6, \quad A = M_{Cu}/M_{Ce} = 0.45$$
$$\text{Al-Ce-Si system,} \quad i = 1, \quad j = 1, \quad A = M_{Si}/M_{Ce} = 0.20$$
$$\text{Al-Ce-Ni system,} \quad i = 7, \quad j = 3, \quad A = M_{Ni}/M_{Ce} = 0.42$$

Introducing the three sets of parameters into Eq. (10.11) gave:

## 10.1 Role of Rare Earths in Ni, Cu, and Al Solutions

$$-lg\,K'_{CeCu_6} = -lg\,K_{CeCu_6} + e_{Ce}^{Cu}([\%\,Cu] + 2.72[\%\,Ce]) \qquad (10.12)$$

$$-lg\,K'_{CeSi} = -lg\,K_{CeSi} + e_{Ce}^{Si}([\%\,Si] + 0.20[\%\,Ce]) \qquad (10.13)$$

$$-lg\,K'_{Ce_7Ni_3} = -lg\,K_{Ce_7Ni_3} + e_{Ce}^{Ni}(7[\%\,Ni] + 1.26[\%\,Ce]) \qquad (10.14)$$

The experimental data of the three systems were processed according to the above three formulae, and the results are shown in the three diagrams.

The experiment points pertaining to lower solute concentration were treated by a least-squares regression method, yielding:

$$-lg\,K'_{CeCu_6} = 10.53 - 9.22([\%\,Cu] + 2.72[\%\,Ce]) \quad (\gamma = 0.95) \qquad (10.15)$$

$$-lg\,K'_{CeSi} = 2.18 - 1.87([\%\,Si] + 0.20[\%\,Ce]) \quad (\gamma = 0.95) \qquad (10.16)$$

$$-lg\,K'_{Ce_7Ni_3} = 16.75 - 6.54(7[\%\,Ni] + 1.26[\%\,Ce]) \quad (\gamma = 0.98) \qquad (10.17)$$

At 850 °C:

$$lg\,K_{CeCu_6} = -10.53, \quad K_{CeCu_6} = 2.95 \times 10^{-11}.$$

At 800 °C:

$$lg\,K_{CeSi} = -2.18, \quad K_{CeSi} = 6.61 \times 10^{-3};$$
$$lg\,K_{Ce_7Ni_3} = -16.75, \quad K_{Ce_7Ni_3} = 1.77 \times 10^{-17}.$$

The three reactions may be rewritten as follows:

$$[Ce] + 6[Cu] = CeCu_{6(s)}$$
$$[Ce] + [Si] = CeSi_{(s)}$$
$$7[Ce] + 3[Ni] = Ce_7Ni_{3(1)}$$

The standard free energy of each product in Al solution was:

$$\Delta G_{CeCu_6}^{\ominus} = -2.303RT\,lg(1/K_{CeCu_6}) = -226.4 \ (k\,J\,/\,mol) \qquad (10.18)$$

$$\Delta G_{CeSi}^{\ominus} = -2.303RT\,lg(1/K_{CeSi}) = -44.8 \ (k\,J\,/\,mol) \qquad (10.19)$$

$$\Delta G_{Ce_7Ni_3}^{\ominus} = -2.303RT\,lg(1/K_{Ce_7Ni_3}) = -344.1 \ (k\,J\,/\,mol) \qquad (10.19)$$

In this way, the reaction products and related thermodynamic analysis of Ce and other elements in Al solution were obtained.

## 10.2   Role of Rare Earths in Mg Alloy Solutions

Luo and Tang (1995) carried out thermodynamic analysis of the role of rare earths in Mg alloys. Their results showed that rare earths react with oxide inclusions in Mg solution to produce rare earth oxides, thereby removing the oxide inclusions. They also react with hydrogen in Mg solution to produce rare earth hydrides, thereby mitigating or eliminating the harmful effects of hydrogen.

## 10.3   Properties of Rare Earth Alloys

Various alloys for metallurgical and engineering applications have properties tailored to their needs. The composition, structure, and treatment mode of an alloy determine its properties. The interplay between these influences is governed by the fundamental relationships and change laws of thermodynamics, dynamics, solid-state physics, solid-state chemistry, chemical physics, and other basic disciplines. Xiao (1985) expounded in detail the compositions, structures, properties, processes, and energetic parameters of alloys in terms of basic and applied sciences.

(1)   Thermodynamics

The expression of the first law of thermodynamics was:

$$dU = \delta Q + PdV - \varepsilon dZ$$

The first laws of thermodynamics for various types of work and states has been summarized in a table, and the items include system (process), strength properties, extensive parameters, work ($W$), and expression of the first law of thermodynamics, for example, when system (process) is chemistry (expansion), the strength properties is pressure $P$ ($dyn/cm^2$), extensive parameters is volume $V$ ($cm^3$), work ($W$) is $- \int PdV$ (erg), and expression of the first law of thermodynamics is $\delta Q = dU - PdV$; when system (process) is magnetic material (magnetization), strength properties is magnetic field strength $H$ (Oe or A/m), extensive parameters is magnetization $J$ (emu/g), work (w) is $\int HdJ$ (erg), and expression of the first law of thermodynamics is $\delta Q = dU - HdJ$. In it, 1 dyn $= 10^{-5}$ N, 1 Oe $= 79.5775$ A/m.

(2)   Chemical thermodynamics

Chemical thermodynamics must be invoked when the chemical properties of a system are changed. The study of alloy structure not only involves physical changes of the alloy, but also chemical changes. For example, the formation and transformation of

## 10.3 Properties of Rare Earth Alloys

alloy structures and related processes (smelting, heat treatment, etc.) involve a range of physical and chemical changes.

In chemical thermodynamics, the amounts of the components are most commonly expressed as matter amount and mole fraction. If a system contains K components, the matter amount is $n_1, n_2, \ldots, n_i, \ldots, n_k$, and the total matter amount in the system is N, then:

$$N = n_1 + n_2 + \cdots + n_i + \cdots + n_k = \sum_{i=1}^{k} n_i$$

### (3) Chemical potential

The terms "position" and "potential" are commonly used to describe energy or matter flows. When discussing the thermodynamics of a material transfer, the concept of chemical potential is introduced to illustrate the meaning of free energy G. If a study concerns a homogeneous system with K components and a single phase, then:

$$G = f(T, P, n_1, n_2, \ldots, n_k),$$

so,

$$dG = \left(\frac{\partial G}{\partial T}\right)_{P,\sum n_i} dT + \left(\frac{dG}{dP}\right)_{T,\sum n_i} dP + \sum_{i=1}^{k} \left(\frac{\partial G}{\partial n_i}\right)_{T,P,\sum n_j} dn_i$$

In this expression,

$$\left(\frac{\partial G}{\partial n_i}\right)_{T,P,\sum n_j}$$

means at constant temperature and pressure, except for i group elements, the matter amount in other components is constant; an incremental increase in i group element $(dn_i)$ causes a change in G (dG), by the derivative definition:

$$\frac{dG}{dn_i} = \lim_{\Delta n_i \to 0} \left(\frac{\Delta G}{\Delta n_i}\right), n_i + \Delta n_i \approx n_i$$

This partial derivative varies only with the composition and is independent of the total system amount. It is termed the chemical potential $(\mu_i)$ of component i, as it represents the change of constant temperature isobaric position caused by a change in chemical composition.

$$\mu_i = \left(\frac{\partial G}{\partial n_i}\right)_{T,P,\sum n_j}$$

A difference in chemical potential can explain not only the normal diffusion phenomenon, but also the "uphill diffusion" phenomenon of material flowing from a low to a high concentration.

(4) Electron concentration factors and Fermi energy

With regard to electron concentration factors and Fermi energy, the binding energies and lattice constants of metals were first considered. For the calculation of copper, the repulsive reaction between nearest-neighbor copper ion shells must be considered. It is precisely because this repulsion energy of the body-centered cubic structure was higher than that of the face-centered cubic structure that pure copper adopted a face-centered cubic structure. When calculating internal energy, three terms were considered:

$$U = \varepsilon_O + \varepsilon_F + \varepsilon_R$$

In the formula, $\varepsilon_O$ is the lowest energy of the electronic state, $\varepsilon_F$ is the Fermi energy, and $\varepsilon_R$ is the repulsive energy. The energy difference between the two crystal structures of copper arises from:

$$\varepsilon_O^B - \varepsilon_O^F \approx -10^{-3}$$
$$\varepsilon_F^B - \varepsilon_F^F \approx -10^{-3}$$
$$\varepsilon_R^B - \varepsilon_R^F \approx -10^{-3}$$

The B and F superscripts of the energy terms denote the body-centered and face-centered cubic structures, respectively. The Fermi energy is the chemical potential of electrons.

The chemical potential $\mu_i$ of a component was:

$$\mu_i = \left(\frac{\partial U}{\partial n_i}\right)_{S,V,\sum_{n_j}} = \left(\frac{\partial H}{\partial n_i}\right)_{S,P,\sum_{n_j}} = \left(\frac{\partial F}{\partial n_i}\right)_{T,V,\sum_{n_j}} = \left(\frac{\partial G}{\partial n_i}\right)_{T,P,\sum_{n_j}}$$

The magnitude of work W reflects the Fermi energy level $\varepsilon_F$. Electron flow occurs when two metals of different W are in direct contact. Electrons flow from metal A with high Fermi energy to metal B with low Fermi energy, creating a deficit of electrons, i.e. a positive charge, at the surface of A, and an accumulation of electrons, i.e., a negative charge, at the surface of B. This electron flow equalized the electronic potentials of the two metals, that is, $\varepsilon_F$ became balanced. This is the working principle of a thermocouple.

(5) Stress, strain, and strain energy

A large number of process problems of alloys involve stress, strain, and strain energy. The processes of casting, welding, and quenching of a hot surface can lead to cracking

10.3 Properties of Rare Earth Alloys

due to thermal stress and phase-change stress. The properties of alloys, especially mechanical properties, involve mechanical processes, which are necessarily affected by stress, strain, and strain energy. For example, toughness is a comprehensive representation of strength and plasticity, which characterizes the ability of an alloy to deform and the extent to which fracture processes absorb strain energy.

Strain energy causes the atoms in an alloy to be displaced from their equilibrium positions with minimum energy, so that the internal energy of the alloy is increased, that is, $\Delta \upsilon > 0$. An object in mechanical equilibrium can be subjected to two types of external forces. The first is a surface force, which acts on the surface of an object, such as the pressure of a gas, liquid, or solid in contact with the object. The second is a volume force, which acts on all elements of the object, such as gravity, centrifugal force, etc. Force and stress not only have size, but also direction, and as such are vectors. Both force and stress vectors can be decomposed, and the sum of the component actions is equal to the action of the resultant force. The most common approach is to decompose the vector into components parallel to the coordinate axis in rectangular coordinates.

The stress on each plane of the minimum element in the elastic body can also be decomposed into stress components along the x, y and z directions. Acting on all vertical stresses across different element areas through known points, the maximum and minimum are principal stresses, all vertical stresses are full stresses.

In the simplest case of unidirectional tension, Hook's law states that stress is proportional to strain, that is:

$$\sigma_x = E\varepsilon_x$$

Connecting the strain ($\varepsilon_x$) corresponding to positive stress by elastic measurement, the strain energy of micro volume $dxdydz$ is:

$$d_w = \frac{1}{2}\left(\sigma_x d_y d_z\right) \cdot \left(\varepsilon_x d_x\right)$$

The strain energy per unit volume, $w_0$, is given by:

$$w_0 = \frac{1}{2}\sigma_x \varepsilon_x$$

$E$ is also called the Young's modulus.

(6) Research methods

The changes of various properties and processes caused by the segregation of solute atoms at grain boundaries can be used to study the phenomenon of grain boundary adsorption. Some of the research methods involve direct observation, whereas others rely on information inferred through existing theories, and details can be found in the relevant literature. The commonly used research methods are as follows.

- Auger energy spectrum method—the energy spectrum of Auger electrons emitted from grain boundaries allows direct and quantitative determination of the composition of these grain boundaries.
- Field ion microscopy—direct observation of solute protons segregated at grain boundaries.
- Electron probe method—chemical composition along a crystal boundary section can be obtained by the electron probe method.
- Study on the composition and concentration of grain boundary region by radioisotope labeling.
- Neutron diffraction—neutron radiation is applied to study the distribution of atoms capable of activating and releasing rays in grain boundaries.
- High-resolution transmission electron microscopy (TEM) method—monitoring energy loss through the inelastic scattering effect of electrons penetrating the specimen.
- Fluorescence—application of fluorescence effect.
- X-ray microscopy—differences in the ability of elements in alloys to absorb X-rays.
- Optical metallography method—the composition in a grain boundary segregation zone is different from that in the crystal, and this is observed by chemical erosion.
- Changes in lattice constants—solute segregation at grain boundaries.
- Electrode potential method—determination of cross-section electrode potential along grain boundaries. Grain boundary segregation is studied by means of calibration curves.
- Internal friction method—grain boundary segregation is inferred from the change of solute or grain boundary internal consumption in the crystal.
- Surface energy method—measurement of the thermal erosion angle of an alloy by means of an interferometer.
- Resistance method—grain boundary segregation is inferred from the change of grain resistivity.
- Ultrasonic method—monitoring anomalous absorption, which results from resonance when the average grain size is equal to the half-wavelength of the ultrasonic wave.

## 10.4 Effect of Rare Earths on Non-ferrous Alloys

Non-ferrous alloys incorporating rare earths show good performances, as described in the following. Rare earth-magnesium alloys have become an important direction in the development of magnesium alloys. Alloys based on magnesium with other elements are characterized by low density ($1.8 \text{ g/cm}^3$), high specific strength, large specific modulus of elasticity, good heat dissipation, good anti-seismicity, greater ability to withstand impact load than aluminum alloys, and good corrosion resistance to organic matter and alkalis. At present, the most widely used is magnesium–aluminum alloy, followed by magnesium-manganese alloy and magnesium-zinc-zirconium alloy. These are mainly used in the aviation, aerospace, transportation, chemical, and rocket industries, and are the lightest practical metals. The density of magnesium is about two-thirds of that of aluminum and a quarter of that of iron. Such alloys also offer high strength and rigidity.

Luo and Zhang (1994) studied the effect of heat treatment on the properties and microstructure of Mg-Nd-Zr extruded alloy:

The studied material was Mg-5Nd-0.5Zr. It was extruded at 350 °C, and then treated for 10 h at 170 °C, 1.5 h at 510 °C, quenched, and then 10 h at 170 °C again. The processed samples were used for mechanical properties and phase analysis. The results showed that the strength increased markedly after the aging treatment, and that the tensile strength increased after the quenching and aging treatment; the yield strength and plasticity decreased following both heat treatments; rare earth phase is $Mg_{12}Nd$ phase, and is small diffuse phase; and many sub-crystals with dislocation were observed.

## 10.5 Application of Rare Earth-Magnesium Alloy Bone Grafting Materials

In clinical medical procedures, stainless steel and titanium alloys are used as fracture fixation parts, but bone tissue recovery must be followed by a second surgical operation to remove plates and pins, because long-term presence of these alloy materials in the body may cause ulceration and so on. With rare earth-magnesium alloys, however, the magnesium-based material has good compatibility with the matrix, and can promote rapid fracture healing with no adverse effects. Lu et al. (1994) reported the results of some studies in this field.

With increasing Nd content in an alloy, the tensile strength at room temperature increased. Artificial aging increased the tensile strength by 30 MPa compared to the extruded state, and quenching further increased it by 20 MPa, but the yield strength concomitantly decreased. The rare earth compound in the alloy was identified as $Mg_{12}Nd$. Extrusion profiles showed that $Mg_{12}Nd$ was dispersed in the crystal and at grain boundaries, and the content of this phase increased with aging.

Rare earth-magnesium alloys account for more than half of all of the magnesium alloys in current use. Generally, rare earth additions to magnesium alloys are only 0.1–0.6%. A series of rare earth magnesium alloys with high strength, heat resistance, and corrosion resistance has been developed for use in motor vehicles, high-speed trains, and military products. For example, Shanghai Jiao Tong University and Guangzhou Institute of Nonferrous Metals undertook the "high-performance rare earth heat-resistant magnesium alloy research and its application in the electronic information industry".

New magnesium alloy calendering materials with the same normal temperature formability as aluminum alloys have been developed at the Japan Institute of Industrial Technology and Kyoto University. These new alloys incorporate trace cerium, etc. In Mg–Al alloys, hot rolling produces a kind of aggregate structure, which is very different from that of general magnesium alloy. The crystalline structure of ordinary magnesium is anisotropic, and calenderability is lower than those of aluminum and iron. In the new magnesium alloys incorporating trace rare earth elements, the magnesium columns are able to slide, and the material can be punched at room temperature.

In 2004, researchers in the United Kingdom demonstrated a new magnesium alloy specifically for use in the aerospace industry, which combines the advantages of various alloys. With excellent casting properties, it can maintain mechanical characteristics at 150–200 °C. Its composition range is:

$$(0.2 - 0.5\ Zn) - (2.6 - 3.1\ Nb) - (1.0 - 1.5\ Gd) - (0.6\ Zr) - Mg$$

Gd shows greater solid solubility than Ce in Mg, and can be completely fixed by heat treatment, resulting in homogenization and good precipitation hardening.

Zhou et al. (1994) studied the effect of rare earth incorporation in the Al-Cu-Mg system of 2024 alloy on aging properties. 2024 alloy is generally used in a naturally aged state, mainly as aircraft skin, panels, etc. When profiled, it can be used in aircraft compartments, wing ribs, etc. However, the stress corrosion tendency of 2024 alloy is high, especially in its naturally aged state. The effect of rare earth addition on the stress corrosion sensitivity of 2024 alloy has been studied. The test alloy was derived from the Southwest China Aluminum Processing Plant. Ingots were hot-extruded into round bars and then pressed into pieces.

Test alloy sheet was heated at 495 °C in a salt bath furnace for 24 min, and then immediately immersed in water for quenching. Microhardness was measured in the naturally aged state, and the morphology of broken products was observed by tensile tests and SEM. The results showed that the strength and plasticity of the materials were improved by adding rare earths. The toughness was uniform, the inclusions were small, and the stress corrosion tendency was reduced. The optimal level of rare earth addition was found to be 0.2–0.3% (mass fraction).

The new high performance super aluminum alloy with multiple dispersion phases of rare earths exceeds the comprehensive performance of high toughness and corrosion resistance of general super aluminum alloy, which is of great significance for the development of aerospace, rail transit, and construction machinery. A single amorphous aluminum-based bulk material with high fracture strength was prepared by direct melt pouring at Shenyang National Materials Science Laboratory and John Hopkins University. The nanocomposite ribbons formed by the aluminum-based amorphous alloys showed ultra-high fracture strength and good plasticity. A 1 mm diameter aluminum-based metallic glass bar was obtained from an Al-Ni-Co-Y-La five-element alloy system.

The internal structure of glassy matter is in a disordered state. A team in the United States, after processing a metallic glass sample at high pressure, found a highly ordered single-crystal structure inside it. Generally no more than a few atomic spacing, "short-range order" refers to alignment over a few atoms; "long-range order" refers to the ordering in ordinary crystals. Cerium-aluminum alloy metallic glass with long-range order was fabricated and placed under a pressure of 25 GPa. "De-glassing" occurred, forming a face-centered cubic crystal structure, which remained stable as the pressure was reduced to ambient. Thus, the application of high pressure provides a method for the preparation of single-crystal materials from metallic glass.

## 10.6 Application of High Pressure to Form New Alloys

Mixing different metals can produce alloys with superior properties to those of the individual components, but not all metals can form alloys because of the different sizes of their atoms. An international research team found that by applying high pressure to atoms, alloys that could not be made under normal conditions could be made, thereby developing new materials. Ce and Al can form compounds, but do not readily form alloys, because Ce atoms are 28% larger than Al atoms, and the electronegativity of Ce is low. Researchers observed the onset of alloy properties when conducting high-pressure tests on Ce-Al metallic vitreous samples. High pressure removes some of the outermost electrons from Ce, leaving them delocalized. The delocalization of electrons not only changes the electronegativity of the Ce atoms, but also reduces their size by about 15%. The Ce and Al atoms can then form a single-crystal alloy structure at 25 GPa (about 250,000 atm).

Alloys formed by the multi-dispersion phase of rare earths have great significance in the aerospace, rail transit, and construction machinery industries.

1. Al-Li alloys

Al-Li alloys have attracted great attention in the aerospace industry because of their high specific strength (fracture strength/density), high specific stiffness, and low relative density. As such, they are attractive as modern aircraft skin materials, potentially reducing the mass of a large passenger aircraft by 50 kg. For example, for a Boeing 747, an annual saving of $2000 could be made for every 1 kg reduction in mass. Titanium alloy is lighter than steel, corrosion resistant, non-magnetic, and strong, and as such is an ideal material for aviation and naval ships.

To further improve the properties of existing Al-Li alloys and to reduce production costs, new materials are being sought. Some research on Al-Li alloys containing rare earths has been carried out, and several such studies are briefly described below.

Wang LM and Zhao MS (1997) studied the effects of rare earth elements on unidirectional solidification structures in Al-Li alloys. They used a corundum crucible to smelt rare earth Al-LI alloys in a resistance furnace. Al, Al-Li, and Al-RE intermediate alloys were added under KCl-LiCl-LiF molten salt protection. At 704–760 °C, the mixture was stirred with a corundum tube, kept at this temperature for about 10 min, then allowed to cool and poured into a graphite mold. It was then placed in a one-way solidification furnace with the temperature controlled at 720 °C. After stabilization, cooling water at a certain flow rate and temperature was passed through the bottom of the graphite mold to induce unidirectional growth of the alloy. The microstructure and properties of rare earths in the alloy were determined by SEM.

The experimental results showed that the rare earth elements had a significant effect on the microstructure of unidirectional solidification of Al-3Li. When the rare earth content was 0.15–0.2% (mass fraction), the micro-longitudinal dendrite spacing decreased significantly, and the rare earths were mainly distributed at dendrite boundaries of the alloy. The supercooling effect of rare earth alloy solidification process is an important reason to affect microstructure.

Ni et al. (1994) studied the effect of cerium on the physical short-crack propagation resistance of Al-Li alloys. Such alloys have excellent fatigue long-crack propagation resistance, but poor resistance to short-crack propagation, thus limiting their broad industrial application. Aircraft manufacturers in Europe and the United States are seeking to improve the fatigue short-crack propagation resistance of Al-Li alloys as a prerequisite for their industrial use. From the point of view of purification and micro alloying of rare earths, theoretical discussions have been presented on improving the intrinsic crack growth resistance of Al-Li alloys, so as to improve their short-crack growth resistance and reduce their short-crack effect.

An in-house-constructed 2 mm thick Al-Li alloy plate was used to study two-dimensional short cracks therein. The threshold for formation of short cracks of different lengths was measured, and the properties were studied. The results showed that the plane slip and textural strength of Al-Li alloy make its crack propagation paths zigzag and roughen the crack area. An extremely strong crack deflection tendency imparted the alloy with a high crack tip shielding effect, and the intrinsic crack growth resistance was low. Overall, the alloy exhibited a strong short-crack effect.

Trace rare earths can significantly improve the intrinsic crack growth resistance of 2090 alloy. Research has shown that coplanar slip is the main reason for the poor plasticity and toughness and low intrinsic crack-growth resistance of Al-Li alloys. The main functions of trace rare earth Ce in Al-Li alloy are twofold. On the one hand, trace Ce can strengthen grain boundaries, the external strengthening and external toughening effect of the alloy is obtained. On the other hand, Ce and Al can form $Al_3Ce$ as an intermetallic compound, which has a cubic structure with a similar atom arrangement as the most densely arranged planes of Li. Ce does not improve the intrinsic crack-growth resistance and crack-closure effect in 8090 alloy; rather, it has an adverse effect.

2.  Zn-based alloys

High strength, wear-resistant, Zinc-based alloy is a new bearing alloy, which has the advantages of light weight and high wear resistance. Rare earths have a good effect on the wear resistance of such alloys.

Xiang et al. (1994) studied the effect of rare earths on the mechanical properties of high-strength, wear-resistant, Zinc-based alloys. Their experimental results showed that the rare earths affect the tensile strength, elongation, hardness, impact toughness, shear strength, and compressive strength of such alloys. Specifically, the tensile strength, shear strength, and compressive strength were improved for alloys containing 0.09–0.45% rare earths. No obvious change in alloy hardness was discerned. A certain amount of rare earth can refine the grains. When adding rare earths, a bulk rare earth-rich phase was formed at grain boundaries, and microstructure shrinkage of the alloy increases. This had an adverse effect on the material properties, so the amount of rare earths added needed to be strictly controlled.

Rare earths also exerted a good effect in noble metal-based alloys within a certain addition range.

# References

A magnesium alloy which can be punched at room temperature has been developed by KYU. http://www.cre.net/show.php?contentid=363. 25 Sept 2008

A single crystalline cerium-aluminum metallic glass has been fabricated. http://www.cre.net/show.php?contentid=97269. 25 July 2011

Du T, Han QY, Wang CZ (1995) Physical chemistry of rare earth alkali soil and its application in materials. Science Press, Beijing, p 151

High specific strength aluminum-based bulk metallic glass. http://www.cre.net/show.php?contentid=84310. 11 June 2009

Hunan institute of rare earth metal materials was successfully recognized as the academician workstation of Hunan province. http://www.cre.net/show.php?contentid=113725. 22 May 2014

Lu LQ, Li CY, Lu Guo (1994) Application of rare earth magnesium alloy bone bonding materials. In: Proceedings of the third annual conference of the Chinese Society of Rare Earths,vol 3, pp 9–15

Luo ZP, Tang YL (1995) Thermodynamic analysis of the role of rare earths in magnesium alloy solutions. J Chin Soc Rare Earths 13(2):119

Luo ZP, Zhang SQ (1994) Effect of heat treatment on properties and microstructure of Mg-Nd-Zr extruded alloys. J Chin Soc Rare Earths 12(2):183

Ni YF, Chen Z, He M (1994) Effect of cerium on the physical short crack propagation resistance of al-lithium alloy. Chin Rare Earths 15(4):37–41

Research on high performance rare earth heat-resistant magnesium alloy and its application in electronic information industry. http://www.cre.net/show.php?contentid=108245. 13 May 2013

Scientists use high pressure to make new alloys, http://www.cre.net/show.php?contentid=58598. 27 Mar 2009

Sun YY, Du T (1994) Basic thermodynamic study on the application of Ce in aluminum alloy. In: Proceedings of the third annual conference of the China Rare Earth Society, p 77

Wang L, Zhao M (1997) Synthesis of $\gamma$ LiAlO$_2$ ultra fine powder by citrate method. J Chin Ceram Soc 25(5):618–621

Xiang HY, Gong BK et al (1994) Effects of rare earths on mechanical properties of high strength wear-resistant zinc-based alloys. In: Proceedings of the third annual conference of the China Rare Earth Society, vol 3, pp 38–42

Xiao JM (1985) Alloy energetics—relationship, calculation and application of alloy energy. Shanghai Science and Technology Press, Shanghai

Zhou XP, Wang SL, Xu YH et al (1994) Effects of rare earths on the aging properties of 2024 alloys. In: Proceedings of the third annual conference of the Society of Rare Earths, vol 3, pp 19–21

# Chapter 11
# Application of Rare Earths in Agriculture, Forestry, and Animal Husbandry

## 11.1 Brief Description

In the early 1980s, the author went to Baotou to attend China's National Rare Earth Conference, and found many small, light-red or light-yellow wild chrysanthemums growing in grassland soil at the edge of a local mine. Their petals were solid, bright, and plastic-like. A few were picked and placed in a vase, whereupon they lasted for years without decay. It transpired that such flowers absorb rare earth elements such as neodymium (Nd) and praseodymium (Pr) from soil, which enhances their vitality.

Later, students at an agricultural university were provided with a small amount of lanthanum chloride, with which they performed physiological experiments on soybean. The experiments involved spraying with different doses at the seed, seedling, flowering, and pod stages. The results showed differences in plant morphology and pod conditions between untreated and treated specimens. This suggested that rare earths may have important effects on crop physiology.

The application of rare earths in agriculture has been reported since the 1930s. Scientists in the former Soviet Union and Bulgaria reported that the application of appropriate amounts of rare earth elements can enhance plant growth. China began to promote the application of rare earths in agriculture, forestry, and animal husbandry in the 1970s.

## 11.2 Rare Earths in Agriculture, Forestry, and Animal Husbandry

1. Rare earths increase the yield and sweetness of watermelon

Water-soluble rare earth nitrates (light rare earths, such as lanthanum, cerium, praseodymium, neodymium, and samarium as major component) have been deployed to treat watermelon in Fumin County, Yunnan, which has a typical subtropical red

© Science Press 2023

C. Wang, *Theory and Application of Rare Earth Materials*,

https://doi.org/10.1007/978-981-19-4178-8_11

183

soil. The yields of different watermelon varieties were increased by 1–20% or more, and their sugar contents were increased by more than 1%. Similarly, small-seed watermelon sprayed with aqueous rare earths on sandy tidal soil in the Western suburb of Tianjin typically showed an increase in production of more than 20% and an increase in sugar content of about 0.75%.

The question arises as to why watermelons sprayed with rare earths show increased production and greater sweetness (Gao 1983). The results of radioisotope tracing have indicated that rare earths can promote phosphorus absorption. Phosphorus is one of the indispensable nutrient elements in crop growth and development, and it participates in physiological and biochemical processes in various ways. The synthesis, hydrolysis, and transfer of carbohydrates are all directly related to phosphorus. The sugar present in watermelon is synthesized and gradually accumulated due to interactions with phosphorus. In this way, appropriate deployment of rare earths can make watermelon sweeter.

2. Effect of cerium on maize seedling growth and enzyme spectrum

Tang and Li (1983) studied the effects of different concentrations of cerium chloride on the growth of maize seedlings and the peroxidase and esterase isozyme profiles in their roots.

The results showed that a low concentration (0.5 ppm) of cerium chloride stimulated maize seedling growth, whereas a high concentration (5.0 ppm) impaired growth. Within the inhibited roots, the numbers of isozyme bands of both enzymes were suppressed. Isozymes can catalyze similar enzyme reactions with different molecular structures, and their synthesis is controlled by genes, so changes in the isozymes are related to metabolic processes in plants. The effect of cerium on maize growth is exerted through internal physiological activities.

3. Acute toxicity of rare earth nitrates

Ji and Cao (1983) studied the acute toxicity of rare earth nitrates. Their results provide a benchmark for the assessment of rare earth toxicity. The experimental subjects were mice of varying sizes. The mice showed transient excitement, restlessness, and increased activity after exposure to poison, promptly followed by mental depression and general weakness. They returned to normal after 24–48 h (low doses) or after 48–96 h (high doses). In terms of toxicity, there are no specific differences between the nitrates and oxides of mixed rare earths and individual rare earths. Rare earth nitrates can irritate the skin and mucous membranes, so protective measures should be taken in their application.

4. Absorption, operation, and distribution of rare earth elements in crops

Using a rare earth radioisotope tracer method, Qi et al. (1984) observed the absorption, operation, and distribution of rare earth elements in four different crops, namely maize (Shandan 9), cotton (Heishan Cotton 1), soybean (June bracts), and cole cabbage. The rare earth radioisotopes [141]Ce, [147]Nd, [152]Eu, and [46]Sc were added in the form of a mixed nitrate solution.

11.2 Rare Earths in Agriculture, Forestry, and Animal Husbandry

The experimental results showed that the uptakes of these four rare earth elements in hydroponic corn seedlings can be very rapid; with maximum absorption levels achieved within 1 min. Over a period of 72 h, the absorption level in the seedlings gradually decreased. The uptake of Ce by corn seedlings was the largest, amounting to 2.5 times the sum of the other three elements. Except for high Sc uptake by Canola stems, the level of Ce absorption remained the highest in all parts of the other crops.

The amount of rare earths absorbed by corn after 7 days was not much higher than that after 2 days, while the first absorption is larger. The other three crops continued to absorb considerable amounts of rare earths after 2 days and, as the plants grew, and significant amounts were transferred to fruiting organs and stems; The initial 2 days have slow absorption, the next 5 days, and the amount of rare earths in cotton running to bolls and buds was 20 times higher than previous 2 days; Cotton plants mobilized the entire dose of rare earth to their fruiting organs. Canola transported Sc to its stems for storage in the first two days. These experiments showed that crop leaves only take up a small fraction of the rare earths.

5. Effect of rare earths on rubber tree

Wang and Chen (1985) studied application techniques and the effect of rare earth elements on Brazilian rubber trees in 1985. Their first experiments were carried out on a plot of 10.67 ha on Hainan Island. The study involved 4425 rubber trees planted in 1957; 1:1 scale design, and take about 0.67 ha as a treatment area. A second test site was set up in Lingshui on Hainan Island in 1981, covering 2 ha. In 1982, a third experimental site was established on the Leizhou Peninsula of western Guangdong, covering an area of 66.7 ha and encompassing 30,000 trees. In order to explore the effects of application in different rubber regions and with different rubber strains, the study was extended to rubber planting areas in Guangxi, Yunnan, and Fujian. The rare earths were applied at 299.85–749.63 g per hectare, either in purely aqueous solution or mixed with ethephon carrier, 5–7 times a year. The treated trees were compared with untreated controls.

The results showed that the amount of rubber produced was increased by 14.86%; the total sugar content in the latex was about 0.56 mg/mL higher than that in the control, and the dry glue content was increased by 0.2–2%. The rubber produced conformed to the national first grade standard. However, it was also found that if the rare earth concentration was too high, premature coagulation could occur.

6. Effects of rare earths on the growth of rhizobia in legumes

Zhang et al. (1988) carried out a study to investigate the effect of rare earth compounds on the reproductive state of *Rhizobium* strains and to identify the optimal dose. The test materials were alfalfa, red clover, white clover, and Shatavan rhizobia. Rare earths were deployed as their chlorides, $RECl_3$, at levels of 0, 100, 200, 300, 500, and 1000 ppm.

The results showed that 100–300 ppm rare earths promoted rhizobia proliferation, and the effect on alfalfa rhizobia was most obvious. When it is 500 ppm, the proliferation of four rhizobia was inhibited, while the growth of all four rhizobia was inhibited when the rare earth concentration was 1000 ppm. The weight of fresh alfalfa, the numbers of the respective rhizoma, and the nitrogen-fixing activities of the rhizoma were significantly higher than those of alfalfa rhizoma without rare earth treatment.

7. Effect of rare earths on somatic cells of Chinese wolfberry

Yang and Du (1994) reported the effects of rare earths on the somatic embryo induction process of Chinese wolfberry. Seeds from Ningxia province were sterilized and cultured in a medium. After germination and seedling formation, leaves were cut and inoculated into a medium to induce callus formation. The embryonic callus was transferred to rare earth solutions of different compositions and concentrations. After culturing for 15 days, somatic embryos at different developmental stages were examined under a microscope. The results showed that various rare earth combinations have different effects on embryonic induction.

8. Effects of lanthanum-rich rare earths on spring wheat

Of the rare earths used in Chinese agriculture, 97% are light rare earths, of which $La_2O_3$ accounts for 25–28.5% and $CeO_2$ 49–51%. Due to the industrial utilization of cerium- and lanthanum-rich rare earth mixtures and the high lanthanum abundance in southern China, their effects should be further studied. Xie et al. (1994) studied the biological effects of lanthanum-rich rare earths on spring wheat.

The soil used in culture and pot experiments was thin-layer black soil from the second terrace of the Songhua River. Field experiments were conducted in a meadow on the Sanjiang Plain. In pot and field experiments, rare earth solutions were sprayed on wheat leaves at the trifoliate stage, and the physiological indices were measured from the tillering to the jointing stage.

The fertility index exceeded that of the blank control for all treatments with rare earth solutions. The fresh weight, leaf area, total chlorophyll content, and root volume increased. Lanthanum-rich rare earths significantly increased chlorophyll *a* content.

9. Grey correlation analysis of the application of rare earth and soil elements to cotton

Zhang et al. (1994) carried out a grey correlation analysis of the application of rare earth and soil elements to cotton. In Kazuo County, rare earths have been applied to cotton plants since 1989. The methods include seed dressing, spraying, and soaking. Results have shown increased leaf area, extended leaf life and functional period, photosynthetic rate, and single boll weight. The cotton fiber developed well, the maturity coefficient was high, and skin flower color and quality were improved.

Fertility tests were conducted on 15 soil samples from 5 towns. The results indicated that rare earth application to cotton affects soil characteristics, most notably pH > total potassium > total phosphorus, with secondary factors being total nitrogen > hydrolyzed nitrogen > organic matter. The rare earths have little effect on $K_2O$ or $P_2O_5$.

10. Rare earth application to white gourd, potato, and cabbage

## 11.2 Rare Earths in Agriculture, Forestry, and Animal Husbandry

Fu et al. (1994) applied rare earths to white gourd, potato, and cabbage, and observed significantly improved yields. Specifically, the average yields of white gourd, potato, and cabbage were increased by 17.4%, 12.4%, and 7.1%, respectively. By applying mixed rare earths or La, Ce, Pr, or Nd individually, it was shown that rare earths can promote nutrient absorption and root growth, and increase chlorophyll content.

11. Rare earth application to Chinese fir

Lin et al. (1994) applied rare earths to Chinese fir, a fast-growing tree species in southern China, to assess their growth-enhancing effects. The study was conducted on a farm in Gujiang county, northern Guangdong province, which has sandy shale red soil. Firs were planted at 37,481 per hectare in the experimental forest, 5 rows × 6 plants per plot.

The seedling stage of Chinese fir was treated with rare earth solution by soaking, leaching, and spraying. The results showed that these methods of rare earth application could all promote growth. Compared with untreated controls, seedling height increased by 21.2–32.7%, 14.4–28.3%, and 13.0–28.6%, respectively, and seedling diameter also increased.

12. Effect of lanthanum on the phosphorus absorption and morphology of rice

Xie et al. (2004) studied the effects of lanthanum on rice growth, its root, stem, and seed grain phosphorus content and absorption, and in vivo morphology by root-separated nutrient solution culture. The results showed that low concentrations of lanthanum (0.05–1.5 mg/L) led to increased rice yields, whereas high concentrations (9–30 mg/L) had a negative effect. Lanthanum significantly increases the uptake of phosphorus by roots, such that both complexes and inorganic phosphorus are increased in stems and roots, and to promote the activity, so that increase phosphorus absorption in grain.

13. Effects of rare earths in soil on the morphological distribution and transformation of effective nitrogen

Ding et al. (2004) studied the effects of rare earths on morphological distribution and transformation of available nitrogen in chloasma by adding them to dry soil saturated soil. The investigated available nitrogen comprised inorganic mineral nitrogen and relatively simple organic nitrogen; that is, the sum of $N H_4^+$-N, $N H_3^-$-N, amino acids, amides, and hydrolytic protein nitrogen. A good correlation between available nitrogen and crop growth was noted; it can reflect the seasonal supply of nitrogen in the soil, so the focus of the study is effect of rare earth application on different types of available nitrogen in a pure soil system.

The results showed that ammonium nitrogen and hydrolytic nitrogen content increased under dry conditions, while the content of nitrate nitrogen decreased. With an extension in culture time, the effect of the rare earths gradually increased. An increase in rare earth content increased the mineral nitrogen content in two weeks, but no significant effect was found after eight weeks. The content of available nitrogen increased when the rare earth content was low and decreased when the content was high. Under the condition of saturated conditions, the ammonium and hydrolyzed

188    11  Application of Rare Earths in Agriculture, Forestry, and Animal Husbandry

nitrogen contents in soil are little affected by low rare earth contents, but they decrease at high rare earth contents. Again, with increasing culture time, the effect of the rare earths gradually increases.

14.   Rare earth applications in pollution ecology

Zhou et al. (2004) have reviewed research progress in the use of rare earths to promote plant resilience, which the aim is to reduce the harmful effects of acid rain, ozone, ultraviolet radiation, pesticides, heavy metals, and other environmental pollution hazards on plants. They proposed that the mechanism of rare earth-induced plant resilience to environmental pollution relates to the regulation of endogenous metabolic networks in plant cells, the maintenance of normal physiological functions, and the enhancement of the tolerance range of plants to environmental pollution. These authors also analyzed the main problems associated with such efforts.

Heavy metals in soil not only inhibit plant growth, but also pose a risk to human health due to their enrichment in plants. Some researchers have studied the effects of heavy metals (Cd, Hg, Cr, Pb) on rice, cabbage, soybean, wheat, pak choi, and spinach, and the possible protective role offered by rare earths. It was confirmed that La, Ce, Pr, and Nd can significantly reduce the enrichment of heavy metals in plants, thereby alleviating the decline of biomass, improving quality, and reducing the uptake of heavy metals in humans and animals.

Rare earths can also reduce the harmful effects of ozone on plants.

Pesticide residues in agricultural products can endanger human health and pollute the environment. Research has indicated that the application of rare earth compound fertilizer can greatly reduce pesticide residues in rice, citrus fruits, cantaloupes, Chinese cabbage, hollow vegetables, pakchoi, tomatoes, tea tree, and rape, while increasing crop yields and improving quality.

Researchers believe that the role of rare earths in inducing plant resilience and alleviation of pollution damage is based on the regulation of plant biochemical metabolic networks, which is manifested in the following aspects.

(1)   Modification of the reactive oxygen species protection system

The damaging effects of acid rain, heavy metals, ozone, and UV-B radiation are associated with strong free radical reactions in plants. REs (La, Ce) can directly inhibit the excessive production of free radicals induced by adversity. In this way, the pollution stress effect arising from free radical damage reactions can be reduced.

(2)   Modification of photosynthesis

A number of studies have shown that under the stresses of acid rain, heavy metals, ozone, and radiation, photosynthesis is inhibited, the chloroplast state becomes abnormal, chlorophyll acid ester reductase is inhibited, and photosynthetic efficiency is reduced. Rare earths can inhibit such damaging processes, such that photosynthesis is maintained.

The plasma membrane is susceptible to pollution damage, but it is also the primary site of rare earth biological action. For example, rare earths can stabilize membrane

## 11.2 Rare Earths in Agriculture, Forestry, and Animal Husbandry

structure, influence the transmembrane proton gradient and membrane potential, and modify mineral nutrition.

Many other enzymes are involved in the plant biochemical metabolic network. Pollutants can inhibit the activities of root dehydrogenase, polyphenol oxidase, lactate dehydrogenase, amylase, nitrogenase, nitrate reductase, protease, deoxynuclease, and ribonuclease, thus hindering the carbon, nitrogen, and secondary metabolism networks of plants. Early results from many rare earth agricultural studies show that they have a regulatory effect on these enzymes.

15. Migration and accumulation of rare earth elements in the soil-rice system

Wang et al. (2006) studied the migration and absorption accumulation characteristics of rare earths in a soil-rice system. Because of the complex environmental factors in paddy fields, the biomass of rice plant samples was observed systematically at different growth stages following the application of different concentrations of rare earths to rice plants. The contents of 14 rare earth elements in soil and various organs of rice plants were simultaneously measured by inductively coupled plasma mass spectrometry (ICP-MS), and compared with those in potted winter wheat.

The soil of potted winter wheat was the Xiashu loess (Nanjing, Jiangsu Province). The developed yellow–brown soil (pH 6.81, organic matter 1.14%) used for potted rice was obtained from the same source. Wet, yellow–brown soil (pH 6.48, organic matter 1.42%, water content 10–20%) was used to mimic a paddy field. All soils were treated with mixed cerium-rich rare earth chlorides.

Samples were taken at the seedling and mature stages. Seedling stage samples were taken aboveground and from the root. Mature stage samples were taken from the root, stem, leaf, ear, glume, and grain. All samples were digested in $HNO_3$-$HClO_4$-HF, and the contents of various rare earth elements were determined by ICP-MS. The quality control is carried out in sampling and analysis process.

The experimental results showed that rare earths were readily taken up by the roots, leaves, and stems of rice. Their accumulation in these plant parts increased rapidly with increasing concentration of the applied rare earths, but they did not accumulate in grains. When the amount of rare earths applied exceeded 400 mg/kg, the growth of the rice plants was significantly impaired. This threshold is much lower than that for dry-wheat.

An obvious positive abnormality in the Tb content of mature rice root was noted. An enhanced distribution of the heavy rare earths Eu and Tb in above-ground organs was also evident. When the rare earth concentration is high, the main components have a significant influence on the distribution pattern of rare earths; have influence to stems and leaves; but not significant influence in grains.

The accumulation abilities of rare earths from soil were ranked as root > leaf > stem, spike, shell > grain. At the root, the absorption capacities of various rare earth elements from the soil are approximately the same, with only Tb showing selective absorption.

The absorption coefficients of most rare earth elements in grain decreased with increasing Nd, Eu, and Tb concentrations.

## 16. Advances in research on rare earth superaccumulation

Superaccumulating plants are wild plants capable of overaccumulating certain chemical elements. Such plants may be used in the phytoremediation of environmental pollution, the extraction or recovery of rare chemical elements, and geological prospecting. The application of rare earth superaccumulating plants and phytoremediation has been the subject of much ecological and environmental research. Wei et al. (2006) outlined the environmental and ecological characteristics of rare earth elements and their superaccumulating plants, including the plant species, their geographical distribution, the distribution of rare earth elements in vivo, differentiation characteristics and influencing factors, the physiological and biochemical mechanisms of absorption and enrichment of rare earth elements, and development prospects for remediation applications.

(1) Environmental pollution by heavy metals and rare earth elements, and its remediation by superaccumulating plants

The term "heavy metals" refers to those with atomic density greater than 5 g/cm$^3$, including Cd, Cr, Hg, Cu, Pb, Zn, and Ag. Because As and Se are similar to heavy metals in terms of toxicity and other properties, they are generally also included in this category. Cu and Zn are essential trace elements for animal and plant metabolism, butthey are toxic above a certain threshold level. Cd and Pb are considered as nonessential elements for animals and plants, and they are highly toxic as they can interact with S and N atoms on the side chains of amino acids.

Although there have been many studies on the biological and toxicological effects of rare earths, there is no unified conclusion. The consensus on rare earths is that they can promote various physiological activities in organisms at a lower dose range, but can inhibit them after exceeding a certain threshold. In response, so-called "rare earth harmlessists" believe that the effect of rare earth elements on organisms is similar to that of essential metals for life. Conversely, scholars who regard rare earth elements as more toxic believe that the above effect is a reflection of a hormesis effect of rare earth elements; that is, a toxic excitatory effect. Toxic excitatory effect refers to the exposure of organisms to a low-level physical and chemical factor, such as radiation or toxic substances.

(2) Environmental pollution and phytoremediation of heavy metals and rare earths

Rare earth elements have been identified as major environmental pollutants. A large number of studies have shown that excessive occupational exposure to rare earths and environmental exposure can lead to various toxic effects or lesions, such as damage to the respiratory system and the blood circulation system. Of particular concern is research which implies a low IQ of children in rare earth-contaminated areas. Related problems have attracted much attention in ecological and environmental studies.

Plants with above-ground rare earth levels of more than 1000 μg/g can be classed as rare earth element superaccumulating plants. Several types of plant have been identified as overaccumulating rare earth elements, namely pilose pecan, pecan, ferns, and mango. Most of the rare earth element superaccumulating plants are ferns, which have some pertinent characteristics. They are highly variable, widely distributed, multiply in many ways, and have a strong adaptability to various extreme harsh environments.

17. Effect of some light rare earth elements on Triticum durum

Aquino et al. (2009) studied effect of some light rare earth elements on seed germination, seedling growth and antioxidant metabolism in Triticum durum. Rare earth elements (REs) enriched fertilizers have been commonly used in China since the 1980s, thus inducing a growing concern about their environmental impact in agriculture. For example, the effect of some light REs nitrate mixture and $La^{3+}$ nitrate on seed germination, seedling growth and antioxidant metabolism in Triticum durum was investigated with the aim of clarifying the potential benefits or damages of REs on plants. Seed pre-soaking for 8 h with $La^{3+}$ and REs nitrate inhibited seed germination at low concentrations (0.01 and 0.1 mM), while pre-soaking for 2 and 4 h already inhibited seed germination when higher concentrations (1 and 10 mM) of $La^{3+}$ and REs nitrate were used.

In roots $La^{3+}$ and REs nitrate treatments induced an increase in ascorbate (ASC) and glutathione (GSH) contents. In shoots only $La^{3+}$ nitrate induced an increase in the ASC content whereas GSH decreased following both $La^{3+}$ and REs nitrate treatments. An increase in ASC peroxidase activity was observed in shoots and roots, while catalase did not change in roots and slightly decreased in shoots. The possible role of the increase in some antioxidants as indicators of stress caused by lanthanide treatments is discussed.

## 11.3 Study of Rare Earth Bio-inorganic Chemistry

Li et al. (2004) discussed issues raised by the application of rare earths in agriculture and medicine, such as transcellular membrane transport of rare earths, the distribution of species after rare earths are ingested by animals, the role of rare earths within cell membranes, their effect on hemoglobin structure and oxygen loading function, and their effect on bone structure. As early as the 1940s, the biological effects of rare earths were raised with the development of nuclear technology. Since rare earths are similar to calcium, they can be used as structural probes of calcium-containing proteins.

1. Transcellular membrane transport of rare earths

The initial target sites of rare earths entering animals are various cells. For example, 90 days after feeding rats with different doses of $LaCl_3$, the La content in the erythrocyte membrane and cytoplasm of all groups was higher than in a control group. These animal studies confirmed that rare earths could penetrate the erythrocyte membrane. In vitro cell experiments showed that after incubating red blood cells in $5.6 \times 10^{-6}$ mol/L $EuCl_3$, the Eu could be detected in cell membranes and cytoplasm.

The process of rare earths entering red blood cells has been tracked by laser confocal microscopy. The rare earth ion transmembrane time ($t_a$) was found to be linearly related to the extracellular rare earth concentration C in the range $1 \times 10^{-6}$–$1 \times 10^{-5}$ mol/L. Experiments showed that the diffusion of rare earths into cells is one of the important pathways. Rare earths can also be transported into cells in the form of complex anions through anion channel carriers of red blood cells, and can use ion carriers to enter cells through $Na^+$-$La^{3+}$ exchange. The mechanism by which rare earths enter cells across membranes is related to cell type, rare earth species, concentration, and reaction conditions. In many cases, several mechanisms operate simultaneously.

2. Distribution of rare earths in animal cells

Rare earths entering the blood can form insoluble components, soluble macromolecular (protein) complexes, and soluble small molecular complexes, leaving only a very small proportion of free rare earth cations. Their effects and cytotoxicities vary greatly. Meng L et al. used ultrafiltration to prove that most rare earths in normal blood bind to proteins with relative molecular weights > 50,000. Three protein peaks containing rare earths were obtained by rapid protein chromatography, corresponding to complexes of the rare earths with ferritin, immunoglobulin, and albumin. The compound can enter cell through transport of ferritin the cell surface.

3. Effects of rare earths on cell membranes

The effect of rare earth and erythrocyte membrane is to determine the hemolysis caused by its destruction of cell membrane. In vitro experiments have been carried out to determine the hemolysis caused by destruction of cell membranes. The critical solution concentration of rare earth was evaluated as about $1 \times 10^{-6}$ mol/L, only 1% of the critical hemolysis concentration of $Ca^{2+}$. However, if rare earth cations complexed by ethylenediamine tetraacetate (EDTA) are added, hemolysis is reversible within a certain period of time. Zheng YJ et al. cultured human plasma with $LaNO_3$ (<1 × $10^{-7}$ mol/L) and found that cell surfaces had a spinous bulge, even though adding $3.98 \times 10^{-7}$ mol/L lanthanum citrate, situation has been reduced, yet make red blood cells expand and bubble.

4. Effects of rare earths on mitochondria and DNA

An important task in rare earth cell chemistry is to study the effect of rare earths on mitochondria and lymphocytes, and their interactions with DNA. After incubation of liver cells with rare earth ions in rats, it was shown that rare earths can enter the nuclei of hepatocytes and can affect the nuclear structure. However, there are inconsistent reports on the interactions of rare earths with DNA. Some authors have concluded that low-dose rare earths inhibit DNA damage in cells caused by $Ni_2O_3$ and asbestos, but high concentrations (100–200 $\mu$g/mg) of yttrium oxide and nitride oxide can cause damage to lymphocyte DNA. Zhao et al. studied the interaction of $Ce^{3+}$ with salmon essence DNA by FTIR spectroscopy. The phosphodiester bond was destroyed and the hydrogen bond between bases was cleaved. Wang et al. used

$Ce^{4+}$ complexes with fucoidan starch to bind to plasmid DNA, and observed that a superhelical ring of DNA was cleaved into linear DNA.

5. Effects of rare earths on hemoglobin

The effect of rare earths on the oxygen-carrying capacity of hemoglobin from human red blood cells has been investigated. When the partial pressure of oxygen, $p_{O2}$, is $< 666.6$ Pa, the normal oxygen saturation of hemoglobin is $< 5\%$. In the presence of $Ce^{3+}$, $Pr^{3+}$, or $Tb^{3+}$ ($1 \times 10^{-5}$–$1 \times 10^{-4}$ mol/L); however, the oxygen saturation increased to 20–30%. This shows that hemoglobin does not release oxygen so easily in the presence of rare earths due to a modification of its structure.

6. Effects of rare earths on bone cells

When rats were fed with 2 mg $La(NO_3)_3$/kg body weight for 6 months, significant leg bone La accumulation was noted. Electron microscopic observation showed that the bone trabeculae became thin, flimsy, broken, and extensively pitted, and the leg bone structure changed, showing the hormesis effect of "low promotion-high inhibition".

7. Accumulation, metabolism, and distribution of rare earths in rat liver

Lu et al. fed rats with different doses of mixed rare earth nitrates for 6 months. Levels of rare earth accumulation in rat liver were in the order La > Ce > Pr, but no significant accumulation was observed in serum. More than six, rare earth-containing components were obtained by stratified separation of the proteins in the water-soluble component of the liver. According to the results, even if the rare earth dose is less than 2 mg/kg, long-term oral administration will cause the accumulation of rare earths in the liver. The effect on bone cells and liver indicates that re-direction of doses of rare earths with more sensitive biomarkers is an urgent issue in toxicology.

## 11.4 Use of Rare Earth Compounds as Drugs

Some rare earth compounds have specific pharmacological effects, such as anti-inflammatory, bactericidal, anti-coagulation, anti-arteriosclerosis, and anti-tumor properties. Many rare earth compounds can be used directly as drugs, but they need to be taken appropriately.

Take Huo et al. (2014) research as example, they synthesized a new 2,7-diphenylamine ethyl naphthalene and its complex with cerium picrate. The interaction modes of the complex with some albumins such as Human Serum Albumin (HSA), Bovine albumin (BSA) were investigated by fluorescence spectroscopy in a buffer solution at pH 7.30. The experimental results showed that fluorescence quenching of the various albumins by the complex was static quenching, the binding constants, binding sites, and related thermodynamic parameters ($\Delta H < 0$, $\Delta S > 0$, $\Delta G < 0$) of the complex with the proteins were calculated at different temperatures, which showed that the binding mainly relied on electrostatic forces.

## 11.5 Glue-Powder-Modified Pavement Asphalt

Zhang et al. (2014) prepared rare earth compound additives and added them to rubber powder to modify pavement asphalt. The aim was to study their characteristics in the dry climate with large temperature variations found in northern China.

The action mechanism is that the polymer material under the action of light, heat, oxygen or mechanical wear, the internal release of the free group and the loss of the original performance, especially in the long-term illumination, can lead to molecular chain disconnection or oxidation of the polymer. The numerous vacant orbitals of rare earth atoms or ions can capture unstable free radicals and render them inactive. Rare earth 4f electrons have strong coordination ability and can form stable complexes, some of which can absorb high-energy ultraviolet light and release low-energy infrared heat energy. This offers long-term potential for reducing the damage done by ultraviolet light, making the product suitable for use in areas with long periods of sunlight.

## 11.6 Biological Effect of Rare Earths on Algae

Since the first application of rare earths in agriculture and medicine, there has been debate about their biological effects and their application safety. This is because the biological effects of rare earths are so often characterized by concentration-dependent divergence. Some scholars have studied the effects of rare earths on promoting the growth of beneficial algal species, inhibiting the proliferation of harmful algal species, and mitigation when algae are affected by adverse factors.

Chen et al. (2014) summarized recent research on the biological effects of rare earths on algae. At present, the main research method employed involves the addition of different concentrations of rare earth compounds to algal culture media to compare and study their growth status and physiological characteristics. The aim is to distinguish the role of rare earths from environmental factors at the cell molecular level. Early studies showed that certain concentrations of La and Pr led to nitrogen-fixing *Azolla anabaena*, which may be used as a feed and green manure in rice fields. La and Pr promote photosynthesis and nitrogen fixation, thereby promoting the rate of photosynthetic incorporation of $CO_2$ into algae.

For harmful algae species, the growth-promoting effect of rare earths will have a negative impact on aqueous ecosystems and human life. The glenodinium in red tide release toxin, and can seriously harm marine ecology. The addition of rare earths can promote the growth and reproduction of these harmful algae. Some researchers fear that the adverse effects of rare earths cannot be mitigated, but others suggest that the addition of complexing agents can provide a good inhibitory effect. Wang DF et al. showed that magnetic chitosan/rare earth/clay composite resin particles had a good inhibitory effect on red tide diocystis.

## 11.7 Application of Rare Earths in Animal Breeding

Rare earth minerals with lanthanum as the main element are often associated with natural radioactive elements. After the application of rare earths in aquaculture and to forage feed, attention has been paid to whether the specific activities of associated trace amounts of uranium, thorium, and radium may be worryingly high in the products, and whether this has an impact on environmental quality and health. Sun et al. (1994) addressed this issue between 1986 and 1993.

Studies on the radioactivity levels of pork, beef, mutton, milk, broiler chicken, eggs, carp, *Astragalus adsurgens* Pall, Mongolian ice grass, grain amaranth, alfalfa, buckwheat, and mung bean after seed dressing, seed soaking, or leaf spraying with different concentrations of rare earth solutions were conducted. The specific activities of uranium, thorium, and radium in soil after application were also studied, and compared with those of blank samples.

The specific activity levels of uranium, thorium, and radium before and after feeding the rare earth additives to livestock, poultry, and aquatic organisms are summarized in a table. The samples include pork, beef, lamb, milk, egg, chicken, and carp. It can be seen that all levels fluctuate within the normal background range, except for a discrepancy caused by fatness of pork samples. The results are in broad agreement with those from other provinces and cities in China, as well as for similar samples from Inner Mongolia and Baotou.

The specific activities of uranium, thorium, and radium in forage feed before and after application of rare earths are summarized in another table. The samples include prairie milk vetch, mongolian ice grass, grain, alfalfa, siberian wild rye, buckwheat, and mung bean. The methods of application include dress seeds, soaking, leaf spray, presoaking, and contrast.

It can be seen from the test results that the relevant specific activity levels are all at the normal background level, regardless of whether leaf spraying, seed soaking, or seed dressing was performed. The results of this experimental study were lower than those of studies on natural grass (without treating by rare earths) in Baima Township, Tibet, and Taipei, Taiwan. The results for mung bean are consistent with those from studies in Shanxi ($^{238}$U at 0.02 Bq/kg, $^{232}$Th at 0.01 Bq/kg, $^{226}$Ra at 0.25 Bq/kg) and Sichuan.

The specific activities of uranium, thorium, and radium in soil before and after the application of rare earths were not significantly altered, with all of them remaining in the range of typical global values. Based on soil profile analyses of Prairie milk vetch, Mongolian ice grass, and alfalfa, uranium, thorium, and radium are uniformly distributed in each layer of soil; that is, they are independent of depth, and their concentrations are consistent with the world literature on natural radioactive elements in soil. Whether the specific activity of radioactive elements in soil is increased by the continuous application of large doses of rare earths needs further study. Rare earths should be used sparingly in breeding and forage feed in view of their potential harm in excessive amounts.

## 11.8 Other Rare Earth Research in Different Fields

Other rare earth studies in recent years in agriculture, forestry, and animal husbandry include agricultural RE, liquid RE, nano-active RE, RE organic chelates, ammonia nitrogen wastewater treatment technology, new RE materials, application effect of new technology in forestry production (Zhang 2015), cultivation and application of South American chrysanthemum in wasteland contaminated with rare earth ore, characteristics of rare earth distribution in *Radix Pseudostellariae* (roots of *Pseudostellaria heterophylla*) and its planting soil, research on rare earth differentiation in tea trees, review of rare earth differentiation in tea plants, spatial distribution characteristics of light rare earths in surface soil of contaminated irrigation areas (Zhang 2015), rare earth content and evaluation in soil and crops around copper mines and smelters (Jin et al. 2015), and the application effect of herbicidal and insecticidal fertilizer in rice (Tan et al. 2015).

## References

Aquino L, Pinto MC et al (2009) Effect of some light rare earth elements on seed germination, seedling growth and antioxidant metabolism in Triticum durum. Chemosphere 75(7):900–905

Chen AM, Shi QS, Xie XB (2014) Advances in the study of the biological effects of rare earths on algae. Chin Rare Earths 35(4):103

Ding SM, Zhang ZL, Liang T et al (2004) Effects of rare earths in soil on morphological distribution and transformation of available nitrogen. J Chin Soc Rare Earths 22(3):375

Fu MH, Gu ZL, Ding QF et al (1994) Benefits and physiological effects of rare earth on winter melon, potato and cabbage. In: Proceedings of the third annual conference of the China Rare Earth Society, vol 5, p 143

Gao L (1983) Why are watermelons grown with rare earths sweeter? Chin Rare Earths 1:85

Huo LN, Liu W, Yang TL et al (2014) Synthesis, characterization and interaction with albumin of novel cerium bitter acid complexes. Chin Rare Earths 35(4):37

Ji YJ, Cao Y (1983) Acute toxicity studies of nitric acid rare earth. J Chin Soc Rare Earths 1(2):60

Jin SL, Huang YZ, Wang F et al (2015) Assessment of rare earth elements in soil and crops around Jiangxi copper mine and smelter. Environ Sci 36(3):1060–1068

Li RC, Yang XD, Wang K et al (2004) Research progress of rare earth bioinorganic chemistry in China. J Chin Soc Rare Earths 22(1):1

Lin SR, Liao GR, Li SY et al (1994) Application of rare earth on Chinese fir. In: Proceedings of the third annual conference of China Rare Earth Society, vol 5, p 153

Qi TQ, Liu PL, Song CZ et al (1984) Absorption, transport and distribution of rare earth elements in crops. J Chin Soc Rare Earths 2(1):94

Sun BC, Cong RH, Na R et al (1994) Specific activity levels of uranium, thorium and radium in the breeding and forage feed of rare earths. In: Proceedings of the third annual conference of the China Rare Earth Society, vol 5, p 75

Tan FJ, Zhou Y, Wan Q (2015) Effect of herbicide fertilizer on rice. Hunan Agric Sci Monthly J 2:31–33

Tang XK, Li GF (1983) Effect of cerium on the growth of maize seedlings and their enzyme profiles. J Chin Soc Rare Earths 2:56

Wang GH, Chen YC (1985) Effect of rare earth on rubber tree. J Chin Soc Rare Earths 3(4):55

# References

Wang LJ, Hu AT, Zhou QS et al (2006) Characteristics of movement, absorption, and accumulation of rare earth elements in soil-rice systems. J Chin Soc Rare Earths 24(1):91

Wei ZG, Zhang HJ, Li HX et al (2006) Advances in research on rare earth superaccumulative plants. J Chin Soc Rare Earths 24(1):1

Xie ZB, Zhu JG, Chu HY et al (2004) Effect of lanthanum on phosphorus uptake and its morphology in rice. J Chin Soc Rare Earths 22(1):153

Xie HG, Zheng TJ, Zhao SX et al (1994) Biological effects of lanthanum-rich rare earths on spring wheat. In: Proceedings of the third annual conference of the China Rare Earth Society, vol 5, p 124

Yang HM, Du L (1994) Effects of rare earth elements on somatic embryo induction frequency of medlar. J Chin Soc Rare Earths 12(2):186

Zhang YM, Xiang H, Xu XY et al (1988) Effect of rare earth on rhizobia growth of legume forage and its application effect. J Chin Soc Rare Earths 6(3):87

Zhang YX, Sun XH, Sun GJ (2014) Application of rare earth compound auxiliaries in rubber powder modified pavement asphalt. Chin Rare Earths 35(3):109

Zhang QH, Liu XW, Cheng L et al (2015) Spatial distribution characteristics of light rare earths in surface soil of farmland in the southern suburb of Baotou. Tianjin Agric Sci 7:39–47

Zhang DX, Sun LD, Niu DL (1994) Grey correlation analysis of rare earth and soil elements applied in cotton. In: Proceedings of the third annual conference of China Rare Earth Society, vol 5, p 139

Zhang XH (2015) Application of new materials and technologies for rare earth agriculture in forestry production. Anhui Agric Sci 43(7):85–86, 88

Zhou Q, Huang XH, Zhang GS et al (2004) Advances in rare earth applications in pollution ecology. J Chin Soc Rare Earths 22(2):17

# Chapter 12
# Rare Earth Catalysts and Catalytic Activity

## 12.1 Brief Description

More than 90% of chemicals, including ammonia ($NH_3$), nitric acid, and sulfuric acid, are produced in the presence of catalysts. Catalysts are also utilized in the inorganic chemical industry, during processes involving catalytic cracking and reforming of petroleum as well as in the organic chemical industry in the production of methanol, butanol, acetic acid, and acetone. Moreover, liquefaction and gasification of coal, production of synthetic materials in the polymer industry, reduction of pollution associated with automobile tail gas, and utilization of waste require the use of suitable catalysts. The earliest application of rare earth elements also started with catalysis.

Application of catalysts enables the design of compounds with specific properties, optimization of methods and operating conditions for the production of raw materials as well as improvement of product efficiency and quality while reducing environmental pollution (Gomer 1985, Huang 1994). The electronic and molecular structures of materials and their thermal characteristics are key factors affecting the progress of chemical reactions, whereas catalysts increase reaction rates.

The preparation and effective use of catalysts must be considered from the viewpoint of physical chemistry. Regarding related research and applications, the materials studio software developed by the former world leading computer science company, namely American Accelrys (currently BIOVIA), can be used to study homogeneous and multiphase catalysts as well as to characterize, optimize, and design various catalytic systems.

## 12.2 Characteristics and Definitions of Catalytic Activity

In 1976 and 1981, the International Union of Pure and Applied Chemistry (IUPAC) proposed that a catalyst is a substance, which can accelerate the rate of reaction without changing the standard Gibbs free energy. Catalysts enable reactions to proceed via a new pathway involving a series of steps. They participate in reactions; however, return to their original composition following chemical cycling. For instance, during the synthesis of $NH_3$, iron is typically selected as the catalyst.

In the absence of catalysts, the direct combination of $N_2$ and $H_2$ is remarkably challenging and requires a lot of energy. At temperatures below 500 °C and atmospheric pressure, the activation energy for this reaction is 334.6 kJ $mol^{-1}$, and the $NH_3$ yields are low. In contrast, when an iron catalyst is employed, $N_2$ and $H_2$ interact through chemisorption to create chemical bonds. This process involves bond weakening and dissociation. Subsequently, a series of surface reactions results in the formation of $NH_3$. According to Maier (2004) research, the formation of $NH_3$ can be represented by the following reactions:

$$N_{2(g)} \rightleftharpoons N_{2, ad}$$

$$\frac{1}{2}N_{2, ad} \rightleftharpoons N_{ad}$$

$$H_{2(g)} \rightleftharpoons H_{2, ad}$$

$$\frac{1}{2}H_{2, ad} \rightleftharpoons H_{ad}$$

$$N_{ad} + H_{ad} \rightleftharpoons NH_{ad}$$

$$NH_{ad} + H_{ad} \rightleftharpoons NH_{2ad}$$

$$NH_{2ad} + H_{ad} \rightleftharpoons NH_{3ad}$$

$$NH_{3ad} \rightleftharpoons NH_{3(g)}$$

Here, ad means adsorption. The step of rate control of the catalytic reaction is $N_2$ molecules to decompose into N atoms, which activation energy is 70 kJ/mol, much lower than when there is no catalyst, so the reaction rate is greatly improved.

Under the same conditions of 500 °C and atmospheric pressure, 13 orders of magnitude higher than the corresponding homogeneous reaction. However, the free energy of the initial and final states of the reaction is constant regardless of whether a catalyst is involved. Thus, the conversion is the same at equilibrium, i.e., the presence of catalyst does not affect the reaction equilibrium.

## 12.3 Basic Features of Catalytic Activity

Catalytic activity is associated with four basic characteristics:

(1) Catalysts can only accelerate thermodynamically feasible reactions, not those which are chemically impossible. When developing a new catalyst, the system should be first thermodynamically analyzed to determine whether the reaction is feasible.
(2) A catalyst can only accelerate a reaction towards equilibrium without changing the position of equilibrium, i.e., the K constant is not altered. Under known conditions, the $-\Delta G_f^{\ominus}$ and K values of catalytic and non-catalytic processes are the same.
(3) Catalysts are highly selective; therefore, when a reaction has more than one route, different catalysts can accelerate one of the possible pathways in a thermo-dynamically feasible manner. For instance, when syngas is used as the raw material, addition of different catalysts results in the formation of various products, e.g., methanol, methane, synthetic gasoline, or solid paraffin.
(4) A catalyst is a chemical substance, which temporarily intervenes in the reaction, changing its energy state and pathway. It enables reactions to proceed more easily. To achieve high conversions, continuous catalyst recycling is important. Catalysts cannot be used indefinitely. Repeated chemical and physical changes lead to catalyst deactivation.

## 12.4 Composition and Functions of Catalysts

Industrial catalysts are usually not composed of a single substance. Most of them exhibit three distinguishable components:

1. Active components

The active component is the main component of a catalyst. Some catalysts are composed of a single substance (e.g., oxidation of ethylene to ethylene oxide in the presence of a silver catalyst). In other cases, they are composed of two substances (e.g., oxidation of propylene ammonia to propylene cyanide using a molybdenum–bismuth catalyst, in which the active component is composed of molybdenum oxide and bismuth oxide).

2. Carrier

The carrier acts as a dispersant, adhesive, or support for the active catalytic component and as a skeleton for the supported active component. The numerous types of carriers can be divided into two categories, namely low specific surface and high specific surface ones. The carrier types of low specific surface include corundum, silicon carbide, pumice, diatomite, asbestos, and firebrick; the carrier types of high specific surface include alumina, $SiO_2/Al_2O_3$, laterite, carclazyte, magnesium oxide, and silica gel.

The carrier affects the activity and selectivity of the catalyst as well as its thermal stability and mechanical strength. Carriers also determine the transfer characteristics of catalytic processes.

The carrier has the following functions:

(1) Providing a large active surface and suitable pore structure as well as maintaining high dispersion of active components.
(2) Enhancing the mechanical strength of the catalyst to ensure that its shape remains unchanged. For example, industrial catalysts have certain mechanical strength requirements, which are achieved through the selection of suitable supports.
(3) Improving the conductivity of catalysts. The carrier should generally exhibit large specific heat capacity and good thermal conductivity, so that the reaction heat energy can be transferred in or out efficiently and to prevent sintering or deactivation of the catalyst by local overheating.
(4) Reducing the content of active components. When noble metal catalysts, such as Pt, Pd, and Rh, are used, the application of a carrier to highly disperse the active components can reduce the necessary amount of precious metals.
(5) Provision of additional active sites. Typically, supports display no catalytic activity to prevent unnecessary side reactions. Nonetheless, carriers such as $\gamma$-$Al_2O_3$ contain acidic active sites on their surfaces, which exhibit different properties. If not treated at the time of preparation, there is the possibility that such carriers could induce undesirable side reactions. The design of a catalyst must carefully consider the carrier function.
(6) Overflow phenomena and strong interactions. Solid catalysts can be chemisorbed to induce new activity or undergo a chemical reaction. The catalyst changes during reactions, which is affected by the temperature, catalyst composition, type of adsorbed species, and catalytic environment.

3. Cocatalysts

Cocatalysts are materials added to catalysts in small amounts. They are not catalytically active; however, they are able to alter the chemical composition, ionic valence state, acidity and alkalinity, crystal structure, surface and pore structure, dispersion state, and mechanical strength of the catalyst to enhance the performance, selectivity, stability, and lifetime of the catalyst.

The selection and research for suitable cocatalysts is important in the field of catalysis. According to different mechanisms of action, cocatalysts are generally divided into two types, i.e., structural and electronic. The key function of the structural cocatalysts is to improve dispersion and thermal stability of the active components, whereas the role of electronic cocatalysts is to modify the electronic arrangement of the catalyst and enhance its selectivity. Cocatalysts also provide a support function. For example, when active component or carrier is $Al_2O_3$, cocatalysts include $SiO_2$, $ZrO_2$, P, $K_2O$, HCl, and MgO, and the functions include promoting thermal stability of the carrier, slow the coking of active components and lower the temperature, promoting acidity of active components, and spacing active components to reduce sintering.

## 12.5 Adsorption and Multiphase Catalytic Reactions

1. Reaction steps of multiphase catalytic reactions

In multiphase catalytic reactions, the conversion of reactants to products typically involves the following steps (Bechesfedt 2007, Xu and Wang 2010):

(1) Diffusion of the reactant molecules from airflow to the catalyst surface and pores;
(2) Adsorption of the reactant molecules on the catalyst surface;
(3) Adsorption of the reactant molecules on the catalyst surface or chemical reaction with the gas molecules;
(4) Desorption of the reaction products from the catalyst surface;
(5) Diffusion of the reaction product on the surface and inside the pore surface of the catalyst into the reaction airflow.

2. Physical and chemical adsorption of reactant molecules

When the reactant molecules reach the surface of the catalyst, the physical adsorption first occurs due to Van der Waals forces, and the adsorption heat is 5–40 kJ/mol. Subsequently, chemisorption involves various chemical forces and follows one of the most favorable energy pathways. Bonds formed during adsorption can be covalent, coordination, or ionic.

Molecules with a lone pair of electrons or $\pi$ electrons can be non-dissociated by chemisorption through rehybridization of the relevant molecular orbitals. For instance, the chemical adsorption of ethyleneis carried out by rehybridization of $\pi$ electron molecular orbitals. The combined state of the carbon atoms prior to adsorption is $sp^2$. In the adsorption bond, the hybrid state changes to $sp^3$. Adsorption bonds are formed during CO chemisorption. An $\sigma$–$\pi$ bond structure is found in CO adsorption on noble metals, such as Rh, Pd, and Pt.

Non-transition metals, the valence layers of which are the s or p electrons, exhibit weak adsorption ability. The angular wave function (dashed line) and angular distribution of the electron cloud (solid line) of the s, p, and d states are schematically illustrated in a diagram.

Additionally, the angular orbit distribution of the f states is demonstrated in another diagram.

During catalytic reactions, the important function of metals, particularly transition metals, is to dissociate molecules and activate them. $H_2$, $O_2$, $N_2$, and CO are examples of important diatomic molecules. The adsorption bonds formed between them and the transition metal surface provide thermodynamic stimulus for the atomization of the molecules and the atomic activation required for the reaction to proceed. The strength of chemical adsorption of metals to gas molecules follows the order of:

$$O_2 > C_2H_2 > C_2H_4 > CO > H_2 > CO_2 > N_2$$

## 12.6 Metal Oxide Catalysts and Their Catalytic Activity

1. Composite oxides

In a composite oxide, at least one of the components is a transition metal oxide. The examples of such materials include $V_2O_5-MoO_3$, $Bi_2O_3-MoO_3$, $V_2O_5-MoO_3-Al_2O_3$, and $MoO_3-Bi_2O_3-Fe_2O_3-CoO-R_2O-P_2O_5-SiO_2$. Depending on the conditions, components might interact with each other.

In composite oxides, some components act as main catalysts, while others are cocatalysts or supports. Key catalyst components are active on their own (e.g., $MoO_3$ in $MoO_3-Bi_2O_3$); however, the activity of the cocatalysts is enhanced by the addition of the catalyst (e.g., $Bi_2O_3$). The functions of the cocatalyst components include the adjustment of the formation of a new phase as well as the regulation of the electron migration speed. Moreover, cocatalysts regulate the formation of the active phase. Based on their role in the improvement of the catalyst performance, co-catalysts can be classified into structural and anti-sintering ones. They enhance the mechanical strength and promote dispersion.

2. Catalytic performance of the spinel structure

Spinel is a composite oxide displaying a unique structure. A number of metal oxides with spinel structures are commonly used as catalysts for the oxidation and dehydrogenation reactions. Spinel-type catalysts are typically denoted as $AB_2O_4$. They are typically utilized for the oxidative dehydrogenation of hydrocarbons and catalytic combustion of methane, among others. Furthermore, $MgFe_2O_4$ and its analogs can be employed for oxidative dehydrogenation of butene to prepare butadiene.

3. Catalytic performance of perovskites

The lattice structure of perovskites is similar to that of mineral $CaTiO_3$. They are classed as oxides and are generally denoted by $ABX_3$. A sites are usually rare earth or alkaline earth ions, while B sites are transition element ions. Typically, A is a large cation, B is a small cation, whereas X is $O^{2-}$. In the unit cell of an ideal perovskite-type structure, A is located at the center of the cell, and B is at the vertex of the cube. Nonetheless, very few perovskite-types exhibit accurate cubic structures at room temperature. In contrast, they might display such structures at high temperatures.

The catalytic oxidation performance of perovskite oxides was first reported in 1952. Further work oncatalysis followed in 1970s. Compared to Pt-based catalysts, $La_{0.8}Sr_{0.2}CoO_3$ has been reported to show high catalytic activity in the electrochemical reduction of oxygen. Meanwhile, it has been found that Co and Mn perovskites are active gas-phase reaction catalysts for the hydrogenation and hydrogenolysis of cis-butene-2 as well as gas-phase oxidation of CO and reduction-induced decomposition of $NO_x$. Hence, it is considered that perovskite oxides could also be employed as catalysts for the combustion and treatment of automobile tail gas.

Extensive work has been conducted on utilizing perovskite oxides as catalytic combustion catalysts. Catalytic combustion catalysts related to CO, $C_1$-$C_4$ hydrocarbons, methanol, and $NH_3$ are summarized in a table. For example, when catalytic

# 12.6 Metal Oxide Catalysts and Their Catalytic Activity

reaction is $CO + O_2 \rightarrow CO_2$, the catalysts include $LaBO_3$ (B is 3d transition metal), $LnCoO_3$ (Ln is rare earth lanthanide metals), $BaTiO_3$, $La_{1-x} A_x' CoO_3$ (A$'$ is Sr, Ce), $La_{1-x} A_x' MnO_3$ (A$'$ is Pb, Sr, K, Ce), $La_{0.7} A_{0.3}' MnO_3 + Pt$ (A$'$ is Sr, Pb), $LaMn_{1-y} B_y' O_3$ (B$'$ is CO, Ni, Mg, Li), $LaMn_{1-x}Cu_xO_3$, $LaFe_{0.9} B_{0.1}'$ (B$'$ is Cr, Mn, Fe, Co, Ni), and $Ba_2CoWO_6$, $Ba_2FeNbO_6$; when catalytic reaction is $NH_3 + O_2 \rightarrow N_2, N_2O$, NO, the catalyst is $La_{1-x}Ca_xMnO_3$.

All of these catalysts contain a lanthanide at the A site and a 3d transition metal at the B site.

## 4. Examples of perovskite rare earth composite oxide catalysts

The unique structures of perovskites enable their potential applications in various fields. However, only few examples of their use in catalysis have been reported. Guo (1986) described the catalytic activity of some perovskite ($ABO_3$) rare earth oxides.

### (1) Catalytic oxidation reaction

Rare earth metal ions (A) are not active sites, and the catalytic performance can be affected by modifying the electronic state of transition metal ions (B) and oxygen bonding. For instance, the ratio of the two spin states in $NdCoO_3$ and $HoCoO_3$ is the same, and their conductive activation energy is higher than that of other $LnCoO_3$ oxides, which results in high catalytic activity.

Alkaline-earth metals can replace some A ions, which leads to changes in the valence state of the B ions or the appearance of oxygen vacancies. For example, following partial substitution of $La^{3+}$ with alkaline-earth metals, the $Co^{3+}$ ions in $LaCoO_3$ are partly converted to $Co^{4+}$. In addition, oxygen vacancies appear in the crystal. Notably, the $Co^{4+}$ species improve the chemical potential of the lattice oxygen, and the oxygen vacancies are beneficial to the reaction.

### (2) Catalytic hydrocarbon oxidation

Almost all of the catalysts used for CO oxidation can catalyze the oxidation of trace hydrocarbons in the gas. During the process, CO and hydrocarbons are oxidized to $CO_2$ and water by oxygen. For the catalytic $C_3H_6$ oxidation, the order of activity is affected by the alkalinity, which decreases with the increase of the alkaline-earth metal ion radius.

### (3) Methanol oxidation

The methanol oxidation can be described by the following equation:

$$CH_3OH + O_2 \rightarrow CO_2 + H_2$$

The catalysts, which can be used for the methanol oxidation, are listed below:

$LnFeO_3(Ln = Gd, Eu, Sm, Nd, Pr, La)$,
alkaline order is $Gd > Eu > Sm > Nd > La$;

$LnCrO_3(Ln = Gd, Eu, Sm, Nd, Pr, La)$,

the order of activity is $Sm > Nd > La > Ga > Eu$;

$LnMnO_3(Ln = La, Pr, Nd, Sm, Eu)$;

$Ln_2CuO_3(Ln = La, Pr, Nd, Sm, Eu, Ga)$,

the order of activity is $Eu > Gd > Pr \approx La > Sm$.

## (4) Oxidation of ammonia

When $Ca_xLa_{1-y}MnO_3$ is used for the catalytic oxidation of $NH_3$, the products are $NO$, $N_2O$, and $N_2$. Active sites in the reaction are the $Mn^{3+} - Mn^{4+}$ moieties. The amount of $La^{3+}$ and $Ca^{2+}$ affects the reaction selectivity. At 350–400 °C, when $0 \le x < 0.3$, the products are predominantly $N_2O$ and $N_2$. However, when $0.4 \le x \le 1.0$, the amount of $NO$ significantly increases. According to the mechanism of the $NH_3$ oxidation, Vieland points occur in the $La^{3+}$-enriched regions, and $Mn$ exists in a + 3 valence state. $Mn^{3+}$ adsorbs the $O_2$ gas to form atomic oxygen (O). Atomic oxygen is then used to generate $N_2O$, which decomposes to form $N_2$. Moreover, $La^{3+}$ is substituted by $Ca^{2+}$, while $Mn^{3+}$ is oxidized to $Mn^{4+}$. $Mn^{4+}$ further reacts to form $NO$.

## (5) Catalysis of nitrogen and sulfur redox reactions

During $NO$ reduction reactions, trace $NO$ in the gas is reduced to $NH_3$, $N_2O$, and $N_2$ by $CO$ or $H_2$. For effective waste gas treatment, the content of $NH_3$ should be as low as possible.

The catalytic reduction of $NO$ can occur in the presence of the following catalysts:

- $LaMnO_3$, $La_{1-x}M_xMnO_3$ (M = K, Na, Rb), and $Ln_{1-x}Pb_xMnO_3$ (Ln = La, Nd) can catalyze $NO$ reduction in the presence of a mixture of gases containing $NO$, $CO$, and $H_2$. The selectivity of generating $N_2$ is high. For example, in the case of $La_{1-x}Pb_xMnO_3$, 100% conversion of $NO$ to $N_2$ can be achieved at 375 °C.
- Among $La_2Cu_{1-x}M_xO_4$ (M = Li, Al, Ti, Cr, Mn, Fe, Co, Ni, Zn, Zr), materials containing Al and Zr exhibit higher activity. They are used for the reaction between $NO$ and $NH_3$ in the gas phase. When the gas phase contains only $NO$ and $NH_3$, the reaction can be described by $6NO + 4NH_3 5N_2 + 6H_2O$, the $NO$ conversion rate does not decrease. Ku do reported a $La_2Cu_{1-x}Zr_xO_4$ catalyst for the treatment of flue gas (toxic gases containing sulfur and nitrogen) with a lifetime of more than 2000 h. The material was shown to catalyze the reduction of $SO_2$.

The following reactions occur during the treatment of flue gas:

$$2CO + SO_2 \rightarrow 1/2S_2 + 2CO_2$$

## 12.6 Metal Oxide Catalysts and Their Catalytic Activity

$$CO + 1/2S_2 \rightarrow COS$$

$LaCoO_3$, $LaTiO_3$, and $La_{0.5}Sr_{0.5}CoO_3$ have been used as catalysts for the above reactions. Notably, their selectivity is better than that of other metal catalysts.

(6) Catalytic hydrogenation of hydrocarbons and hydrogenolysis

### Hydrogenation

$LnCoO_3$ (Ln = La, Nd, Dy and mixed rare earth metals) materials can be applied as catalysts for the hydrogenation and hydrogenolysis reactions, and their activity and selectivity are comparable. The hydrogenation of *cis*-2-butene in the presence of the aforementioned catalysts takes place at low temperatures. Hydrogenolysis at temperatures over 300 °C results in the formation of methane and ethane.

$LaRhO_3$ is a stable CO hydrogenation catalyst at 225–375 °C. It is noteworthy that the product distribution varies with temperature. Methanol is the main product at temperatures below 225 °C, while ethanol and acetaldehyde are the key products at temperatures between 225 and 350 °C. In contrast, at > 350 °C, ethane is main product. X-ray diffraction (XRD) and Auger electron spectroscopy studies revealed that $Rh^0$ and $Rh^{+1}$ are the active sites on the surface of the $LaRhO_3$ catalyst.

### Hydrogenolysis

$LaCoO_3$ catalyzes the hydrogenation of ethylene to ethane at temperatures below 147 °C. Ethane is subsequently hydrogenated to methane. In the presence of this catalyst, the main product of the $C_3$–$C_5$ hydrocarbon hydrogenolysis is methane. Moreover, the hydrogenolysis behavior of $LaCoO_3$ is similar to that of $Co_2O_3$. $Co^{3+}$ and $La^{3+}$–$O^{2-}$ break the C–C bond, and then dissociate and adsorb hydrogen, respectively.

The hydrogenolysis activity of $LaCoO_3$ is superior to those of noble metal-based catalysts. At an ethylene pressure 10 Torr (1 Torr = $1.33322 \times 10^2$ Pa) and $H_2$ pressure of 100 Torr, $LaCoO_3$ hydrogenolysis rate was determined at $1 \times 10^{13}$ molecules/(s·cm$^2$). In the case of platinum the rate was $2 \times 10^{11}$ molecules/(s·cm$^2$) and for palladium catalysts it was $4 \times 10^{11}$ molecules/(s·cm$^2$). This could be attributed to the synergistic effects of $Co^{3+}$ and $La^{3+}$–$O^{2-}$ in $LaCoO_3$.

(7) Catalysts for automobile exhaust treatment

Perovskites can be used for the reduction of trace nitrogen oxides as well as oxidation of CO and hydrocarbons. They have been considered as alternatives to precious metals for catalyzing vehicle tail gas treatment. Nonetheless, the disadvantages of utilizing perovskites include lower activity and possibility of sulfur poisoning; therefore, trace Pt should also be added in the reactions. Johnson found that Pt existed in the $Pt^{4+}$ valence state on $La_{0.7}Pb_{0.3}MnO_3$, and its activity was considerably higher than that of metal Pt.

(8) Electrocatalysis

LaCoO$_3$, LaMnO$_3$, LaNiO$_3$, or their analogs with partially substituted A and B ions can be used as electrode materials to catalyze the cathodic reduction of oxygen in alkaline solutions. Under working conditions, the above catalysts are more stable than other electrode materials. For instance, the activity of La$_{0.5}$Sr$_{0.5}$CoO$_3$ at temperatures over 170 °C is higher than that of a Pt electrode.

Matsumoto determined that the electrocatalytic redox activity of electrode materials followed the order of La$_{1-x}$Sr$_x$MnO$_3$ > La$_{1-x}$K$_x$MnO$_3$ > La$_{1-x}$Pb$_x$MnO$_3$. Additionally, the activity increased with the increase of $x$.

Considering catalysis of oxidation, hydrogenation, and hydrogenolysis reactions, compared with other catalysts and some noble metals, perovskite rare earth oxides exhibit better activity and toxicity resistance. Moreover, their structures are stable under the redox reaction environment. The active sites of such catalysts are generally transition metal ions. Rare earth metal ions are used to stabilize and improve the selectivity of the transition metal ions.

## 12.7 Molecular Sieve Catalysts and Their Catalytic Activity

Molecular sieves are crystalline aluminosilicates with uniform pore structures, which contain a large amount of crystalline water. Water can be vaporized and removed upon heating; therefore, the materials are also called zeolites. The chemical composition of molecular sieved can be expressed as:

$$M_{x/n}\left[(AlO_2)_x \cdot (SiO_2)_y\right] \cdot zH_2O$$

where M is a metal cation, $n$ is its valence, while $x$, $y$, and $z$ are the numbers of the AlO$_2$, SiO$_2$ and H$_2$O molecules, respectively. More than 40 kinds of natural zeolites have been found, and as many as 120 kinds have been synthesized. The common types include A, X, and Y molecular sieves. Molecular sieves exhibit high activity and selectivity in various catalytic reactions under acidic conditions, which offers potential applications in the refining and petrochemical industry.

(1)   Structural configuration of molecular sieves

The structural configuration of molecular sieves can be divided into four aspects and three different structural levels. The first and most basic structural units are the silicon–oxygen (SiO$_4$) and aluminum–oxygen (AlO$_4$) tetrahedrons, which form the skeleton of a molecular sieve (Diagram-a). Both large and small rings are second in the structural hierarchy of molecular sieves (Diagram-b). The Diagram-a is Si (aluminum) oxygen tetrahedron schematic diagram; The Diagram-b is different size oxygen rings and pore window oxygen rings corresponding to different molecular sieve structure, and Diagram-b also demonstrates the relationship between the pore window oxygen rings and the molecular sieve structure.

The oxygen rings are connected to each other by oxygen bridges, forming a polyhedron, which is third in the structural hierarchy (Diagram-c). The Diagram-c

12.8 Preparation and Application of Industrial Catalysts

includes two kinds of forms, that is, various polyhedral structures, and association of mono-oxygen ring, hydrogen peroxide ring and polyhedron.

Polyhedrons exhibit hollow cages, which are important features of the molecular sieve structure. Cages also vary, e.g., in the A-type molecular sieve structure, $\alpha$ cage is the main hole. In contrast, $\beta$ cage is predominantly used to form the skeleton structure of A, X, and Y molecular sieves, and is the most important hole. The relationships between various cage structures and three different structural levels are summarized up. As demonstrated in the diagram, the cages form molecular sieves with various structures by connecting to each other through oxygen bridges.

Diagram-c shows that (a) correlation of various molecular sieve structures with the first, second and third different structural levels, that is, the four structural levels of the aforementioned molecular sieve structure; (b) $\alpha$-cage structure and $\beta$-cage structure and tetrahedral representation of cage connected by 24 to 4 (t as Si or Al) with oxygen bridges.

- A-type molecular sieve structure
- Structure of X- and Y-type molecular sieves
- Structure of a mordenite molecular sieve (Diagram-d)
- Structure of zeolite socony mobil (ZSM)
- Structure of an aluminum phosphate molecular sieve.

Diagram-d includes (a) double five-membered ring; (b) structural unit of mordenite; (c) layered structure of mordenite; (d) main channel

(2) Catalytic performance and alignment of molecular sieve catalysts

Molecular sieve catalysts display a clear pore size distribution, high surface area (typically up to 600 $m^2/g$), and good thermal stability. They are stable under heat treatment in air up to 1000 °C; therefore, they are widely used as industrial catalysts or catalyst supports. Heterogeneous catalytic processes typically require the consideration of three performance indicators, namely the catalyst activity, selectivity, and operational stability.

## 12.8 Preparation and Application of Industrial Catalysts

Maximum efficiency can be realized only when the catalyst with excellent performance is prepared and used correctly. The activity, selectivity, and stability of industrial catalysts depend not only on their chemical composition but also on various physical properties, such as shape, particle size, phase, density, surface area, pore structure, and mechanical strength. All these characteristics influence the catalytic performance as well as the reaction kinetics and fluid dynamics.

Mechanical strength is an important indicator of industrial catalysts. Modern chemical processes often require pressurization; thus, the mechanical strength of catalysts is particularly significant. If the mechanical strength decreases rapidly and the catalyst is crushed and powdered, the catalytic efficiency will be reduced or even completely diminished. The shape and particle size of industrial catalysts are related to their preparation processes. At present, the preparation methods of solid catalysts used in the industry include precipitation, impregnation, mixing, ion exchange, and melting approaches, which are summarized below.

(1) Precipitation

This method is widely employed for the preparation of non-precious metals, metal oxides, metal salt catalysts, and catalyst supports.

(2) Impregnation

The carriers are immersed in a solution of soluble compounds containing active components (i.e., main and auxiliary catalyst components). After some time, the excess solution is removed, and the catalyst is obtained by drying, roasting, and activating.

(3) Mixing

The mixing method is frequently used during the industrial preparation of multi-component solid catalysts. It involves mechanical mixing of several components. To improve the mechanical strength, a certain amount of binder should be added during the mixing step. The mixing approach can be divided into dry and wet processes.

(4) Ion exchange

The ion exchange method utilizes the presence of exchangeable ions on the surface of the support. The active components are exchanged with the support through ion exchange (usually cation exchange). Following appropriate post-treatment, such as washing, drying, roasting, and reduction, the metal-supported catalyst is effectively obtained.

(5) Fusion

The melting approach depends on fusing the catalyst components at high temperature to obtain a uniform mixture, i.e., a solid solution of alloys or oxides. At the melting temperature, the metals and metal oxides are present in the fluid state, which is beneficial to their uniform mixing and promotes the distribution of the cocatalyst components on the main active phase. The melting manufacturing process is conducted at high temperatures; hence, the temperature is the key controlling factor.

## 12.9 Research and Application of Rare Earth Catalysts

Rare earth elements exhibit unique chemical properties due to their special arrangement of electron layers. As catalysts, they have been applied and evaluated in the

## 12.9 Research and Application of Rare Earth Catalysts

petrochemical industry for the catalytic combustion of fossil fuels, purification of vehicle exhaust gases and toxic gases, and olefin polymerization (Kitto 1992, Liu 2009). Additionally, they have also been utilized in the carbon chemical industry. According to their composition, rare earth catalytic materials can be roughly divided into rare earth composite oxides (e.g., spinel and perovskite types), rare earth precious metals, and rare earth molecular sieves.

Meanwhile, the use of high-tech products has increased the production of wastes containing rare earth elements, therefore, reduce and recycle concept is also being widely adopted.

### (1) Petrochemical catalyst

Fluid catalytic cracking (FCC) is an important step of petroleum processing. Notably, the petroleum rare earth cracking catalysts used for the secondary processing of crude oil are most frequently used in the petrochemical industry. More than a third of gasoline of the world comes from FCC, while in China, 80% of finished gasoline and 35% of finished diesel originate from FCC. More than 300 catalytic cracking catalysts have been developed worldwide. A series of FCC catalysts have also been reported in China. For example, rare earth Y zeolite molecular sieves show good catalytic cracking performance, high gasoline yield, and low dry gas and coke yield. Moreover, composite and rare earth hydrogen molecular sieves exhibit high activity and heavy metal resistance.

To improve the gasoline octane number, China has developed an ultra-stable Y zeolite cracking catalyst, the performance of which is comparable to imported products. Employing this material, the gasoline recovery rate is higher than that of imported catalysts. It has been shown that when rare earth elements are used as cracking catalysts, the catalyst performance and stability are significantly improved. In addition, the cracking conversion of raw oil is considerably enhanced, and the gasoline and diesel oil yield is increased.

The vanadium content in the FCC feedstock oil rapidly increases with the increase in the processing capacity of crude oil. Typically, the vanadium content in heavy oil is higher than 20 $\mu$g/g. During the process of FCC, inexpensive vanadium deposited on the catalyst surface is oxidized to $V_2O_5$. Concurrently, acid regurgitation occurs during regeneration, and aluminum vanadate is formed in the presence of aluminum in the molecular sieves, destroying their structure. Rare earth oxides readily react with vanadium to form stable rare earth vanadates, which can obviously improve the vanadium capacity of the catalyst and protect the sieve structure.

Introducing rare earth elements into molecular sieves can regulate their acidity and aperture distribution. Depending on the type of $RE^{3+}$, exchange amount, and introduction mode, the number of acid centers and strength distribution of molecular sieves can be modified, which affects the performance of the catalyst. Rare earth molecular sieves formed after the exchange of rare earth ions (e.g., La, Ce, and Pr) retain the aluminium skeleton and improve the stability of the molecular sieve structure.

As important cocatalysts, rare earth elements also play key roles in ammonia oxidation of olefins, aromatization of low-carbon alkanes, and isomerization of aromatic compounds. Rare earth element-containing materials, such as Ce/AlPO$_4$ molecular sieves, can be used as main catalysts. They exhibit high activity and selectivity for the catalytic oxidation of ethylene oxide to cyclohexanone and cyclohexanol.

Practiclly, for example, DigitalRefining thinks rare earth metals are an important component of FCC catalysts, as well as being a key raw material for many strategic industries with applications ranging from military devices to electronic components. In addition, they are essential constituents in newly evolving green technologies, such as hybrid cars and wind turbines. However, rare earth metals tend to be concentrated in hard to extract ore deposits, the world's supply comes from only a few sources, the low and zero rare earth FCC catalysts are developed. In the 1990's, a rare earth free stabilized zeolite Y was developed and commercialized.

(2)  Catalyst for vehicle exhausts purification

The purpose of catalytic purification is to convert harmful CO and hydrocarbons into CO$_2$ and H$_2$ as well as to reduce NO$_x$ to N$_2$. Utilizing catalysts for automobile exhaust purification improves the effectiveness of methods used to reduce pollution. Catalysts containing rare earth elements display high activity, good thermal stability, optimal anti-poisoning ability, and long shelf life. Rare earth element-based automotive exhaust purification materials containing a small amount of precious metals have been successfully developed in China and abroad. Ce–Zr rare earth composite oxides used in active coatings are particularly noteworthy and have been successfully used in automotive exhaust purifiers. In the United States, car exhaust purifiers have been manufactured employing CeO$_2$-based mixed oxide catalysts.

Three-effect catalysts have been widely used in developed countries. Such catalysts are composed of three parts, i.e., the catalyst support (cordierite honeycomb support or metal support), active coating (often composed of Al$_2$O$_3$, BaO, CeO$_2$, or ZrO$_2$), and active component (e.g., Pd, Pt, Rh). At present, the solid solution of CeO$_2$–ZrO$_2$ is the most extensively studied. It can enhance the oxygen storage capacity of the catalyst, expand the operating window, and improve the thermal stability of the high specific surface coating as well as dispersion, toxicity, and durability of noble metal components.

A catalyst can exert its maximum efficiency only when it is operating close to the theoretical air–fuel ratio [$m(A)/m(F) = 14.6$], which is usually referred to as the operating window. The actual tail gas composition is often out of the window range. Thus, widening of the operating window is often required to enhance the catalytic efficiency. Studies show that the addition of rare earth oxides reduces the amount of necessary precious metals and effectively broadens the operating window. Rare earth oxides also play roles of accelerators, activators, dispersants, and stabilizers.

In recent years, research into solid solution composites formed by cerium oxide, zirconium oxide, and lanthanum oxide has led to the development of a new generation of catalysts resistant to high temperatures. They display excellent redox performance and oxygen storage capability, high thermal stability, and outstanding catalytic performance at low temperatures. The catalysts also show good oxygen storage and

12.9 Research and Application of Rare Earth Catalysts

discharge properties, which are manifested by the total and dynamic oxygen storage capacity.

Furthermore, China has made several breakthroughs in the purification of rare earth catalysts from gasoline vehicle exhaust. Particular advancements have been achieved in the research concerning are earth-transition metal-micro precious metal catalysts. The catalysts developed for tail gas purification have shown good activity and stability as well as strong anti-sulfur and lead poisoning activities. Employing these materials prolongs the service life of vehicles up to 50,000 km.

Rare earth Ce is the most important component of three-effect catalysts. Its functions include storage and release of oxygen by forming two cerium oxides, i.e., $Ce_2O_3$ and $CeO_2$, and a series of non-stoichiometric oxides. Additionally, Ce facilitates the dispersion of precious metals on the surface of the catalyst and prevents them from interacting with $Al_2O_3$ at high temperatures. It also improves the utilization rate of active components and prevents the aggregation and phase transition of $Al_2O_3$ carriers due to sintering. For instance, it inhibits the conversion of alumina from $\gamma$ to $\alpha$. Lastly, rare earth Ce promotes the conversion of water into gas as well as the steam reforming reaction. La is an auxiliary element in three-effect catalysts. It predominantly improves the thermal stability of $\gamma$-$Al_2O_3$, increases the active component dispersion and the number of active sites, reduces the catalyst initiation temperature, and increases the $NO_x$ conversion rate. La can reduce the role of noble metals and $Al_2O_3$. Similarly to Ce, Pr has two valence states, namely $Pr^{3+}$ and $Pr^{4+}$. Despite a report stating that Pr cannot directly replace Ce as an oxygen storage material, it has been shown that it can release a large amount of oxygen at low temperatures. Undoubtedly, the theoretical research into this rare earth element is valuable for reducing the catalyst starting temperature and improving the activity.

(3) Catalytic combustion of fossil fuels (e.g., natural gas)

At present, the key energy sources are coal, oil, natural gas, and other fossil fuels. The development and application of highly efficient, energy-saving and environmentally friendly combustion technologies is therefore of great significance. Rare earth element-based catalytic combustion is a new method for the complete oxidation of fuel on the catalyst surface. During this reaction, the reactants form a low-energy surface radical on the surface of the high-performance combustion catalyst. To avoid energy loss of visible light, the energy is released by infrared radiation. This process rarely produces pollutants, such as $NO_x$, CO, and HC. Hence, rare earth catalytic combustion technologies have the advantages of high productivity, energy efficiency, and low pollution to the environment.

Rare earth element-based catalytic combustion can be divided into low temperature catalytic combustion ($< 600\ °C$), medium temperature catalytic combustion ($600–1000\ °C$) and high temperature catalytic combustion ($> 1000\ °C$). Low temperature catalytic combustion is mainly used in waste gas purification and low temperature drying, while medium temperature combustion is predominantly utilized in household gas appliances as well as in indoor and outdoor heating systems. Lastly, high temperature combustion is suitable for aircraft engines, natural gas power generators, industrial boilers, and high temperature furnaces.

(4)  Rare earth synthetic rubber catalysts

Rubber can be divided into two categories, i.e., natural and synthetic rubber. Currently, the yield and applications of synthetic rubber exceed those of natural rubber. Catalysts composed of IVB-VIIIB transition metal salts and IA–IIIA metal alkyls play key roles in the production of synthetic rubber. Application of rare earth elements containing 4f electrons in synthetic rubber catalysts began in the 1960s (Liu 2007).

- Composition of rare earth synthetic rubber

Rare earth catalysts, which catalyze the directional combination of diene, are composed of rare earth element-based salts and metal alkyl compounds. For instance, $AlR_3$ alkyl aluminum compounds are commonly employed for this purpose. The binary systems comprise anhydrous rare earth halides, ethers, amine and phosphate ester complexes, and alkylaluminum compounds. Ternary catalytic systems are composed of rare earth carboxylate or acid phosphate esters and halogen-containing reagents.

- Factors affecting activity

It has been confirmed that the catalytic activity of different rare earth elements varies. Specifically, the catalytic activity of light rare earth elements is usually higher than that of heavy rare earth elements. Moreover, different halogens in rare earth synthetic rubber catalysts have a certain effect on the structure of polymeric active polymers. In binary catalytic systems, the activity of different halogens follows the order of Cl > Br > I > F. In contrast, internary catalytic systems, this order changes to Br > Cl > I > F. Typically, in ternary systems, the proportion of the third component to rare earth ions must be balanced. Otherwise, it could affect the catalyst activity and phase state. The X/RE molar ratio in ternary systems is mostly 2.5–3.0, as the catalyst activity is the highest in this range.

- Structure and properties of rare earth synthetic rubber

Rare earth catalytic polymerization of diene to the corresponding polymers is an important reaction because the resulting products exhibit numerous unique structural characteristics and remarkable properties. Rare earth cis-butadiene rubber displays high strength and viscosity. It can be employed to enhance the properties of vulcanizates and is superior to traditional cis-butadiene rubber. Notably, its blend with natural rubber or styrene-butadiene rubber is also more optimal than traditional cis-butadiene rubber.

(5)  Rare earth element-based catalytic materials

Different elements can be substituted by rare earth materials, which can be used as the main or secondary catalytic components (Liu 2007).

The main functions of such materials include:

## 12.9 Research and Application of Rare Earth Catalysts

- The key functions of cocatalysts for supported metal catalysts involve the improvement of the catalytic activity or selectivity by increasing the dispersion of the active surface components. An additional function is to enhance the catalyst stability by preventing the sintering of active components. Lastly, cocatalysts of this type improve the carbon resistance by modifying the acidic and basic moieties on the surface.
- New composite oxides are synthesized through rare earth oxides and other transition metal oxides to obtain catalysts suitable for high temperature oxidation. Such materials are used to enhance the catalytic activity and prevent carbon formation by modifying the acidic and basic moieties on the surface.
- Catalytic materials for rare earth organic compounds.

Many reactions in organic synthesis, such as oxidation, reforming, aromatization, hydrogenation, dehydrogenation, dehydration, hydrolysis, and esterification, employ rare earth oxides or composite oxides as catalysts (Rosynek 1977).

There key reactions utilizing these materials include:

- Hydrocarbon reforming
- Hydrocarbon oxidation and ammonia oxidation
- Oxidative coupling and selective oxidation of methane
- Methanation
- Conversion of light oil and natural gas using water vapor
- Oxidation and transformation of formaldehyde
- Hydrolysis and esterification
- Catalytic materials for rare earth inorganic compounds.

Rare earth oxides or salts are frequently used as catalyst components or cocatalysts to improve the catalytic activity and stability in reactions, such as oxidation of ammonia, synthesis of nitric and sulfuric acids, and medium temperature water gas transformation.

Rare earth composite oxides can also be employed for complete oxidation of hydrocarbons as well as for the oxidation of ammonia to nitric acid. An important process during the synthesis of ammonia and production of hydrogen is the generation of $CO_2$ and $H_2$ to remove CO. During catalytic reactions involving CO and $H_2O$ at 300–500 °C, rare earth elements act as cocatalysts for Fe–Cr catalysts. They play the role of tuning and structural aids to prevent the adverse effects of various additives on the structure and activity of the catalyst. Such materials include $K_2O$, MgO, ZnO, and CaO.

Qian et al. (2014) studied the preparation, characterization, and catalytic activity of nano-perovskite $ABO_3$ (A = La, Y; B = Mn, Co, Ni) in reactions involving NO and CO. Some related state laboratory investigations examined the use of the activation position of doped catalysts from the atomic level to optimize the catalytic effect. For instance, effective uniform oxidation of CO on $CeO_2$ nano-filaments was evaluated.

(8)  Reduce rare earth elements

Silva (2014) studied the determination of rare earth elements in spent catalyst samples from oil refinery is proposed. Their study indicated that oil refinery industry generates huge amounts of waste from the most widely process used in petroleum refining, the fluid catalytic cracking, catalysts containing high concentrations of rare earth elements are used to convert heavier oil into lighter and more valuable products. Gradually the catalyst loses its activity and is rejected as a waste. However, considering the crucial importance of rare earth elements, people hope that the spent catalysts could have a better destination.

# References

Bechesfedt F (2007) Principles of surface physics. Photocopy. Foreign Physics Series 2, Science Press, Beijing
BIOVIA Materials Studio. http://accelrys.com/products/collaborative-science/biovia-materials-stu dio/quantom-catalysis-software.html
DigitalRefining.com. Low and zero rare-earth FCC catalysts. https://www.digitalrefining.com/lit erature/1000193/low-and-zero-rare-earth-fcc-catalysts
Gomer R (1985) Metal surface interactions (trans: Zhang WC et al). Science Press, Beijing
Guo XM (1986) Calcium iron ore type rare earth composite oxide catalyst. Rare Earth 5:46
Huang ZT (1994) Industrial catalysis. Chemical Industry Press, Beijing
Kitto M (1992) Rare-earth elements in refinery cracking catalysts and fuel oils. In: Proceedings of catalytic selective oxidation, INIS, IAEA, vol 24, p 9
Liu GH (2007) Rare earth materials. Chemical Industry Press, Beijing
Liu XZ (2009) Rare earth fine chemicals. Chemical Industry Press, Beijing
Maier J (2004) Physical chemistry of ionic materials, ions and electrons in solid, II. John Wiley & Sons Ltd., Chichester
Qian FB, Wang L, Shen JY et al (2014) Preparation, characterization and catalytic activity of nano-perovskite abox (a =La, y; b=Mn Co, Ni) for NO+CO. Chin Rare Earths 35(3):86
Rare earth elements [catalysis and inorganic chemistry], explore rare earth elements [catalysis and inorganic chemistry] categories, TCI. https://www.tcichemicals.com/TH/en/c/12477
Rosynek MP (1977) Catalytic properties of rare earth oxides. Catal Rev, Sci Eng 16:1
Silva JSA (2014) Determination of rare earth elements in spent catalyst samples from oil refinery by dynamic reaction cell inductively coupled plasma mass spectrometry. J Braz Chem Soc 25:6
Xu GX, Wang XY (2010) Material structure. Version 2, Science Press, Beijing

# Chapter 13
# Rare Earth Hydrides and Hydrogen Storage Alloys

## 13.1 Brief Description

It has long been found that hydrogen can be dissolved in steel and other metal alloys, which can cause stress corrosion and hydrogen embrittlement. Hence, the research on the interactions between different metals and hydrogen has gradually increased. Studies have found that certain metals and their alloys produce hydrides, which can be heated to release hydrogen. Notably, hydrogen absorption and evolution reactions are reversible. This phenomenon can be used to convert different types of energy.

Hydrogen storage alloys have been developed into materials, which can store and transport hydrogen in a clean, pollution-free, simple, and safe manner. In August 1977, an international seminar on hydride as an energy reservoir was held in Geilo, Norway, which was attended by more than 70 researchers.

Lundin (1979) studied hydrogen storage properties and characteristics of rare earth compounds, proposed some applications, potential and realized areas, such as automobiles, buses, industrial vehicles, railroads, storage of converted electrical off-peak energy, power plants, storage of converted wind, solar, or geothermal energy, storage of converted industrial waste heat energy, storage of feedstock for chemical, petrochemical, or other industrial uses, fuel storage for electrochemical systems, storage for a power cycle working fluid, storage for fuel cell application, application for heat pumps for heating, airconditioning, or refrigerative purposes, separation of hydrogen isotopes, hydrogen pump, source for high-purity hydrogen, and getters.

Subsequently, in the spring of 1980, an international seminar on the properties and applications of metal hydrides was held in Colorado. It was attended by 225 researchers, approximately 35% of who were representatives of various industrial institutions. In 1982, a third meeting was held in Japan and similar international conferences have been organized since then.

Hydrogen is a green energy source, which has the advantages of a high calorific value. For instance, burning 1 kg of hydrogen can heat $1.25 \times 10^6$ kJ, which is equivalent to 3 kg of gasoline or 4.5 kg of coke. The hydrogen content in water resources is estimated at 11.1%. Methods, such as electrolysis in the presence of a

© Science Press 2023
C. Wang, *Theory and Application of Rare Earth Materials*,
https://doi.org/10.1007/978-981-19-4178-8_13

certain amount of a salt as well as solar photolysis of water, have been used to produce hydrogen. In the late 1960s, Zijlstra H et al. discovered that samarium cobalt absorbed hydrogen at 2 MPa, and released it at 0.1 MPa. Later on, in the Netherlands, Philips found that $LaNi_5$ absorbed hydrogen to form an alloy hydride, i.e., $LaNi_5H_6$, the hydrogen density (atomic/cm$^3$) of which was comparable to the density of liquid hydrogen (approximately 1000 times higher than the hydrogen density).

Hydrogen storage materials enable safe, reliable, and effective storage as well as transportation of hydrogen. Importantly, the heat and pressure produced during hydrogen absorption and discharge reactions can be utilized (Shiokawa 1993).

## 13.2 Hydride Types and Hydrogen Storage Alloy Components

Except inert gases, nearly all elements can react with hydrogen to produce hydrides. According to the structure and binding force, hydrides can be divided into the following four types:

(1) Salt-like hydrides

Hydrogen, specifically $H^-$, can react with alkali metals and alkaline earth metals. The produced hydrides are in the form of white crystals, which generate a lot of heat and are remarkably stable. Thus, such hydrides are suitable for hydrogen storage.

(2) Metal-like hydrides

Most transition metals can react with hydrogen to produce different types of metal hydrides. Hydrogen reacts with the IIIB–VB transition metals to form metal hydrides, such as $TiH_2$, $ZrH_{1.9}$, $PrH_{2.8}$, $TiCoH_{1.4}$, and $LaNi_5H_6$. Notably, hydrogen exhibits intermediate properties between $H^-$ and $H^+$. It can also react with VIB–VIIIB transition metals. Hydrogen forms solid solutions in the $H^+$ form. Hydrogen atoms enter the metal matrix to form interstitial-type compounds.

(3) Covalently bonded and highly polymerized hydrides

These are hydrides formed by boron or elements near boron in the periodic table (e.g., $B_2H_6$).

(4) Molecular hydrides

Nonmetallic hydrides formed by the reaction of hydrogen with elements in the IIIA–VIIA group of the periodic table belong to this category.

At present, most hydrogen storage materials are hydrides formed by the reactions of hydrogen with transition metals, alloys, or intermetallic compounds. Experiments show that the use of metal hydrides alone is often not suitable for hydrogen storage due to the high heat generation and low dissociation of hydrogen. More practical hydrogen storage materials are polymetallic and intermetallic compounds composed

## 13.4 Thermodynamic Properties of Rare Earth Hydrides

of hydrides and heat-positive endothermic metals (e.g., Fe, Ni, Cu, Cr, and Mo) or hydrides and heat-negative exothermic metals (e.g., Ti, Zr, Ce, Ta, V).

Hydrides of elements in groups IA–VA and transition metals (groups IIIB–VB) have a larger heat of formation and are called strong bonded hydrides. Such compounds exhibit hydrogen stabilization factors. Group VIB–VIIIB transition metals other than Pd (e.g., Fe, Co, Ni, Cr, Cu, and Zn) do not easily form hydrides due to hydrogen instability factors. They are often denoted as weak bonded hydrides. Strong bonded hydrides control hydrogen storage, while the weak bonded hydrides regulate the reversibility of hydrogen absorption and discharge reactions. For hydrogen storage alloys to absorb hydrogen reversibly, there must generally be at least one element with a strong affinity for hydrogen and at least one element with a weak affinity for hydrogen. Such a system can act as an ideal hydrogen storage material (e.g., $LaNi_5$ containing rare earth elements).

## 13.3 Structures of Rare Earth Hydrides

Rare earth elements combine with hydrogen to form dihydrides ($REH_2$), trihydrides ($REH_3$), and non-stoichiometric hydrides. Examples of rare earth hydrides are summarized in a table. The main items are Group I calcium fluoride (e.g. $LaH_{1.95-3}$, $CeH_{1.8-3}$), Group II calcium fluoride (e.g. $YH_{1.902-2.3}$, $SmH_{1.92-2.35}$), Group II hexagon (e.g. $YH_{2.77-3}$, $SmH_{2.59-3}$), and Group III orthogonal type (e.g. $EuH_{1.86-2}$, $YbH_{1.80-2}$).

Rare earth hydrides can be divided into three groups according to their structures. The first group includes La, Ce, Pr, and Nd hydrides. The dihydrides of the aforementioned compounds exhibit cubic system structures and continuously generate solid solutions with trihydrides. The second group comprises Y, Sm, Gd, Tb, Dy, Ho, Er, Tm, and Lu hydrides, which display $CaF_2$-type structures. Moreover, trihydrides of these compounds have cubic system structures. Lastly, the third group includes Eu and Yb dihydrides, which similarly to alkaline earth metal hydrides belong to otropic structures.

The lattice constants of various rare earth hydrides are shown in a table. The items include hydride, crystal system, and lattice constant/$10^2$ pm. For example, hydride $LaH_2$, its crystal system is face-centered cubic, and lattice constant/$10^2$ pm is $a = 5.663$; hydride $YH_3$, its crystal system is hexagon, and lattice constant/$10^2$ pm is $a = 3.672$.

## 13.4 Thermodynamic Properties of Rare Earth Hydrides

The thermodynamic properties of rare earth hydrides can be generally determined based on the hydrogen pressure measurement of the hydride phase equilibrium (Li et al. 2009). The thermodynamic relation is as follows:

$$\Delta G^{\ominus} = \Delta H^{\ominus} - T\Delta S^{\ominus} = -RT \ln K$$

For instance, a hydride generation reaction can be represented by the following equations:

$$RE_{(S)} + H_2 = REH_{2(S)}$$

$$\Delta G_f = \Delta H_f - T\Delta S_f = RT \ln p_{H_2}$$
(f represents to generate)

$$K = 1/p_{H_2}$$

For $REH_2$ hydrides, the thermodynamic data can be obtained experimentally using the above formula. For other types of hydrides, the thermodynamic data can be determined employing other methods. There is a table to summarize the thermodynamic data of rare earth dihydrides, which were calculated based on the decomposition pressure data. To dihydrides, e.g. $LaH_2$, $CeH_2$, $PrH_2$, $NdH_2$, etc., the measured thermodynamic items include: $-\Delta H_f/(4.184\,kJ/mol\,H_2)$, $-\Delta S_f/(4.184\,kJ/mol\,H_2/deg)$, and $-\Delta G_f/(4.184\,kJ/mol\,H_2)$.

## 13.5 Magnetic Properties, Conductivity, and Structure of Rare Earth Hydrides

1. Magnetic properties

Following the formation of hydrides by light rare earth metals (from Ce to Sm), magnetism remains basically unchanged. In contrast, magnetism of heavy rare earth hydrides is lower than that of metals. For instance, lanthanum contains no f electrons; therefore, its magnetism marginally decreases after dihydride formation. The generated $LaH_3$ species is diamagnetic. Moreover, $NdH_2$ exhibits ferromagnetism. The magnetic moment of $EuH_2$ (7.0 B.M.) is close to that of $Eu^{2+}$ (7.94 B.M.), which has seven f electrons with strong magnetism.

2. Conductivity

With the exception of $YbH_2$ and $EuH_2$, $REH_2$ are metal conductors, and hydrogen-deficient $REH_{1.8-1.9}$ species exhibit better conductivity than rare earth metals. The resistance ratios of hydrogen-deficient dihydrides to the corresponding metals are shown in a table, for example, element Y, the $REH_{2-\delta}/RE$ is 0.27; element Ce, the $REH_{2-\delta}/RE$ is 0.49.

Hydrides, which are nearly exclusively composed of trihydrides, change from metal conductors to semiconductors and display typical semiconductor behavior after H/RE > 2.8. Furthermore, hexagonal rare earth trihydrides, $REH_3$, show semiconductor properties.

3. Structures of metal hydrides

Following hydrogen absorption by $LaNi_5$, the position of La, Ni, and H can be clearly seen from the neutron diffraction pattern. Because hydrogen in metal lattices exists in an atomic state, hydrogen storage materials exhibit high hydrogen storage bulk density and good safety.

## 13.6 Principles of Metal Hydride Hydrogen Storage

Hydrogen can react with metals, alloys, or intermetallic compounds to form metal hydrides, which release heat. When heated, metal hydrides release hydrogen, which can be expressed by the following reaction:

$$M_{(S)} + \frac{X}{2} H_{2(G)} \rightleftharpoons Mhx_{(S)} + \Delta H$$

In the formula, $p_1$, and $T_1$ indicate the pressure and temperature required for the hydrogen absorption, respectively, while $p_2$ *and* $T_2$ refer to the pressure and temperature during hydrogen release. M denotes metal alloys or intermetallic compounds and $\Delta H$ Is the reaction heat.

The absorption and release of hydrogen in metals depends on the phase equilibrium relationship between the metals and hydrogen. The factors affecting the phase equilibrium include composition, temperature, and pressure. Phase equilibrium of metal-hydrogen systems can be expressed by composition-decomposition pressure isotherms demonstrated in a diagram.

Hydrogen molecules dissociate into hydrogen atoms on metal surfaces. Subsequently, hydrogen atoms enter the gaps in the metal lattice to form a solid solution, which is called the $\alpha$ phase. This is equivalent to $OA$, while the $A$ point is the maximum amount of hydrogen dissolved in the $\alpha$ phase. When the $A$ point is reached, the hydrogenation reaction begins. As the concentration of hydrogen in the metal increases at a constant hydrogen pressure, the reaction produces a metal hydride, which is called the $\beta$ phase.

Another diagram illustrates the hydrogen composition–pressure isotherms of the $LaNi_5$ alloy during hydrogen absorption and release. The pressure varied when the metal hydride was at a constant temperature, which was caused by reaction hysteresis. Due to the hysteres is phenomenon, the hydrogen absorption curve is not completely coincident and deviates from the ideal state. In $LaNi_5$—hydrogen composition—pressure isothermal curve, the units are $pH_2/MPa$, $W_H/\%$, and $C_H/(mol/L)$.

The composition-pressure–temperature curve of a hydride can illustrate the solubility of hydrogen in the $\alpha$ phase, the type of hydride generated during the hydrogenation reaction, its stability, effective hydrogen storage capacity, and the hysteresis phenomenon.

The composition-pressure–temperature curve, which shows the relationship between the pressure and temperature of hydrogen, enables the establishment of $\Delta H$ and $\Delta S$. Notably, these two values are of great theoretical and practical significance.

## 13.7 Characteristics of Hydrogen Storage Materials

1. Characteristic values of hydrogen storage materials

    (1) Activation

    Activation refers to the formation of a catalyst on the surface of an alloy to dissociate hydrogen molecules into atomic states. This makes the alloy have hydrogen absorption characteristics (Liu 2009). The ease of initial alloy activation and the ability to resist gas impurities (e.g., CO, $O_2$, $H_2O$, $CO_2$, and $SO_2$) are important indicators for evaluating hydrogen storage materials. First, in early stages, the hydrogen absorption rate of a hydrogen storage alloy is slow, and the hydrogen absorption performance is unstable. Second, because the alloy is often exposed to the atmosphere, the surface is covered by a hydrogenated film and adsorbed gases. This leads to different degrees of poisoning. Hence, prior to measuring its performance, the alloy needs to be activated to restore the hydrogen absorption activity.

    The activation treatment generally involves vacuuming the high pressure vessel containing the hydrogen storage material (vacuum degree is higher than 2–10 Pa), and then subjecting it to hydrogen gas at 3.5–6.0 MPa. After a period of hydrogen absorption, vacuuming, hydrogen treatment, and multiple hydrogen absorption and discharge reactions, the particle size and performance of the hydrogen storage material can be restored. Activated hydrogen storage materials typically have no hydrogen absorption gestation period. After 1 min, the hydrogen absorption can exceed 50%, while after 2 min; it can reach more than 95%.

    (2) Saturation hydrogen absorption, equilibrium decomposition pressure, plateau region, and hysteresis

    - Saturated hydrogen absorption: Saturated hydrogen absorption refers to the absorption of the maximum number of hydrogen atoms by 1 mol of an alloy under certain conditions, which is expressed in H/mol or as a proportion of hydrogen atoms H to metal atoms. It is generally expressed by the hydrogen concentration or atomic ratio corresponding to the end of the platform of the composition-pressure–temperature curve. Moreover, the ratio of hydrogen to metal atoms, indicated by the $B$ points in the diagram, is the saturated hydrogen uptake.

13.8 Rare Earth Hydrogen Storage Alloys and Their Optimization

- Balanced decomposition pressure: In the composition-pressure–temperature curve, the pressure of hydrogen corresponding to the platform order is the equilibrium decomposition pressure.
- Platform area: In the composition-pressure–temperature curve, the platform area corresponding to the equilibrium decomposition pressure is generally expected to be wider and have a smaller inclination. Such a small pressure difference enables a large amount of hydrogen to be produced and moved, absorbed, or released.
- Lag: The gas equation of state can be used to calculate the hydrogen absorption of a material by measuring the pressure before and after hydrogen absorption as well as the volume of each part of the system. The equilibrium pressure ($p_{absorption}$) when hydrogen is absorbed to form a hydride is typically higher than that when the hydride is released from hydrogen. The pressure $p_{release}$ difference between the two is called the pressure lag. The hysteresis effect tends to become more significant with the increase of temperature.

2. Requirements for hydrogen storage materials

For hydrogen storage materials to be practical, the following conditions must be met:

- Easy activation, high hydrogen storage capacity, and high energy density.
- The rate of hydrogen absorption and discharge must be fast and reversible.
- Heat of formation is small when hydrogen storage alloys absorb hydrogen.
- Moderate decomposition pressure (0.1–1 Pa) and room temperature are most suitable.
- Good chemical stability, persistence, and stable hydrogen absorption.
- Appropriate mechanical properties, electrical conductivity, heat conduction, and safety.
- Safe during storage and transport.
- Raw materials are widely available, inexpensive, environmentally friendly, and easy to prepare.

## 13.8 Rare Earth Hydrogen Storage Alloys and Their Optimization

1. Rare earth hydrogen storage alloys

Rare earth elements as well as Fe, Co, and Ni can form $REM_5$-type compounds with hexagonal structures and generate orthorhombic hydrides. Among them, the La-Ni series of compounds is the most studied. For example, in a table of La-Ni systems compounds and hydrides, when the compound is LaNi, structure is orthogonal, and corresponding hydride is $LaNiH_{2.6}$, $LaNiH_{3.0}$, and $LaNiH_{3.6}$; when compound is $LaNi_2$, structure is cube, and corresponding hydride is $LaNi_2H_{4.10}$ and $LaNi_2H_{4.5}$.

La, Ce, Pr, Nd, and Mg alloys absorb large amounts of hydrogen. For instance, $La_2Mg_{17}$ forms $La_2Mg_{17}H_{12}$. Moreover, $LaNi_5$ is a commonly used hydrogen storage material, the hydrogen storage capacity of which is greater than 190–200 $cm^3/g$. It rapidly absorbs and releases hydrogen, and can be used repeatedly. $LaNi_5$ reacts with hydrogen of several barometric pressure at room temperature to be hydrogenated into $LaNi_5H_6$. The hydrogen storage capacity of $LaNi_5H_6$ is approximately 1.4% (mass fraction), while 25 °C decomposition pressure is about 0.2 MPa. The decomposition heat is 30.1 kJ/mol $H_2$, which is suitable for operation at room temperature. The $LaNi_5$alloy has the advantages of large hydrogen absorption, easy activation, and moderate equilibrium pressure, little lag as well as fast hydrogen absorption and release. Importantly, the alloy does not easily get poisoned. It has long been used as a candidate material for heat pumps, batteries, air conditioning devices, among others. The biggest disadvantage of the $LaNi_5$ alloy is the large volume expansion of the unit cell during the hydrogen absorption cycle (approximately 23.5%). Additionally, the cost of the pure metal used as the raw material is very high, limiting the large scale application.

To reduce the cost and further improve the performance of hydrogen storage alloys, hybrid rare earth elements denoted by Mm, i.e., La, Ce, Pr, Nd, and Ce (> 45%), have been used instead of costly La to generate the $MmNi_5$ series of alloys at a cost considerably lower than that associated with $LaNi_5$ alloys. Notably, addition of elements such as Mn, Al, Co, Fe, Zr, Cr, and Sn can further reduce costs to form multi-element alloy hydrogen storage materials.

2. Main measures for performance optimization

The base of rare earth hydrogen alloys is composed of A and B elements. Different atomic ratios of A and B can be used to obtain various alloys, e.g., $AB_5$, $AB_2$, AB, and $A_2B$. In all kinds of alloys, A and B represent different metals. For example, A in $AB_5$ alloys consist mainly of La and other lanthanide rare earth elements or mixed rare earth metals and Ca. In other types of alloys, A is composed of different elements, including Ti, Zr, and Mg. Moreover, in all kinds of alloys, B is Ni, which is often replaced by one or more elements, such as Co, Mn, Al, and Fe, forming thousands of different hydrogen storage alloys. Different element substitution leads to different functions of the resulting materials (Liu 2007). The performance of hydrogen storage materials in $AB_5$ rare earth systems can be enhances by A and B composition optimization, i.e.,

(1) Optimization of A side (rare earth) components in $AB_5$ alloys.
(2) Optimization of B side elements in $AB_5$ alloys.

The characteristics of some rare earth hydrides are summarized in a table. For example, when Metal hydride is $LaNi_5H_{6.0}$, hydrogen content/% is 1.4, decomposition pressure/MPa is 0.4 (50 °C), and heat of formation (1 mol $H_2$)/kJ is − 30.24.

Among the commercial multi-element rare earth nickel-based hydrogen storage alloys, the most commonly used alloying elements are Ni, Mn, Al, and Co. The study

13.9 Preparation of Rare Earth Hydrogen Storage Materials

of the synergistic effects between different elements remains the focus of the current research.

To obtain rare earth hydrogen storage materials with excellent properties, in addition to optimizing their alloy components, attention should be paid to the selection and control of the preparation process conditions, melt cooling conditions, alloy crushing rate, and surface treatment methods.

At present, there are two key aspects in the research and application of hydrogen storage materials. The first involves the use of other metals to form mixed rare earth hydrogen storage materials. The other is achieving unique properties of non-stoichiometric alloys, composite alloys, nano alloys, and amorphous alloys, which is currently a hot spot in the research concerning hydrogen storage materials. The production of hydrogen storage alloys in some developed countries mainly involves rapid solidification processes. The electrochemical capacity of $AB_5$ hydrogen storage alloys has exceeded 340 mA·h/g, and their overall performance is good. However, at present, they do not meet the requirements of power batteries and fuel cells. Thus, hydrogen storage materials with better hydrogen storage performance are actively being investigated by numerous researchers (Sakai et al. 1992).

## 13.9 Preparation of Rare Earth Hydrogen Storage Materials

1. Alloy smelting

The main processes involved in alloy smelting are batching, smelting, pouring, crushing, and activating hydrogen storage materials.

According to the composition of hydrogen storage alloy ingredients, for all kinds of metals with purity greater than or equal to 99%, smelting in an electric arc furnace or vacuum induction furnace ($5 \times 10^{-2}$ Pa) proceeds by filling it with argon gas and then emptying it three times. After heating up, the alloy is melted and homogenized, and the molten alloy is poured into a water-cooled copper mold under argon protection.

Following coarse crushing, the alloy ingot is ground by a mechanical method under argon protection. Otherwise, the alloy composition of the hydrogen storage material prepared before hydrogen absorption in a high pressure vessel is not uniform, which will affect the hydrogen absorption performance. The alloy can be treated uniformly at a higher temperature (e.g. 1100 °C) for a certain time. It is noteworthy that the alloy casting conditions affect the hydrogen absorption and electrochemical performance. Fine grain microstructure is beneficial to enhancing the electrochemical cycling stability.

2. Chemical synthesis

During the preparation of lanthanum-rich $Ni_5$ alloys, the lanthanum-rich mixed rare earth fluoride and $NiCl_2$ mixture ($n_{lanthanum-rich}: n_{Ni} = 1:5$) are co-precipitated in the presence of an aqueous solution of $Na_2CO_3$. Following dehydration, the materials are heated under hydrogen atmosphere at 750 °C. After hydrogen reduction, the powder is mixed with an appropriate amount of calcium, and the hydrogen thermal reduction diffusion reaction is carried out at 1000 °C. The synthesized compounds are then washed and dried to produce lanthanum-rich $Ni_5$ alloys. Hydrogen storage materials obtained by this method exhibit the good hydrogen absorption properties. There is a large amount of active Ni atoms on the surface of the alloys, and their catalytic activity can promote hydrogenation reactions.

3. Physical vapor deposition

Evaporation, sputtering (e.g., ion beam sputtering), and other methods are used to condense or deposit metal atoms or ions to obtain polished, deoiled, and chemically activated metal sheets in a 3 mol/L $H_2SO_4$ solution. The sputtering targets are typically made of mixed rare earth metals and pure nickel. The prepared alloy films are amorphous or microcrystalline and display excellent electrochemical stability as well as strong resistance to hydrogen embrittlement and pulverization at high current density. For instance, $LaNi_{3.94}SiO_{0.54}$ alloy films prepared by this approach have this advantage. Prior to sputtering deposition, a high-energy ion beam is used to pre-sputter the target for 10 min to clear the oxide film on the metal surface. Example conditions for sputter deposition are as follows: argon ion beam voltage of 2.8 kV, beam current of 80–90 mA, vacuum chamber pressure $9.2 \times 10^{-3}$ Pa. The size and distribution of a hydrogen storage alloy powder also affect its electrochemical properties. Generally, the smaller the particle size, the higher the discharge capacity and the longer the cycle life. When the ratio of average particle size to average particle size is 0.11 and the mass ratio of coarse fine alloy powder is 0.7, the electrochemical capacity of the alloy is the highest.

## 13.10 Storage and Application of Hydrogen Storage Materials

For applications in transport, hydrogen is typically stored in the form of compressed gases or cryogenic liquids (e.g., steel bottle hydrogen). In contrast, metal hydrides store and release hydrogen in the form of solid hydrides. Small hydride storage systems have a commodity supply. In practice, the pressure of hydride storage systems should be several times higher than the platform pressure (e.g., 2–5 times) to rapidly charge hydrogen and adapt to any changes in the ambient temperature. Hydrides exhibit a high volumetric storage density. Voids are required due to the granular character of hydrides and the necessity to adapt to volume expansion associated with hydride formation. Hydrogen storage alloys continuously expand and shrink, resulting in the micronization (Sahlberg et al. 2016). This not only leads to a decrease in the heat transfer efficiency, but also results in the local stress concentration of the device. A microencapsulation processing method has been developed for micronization (Houston 1983).

## 13.10 Storage and Application of Hydrogen Storage Materials

Hydrogen storage materials are predominantly employed for the following applications:

(1) Purification and separation of hydrogen

Selective hydrogen absorption characteristics of hydrogen storage materials enable the preparation of > 99.9999% pure hydrogen. Therefore, these materials can be used in semiconductor devices, electronic materials, large-scale integrated circuits, fiber production. The chemical industry, petroleum refining as well as the pharmaceutical and metallurgical industries have a large number of hydrogen-containing tail gas emissions.

(2) High performance charging batteries

Nickel metal oxide batteries utilize nickel oxide as the positive electrodes and hydrogen storage alloys as the negative electrodes to generate high specific capacity and pollution-free chemical power. Such batteries are widely used in notebook computers, mobile phones, and electric vehicles.

(3) Hydride heat pumps for air conditioning and heating

Hydrogen storage alloys absorb hydrogen with a high reaction heat up to 210 kJ/kg; therefore, they can be utilized for chemical heat storage and chemical heat pumps. Metal hydride heat pumps are devices, which can be heated without fuel combustion and cooled without the use of refrigerants, limiting environmental pollution. Unlike other heat pumps, which use mechanical power, metal hydride heat pumps do not generate any noise or vibration. The current hydride heat pumps can be divided into warming-up, heating, and refrigeration types according to their power.

Hydrogen storage alloys used as heat pump materials must meet the following conditions:

(1) Little lag so the exchange of hydrogen between two different alloys can proceed smoothly.
(2) The platform area of the composition-pressure–temperature curve should have a small inclination and large width, which increases the amount of hydrogen that can be absorbed and releases.
(3) Easy activation and fast hydrogen absorption reaction rate.

The hydrogen storage alloys satisfying the above conditions all contain rare earth elements.

Numerous hydrogen storage alloy pairs have been developed for hydride heat pumps. For instance, automotive air conditioners use $La_{0.6}Mi_{0.4}Ni_{4.7}Cr_{0.3}$ as a high temperature end alloy and $La_{0.2}Mm_{0.8}Ni_{4.35}Fe_{0.35}$ as a low temperature end alloy. Mi refers to a lanthanum-rich mixed rare earth element (La 51.2%, Ce 3.9%, Pr 8.8%, Nd 26.9%), while Mm indicates a cerium-rich mixed rare earth species (Ce 47.4%, La 20.8%, Pr 6.1%, Nd 15.6%, Sm and Y < 0.5%). Solar conversion systems employ $CaNi_5/LaNi_5$ alloy pairs with a net refrigerating capacity of 3500 W at 117 °C/40 °C/8 °C. Rare earth hydrogen storage materials are also used in digestion and inspiratory agents, ultra-low temperature refrigeration materials as well as in isotope separation of hydrogen (Chuang Teng Technology Co., Ltd. 2015).

## 13.11 Experimental Studies on Hydrogen Storage Materials

Experimental and theoretical research can provide the basis for the development of excellent hydrogen storage materials. Some examples of studies on hydrogen storage materials are briefly described below.

Wang et al. (1994) investigated hydrogen storage alloys for automotive hydride air conditioners. The Israeli Institute of Technology designed a bus hydride air-conditioning system using $LaNi_{4.7}Al_{0.3}/MmNi_{4.15}Fe_{0.85}$ alloy pairs and porous block technology; however, the obtained results required further investigation. To determine excellent alloy pairs suitable for automobile hydride air conditioners, Wang applied La-Mi-Ni–Cr-La-Mm-Ni–Cr and La-Mm-Ni–Fe series alloys.

The employed experimental methods are described below.

Raw materials: the total amount of lanthanum-rich rare earth element (Mi) was more than 98%, of which La accounted for more than 45%. Moreover, the total amount of cerium-rich mixed rare earth element (Mm) was more than 98%, of which Ce accounted for more than 45%. The purity of La, Ce, and industrial Fe was 99%, while the purity of electrolytic Ni was 99.9%.

Alloy melting was carried out in a vacuum arc furnace protected by argon ($\geq$ 99.99%). To ensure uniform composition of the sample, the alloy ingot was rolled over and remelted twice.

Alloy activation was performed at room temperature at a hydrogen pressure of 4.0 MPa. After several hydrogen absorption cycles, the hydrogen storage characteristics of the alloy were determined. The hydrogen absorption composition-pressure–temperature curve was determined using the constant volume/pressure differential stage hydrogenation method. In addition, the hydrogen release composition-pressure–temperature curve was established employing the drainage gas gathering method. X-ray diffraction (XRD) and energy spectrum analyses were used to determine the alloy structure and chemical composition. Moreover, the morphology was evaluated by scanning electron microscopy (SEM).

The obtained experimental results are discussed below.

(1) High temperature end alloys: The $La_{1-x}$–$Ni_{5-y}$–$Cr_y$ alloy series, where $x = 0.2$, 0.4, 0.8, and $y = 0.1$–0.5, was analyzed. The alloy platform pressure, effective hydrogen capacity, lag factor, and slope factor were comprehensively investigated. It was determined that the $La_{0.6}Mi_{0.4}Ni_{4.7}Cr_{0.3}$ alloy could be used as a high temperature end material of automobile hydride air conditioners. The main properties of this alloy are superior to those of $LaNi_{4.7}Al_{0.3}$ used by the Israeli Institute of Technology.

(2) Low temperature end alloys: For comparison with $La_{0.6}Mi_{0.4}Ni_{4.7}Cr_{0.3}$, the $La_{1-x}Mm_xNi_{5-y}Fe_y$ ($x = 0.2, 0.4$ and $y = 0.2$–0.8) series of alloys was designed to study the hydrogen storage characteristics. Notably, compared with the $MmNi_{4.15}Fe_{0.85}$ alloy used by the Israeli Institute of Technology, Mm was partially substituted by La to improve the La/Ce ratio of rare earth elements.

13.11 Experimental Studies on Hydrogen Storage Materials

The effective hydrogen reserves of the alloy were increased due to reduced iron replacement and the platform slope was improved.

(3) Matching properties of $La_{0.6}Mi_{0.4}Ni_{4.7}Cr_{0.3}/La_{0.2}Mm_{0.8}Ni_{4.35}Fe_{0.65}$.

Zeng et al. (1994) studied hydrogen storage electrode materials consisting of a $MmB_{5-x}$ system according to the characteristics of good electrochemical catalytic activity and long charge and discharge life at low temperature for alloys with non-stoichiometric ratio. The following experimental methods were used:

Raw materials: mixed rare earth Mm (La 20.84%, Ce 59.14%, Nd 14.95%, Pr 4.35%, Sm < 0.5%, Y < 0.5% mass fraction); lanthanum-rich rare earth Mi (La 46.53%, Ce 5.07%, Pr 37.26%, Nd 9.26%, Sm < 0.5%, Y < 0.5% mass fraction); purity of Ni, Co, Mn, and Al was 99.6%. The alloy was melted in a vacuum medium frequency electric furnace. Considering the alloy loss during the smelting process, when batching, the burning loss of each element was set as Mm, Mi 6%, and Mn, Al 5%. The hydrogen storage performance of the alloy was evaluated utilizing a self-made device for activation and absorption release of hydrogen.

The experimental results demonstrated that the melted alloy exhibited segregation on the micro level. XRD analysis revealed that rare earth elements exceeding the metrological portion in non-stoichiometric alloys were distributed on the $MmB_6$ matrix as the second phase, and the chemical formula was expressed as $Mm(NiCoMnAl)_5RE_y$. Notably, the presence of excess rare earth elements enhanced the hydrogen absorption and discharge properties of the alloy; however, it reduced its electrochemical capacity. Nonetheless, the above methodology and experimental conditions are not exact. Specifically, small amounts of $O_2$ and $H_2O$ were present in the system and had to be purified and removed. The removal method is described in Wang CZ's Metallurgical Physical and Chemical Research Methods (Metallurgical Industry Press 2013).

Furthermore, Lin et al. (1994) conducted a hydrogen absorption kinetics study involving various hydrogen storage alloys. The authors reported the thermostatic hydrogen absorption kinetics of ML-Ni-Co-Mn-Al and ML-Ni-Co-Mn-Al-Cu series of alloys (ML is lanthanum-rich mixed rare earth metals), which showed potential for application in nickel/hydride batteries in the $\alpha + \beta$ phase region. The employed experimental methods were as follows:

(1) The alloy was prepared with lanthanum-rich mixed rare earth element ML (La > 40%, Ce < 10%, Pr ≈ 10%, Nd 30–40%), Ni, Co, Mn, Al, and Cu (purity > 99.5%). Smelting in a vacuum arc button furnace at a stoichiometric ratio (vacuum degree $1 \times 10^{-2}$ Pa) and filling the system with high purity argon four times were performed to obtain a uniform composition. The as-cast alloys were activated by hydrogen absorption at 3.5 MPa and room temperature. Following dehydrogenation, the powder alloy samples were analyzed on a rotating anode X-ray diffractometer. Each alloy phase exhibited a hexagonal crystal structure and no second phase existed.

(2) To determine the hydrogen absorption kinetics, 2 g of the activated alloy was weighed and fully mixed with 4 g of copper powder under Ar atmosphere. The mixture was then loaded into a special double-layer water constant temperature

reactor with water circulating inside and outside. The kinetics of hydrogen absorption was automatically recorded by a pressure sensor and an X–Y function recorder.

The following kinetic rate equation for gas–solid reactions was used for the calculations:

$$\frac{d\alpha}{dt} = k \cdot f(\alpha) \tag{13.1}$$

In the formula, $\alpha$ is the hydrogen absorption fraction of the alloy, $k$ refers to the reaction rate constant, and $f(\alpha)$ depends on the differential form of the reaction mechanism function.

The corresponding integral form is as follows:

$$g(\alpha) = \int d\alpha/f(\alpha) = kt \tag{13.2}$$

The value of $\alpha$ of the hydrogen absorption fraction at time $t$ can be calculated according to the kinetic curve. Applying Eq. (13.2) results in a linear fit to 38 reaction mechanism functions, including chemical reactions, diffusion, and nuclear growth. The reaction mechanism function with the largest correlation coefficient and the smallest standard deviation is determined as the hydrogen absorption reaction mechanism, and the corresponding kinetic parameters can be acquired based on this data.

The results of different kinetic studies are summarized in the dfferent tables.

## 13.12 Yttrium Hydride as a Hydrogen Sensor Reference Electrode

Iwahara (1992, 1995) pioneered the determination of dissolved hydrogen in aluminum and aluminum alloys using proton conductor hydrogen sensors (Yajima et al. 1991). In their method, standard hydrogen is used as the reference electrode. The electromotive force of a hydrogen sensor battery can be obtained employing the following equation:

$$E = \frac{RT}{2F} \ln \frac{p_{H_2(I)}}{p_{H_2(II)}}$$

In the formula, $p_{H_2(I)}$ and $p_{H_2(II)}$ indicate the partial hydrogen pressure on both sides of the battery.

Either of them can be used as the reference electrode. Accordingly, either gas with a known hydrogen partial pressure can be selected as the reference electrode. Standard hydrogen is stored in a steel bottle, which is not convenient. Based on the composition-pressure system phase diagram of the Y-H system, Chen and Wang selected yttrium hydride with an analogous composition as the reference electrode to generate a hydrogen sensor. The hydrogen content in the two grades of aluminum alloys was determined by the above method. A stable value of the battery electromotive force was obtained after 7–10 s (Chen et al. 1995, Wang 2000).

The composition-pressure relationship of the Y-H system has been summarized in a diagram (Yannopoulos and Edwards 1965). In it, the abscissa indicates the atomic ratio (gram atom H number in per gram atom Y), the ordinate presents pressure.

# References

Chen W, Wang CZ, Liu L (1995) Determination of hydrogen activity in fused aluminum alloy by sensing method. J Metals 31(7):B306-310

Chuang Teng Technology Co., Ltd. http://www.neotrident.com/news/detail.aspx?id=931. 2015-12-28

Houston EL (1983) Application of metal hydride with repeated hydrogen charge. Chin Rare Earths (Foreign Rare Earth Application Album) (supplement):98–109

Iwahara H (1992) High temperature protonic conductors based on perovskite type oxides. ISSI Lett 3(3):11–13

Iwahara H (1995) Technological challenges in the application of proton-conducting ceramics. Solid State Ionics 77:289–298

Li M, Liu ZG, Wu JX et al (2009) Rare earth elements and their analytical chemistry. Chemical Industry Press, Beijing, pp 117–121

Lin Q, Li, Ye W et al (1994) Proceedings of the third annual conference of the China Rare Earth Society, vol 4, pp 139–146

Liu GH (2007) Rare earth materials. Chemical Industry Press, Beijing, pp 328–342

Liu XZ (2009) Rare earth fine chemicals. Chemical Industry Press, Beijing, pp 184–197

Lundin C (1979) Hydrogen storage properties and characteristics of rare earth compounds, J Phys Colloques 40(C5):C5-286–C5-291

Sahlberg M, Karlsson D, Zlotea C et al (2016) Superior hydrogen storage in high entropy alloys. Sci Rep 6:36770

Sakai T et al (1992) Rechargeable hydrogen batteries using rare-earth-based hydrogen storage alloys. J Alloys Compd 180(1–2):37–54

Shiokawa J (1993) The latest application technology of rare earths (Trans: Zhai YS, Yu ZH). Chemical Industry Press, Beijing, pp 159–169

Wang CZ (2000) Solid electrolytes and chemical sensors. Metallurgical Industry Press, Beijing, pp 562–565

Wang XH, Chen CP, Yan M et al (1994) Study on hydrogen storage alloy for automobile hydride air conditioner. J Chin Soc Rare Earths 12(2):141–144

Wang CZ (2013) Metallurgical physical and chemical research methods. Version 4. Metallurgical Industry Press, Beijing, pp 74–88

Yajima T, Iwahara H, Koide K et al (1991) Sens Actuators, B Chem 5(91):145–147

Yannopoulos LN, Edwards RK, Wahlbeck PG (1965) The thermodynamics of the yttrium-hydrogen system. J Phys Chem 69(8):2510–2515

Zeng SY, Jiang ZL, Zhang WK et al (1994) $MmB_{5-x}$ hydrogen storage electrode materials. In: Proceedings of the third annual conference of the China Rare Earth Society, vol 4, pp 130–135

# Chapter 14
# Theory and Application of Superconducting Materials

## 14.1 Brief Description

The discovery of superconductivity is closely related to the development of low temperature technology. In 1897, it was determined than oxygen and nitrogen were liquefied at 90 and 77 K, respectively. Moreover, in 1898, it was found that hydrogen was liquefied at 20 K. In 1908, a Dutch physicist Kamerlingh Onnes generated liquefied helium at a temperature of 4.25 K. This remarkably low temperature laid the foundation for his discovery of metal superconductivity 3 years later. In 1911, Kamerlingh Onnes called the sudden drop in metal resistance to zero a superconducting state, and the temperature, at which resistance changed, was denoted as the superconducting transition temperature or critical temperature ($T_C$). He was awarded a Nobel Prize in Physics for his contributions to the research on low temperature material characteristics. Since then, the search for new superconducting materials and the study of increasing the critical temperature has attracted attention of physicists and chemists all over the world.

There are 28 elements, which exhibit superconductivity at atmospheric pressure, of which niobium (Nb) has the highest $T_C$ of 9.26 K. In addition, thousands of alloys and metal-containing compounds can also become superconductors. To date, more than 250 kinds of rare earth minerals have been found. One of the most important minerals is the black rare gold ore [(Y, Ce, Ca) (Nb, Ta, Ti)$_2$O$_6$], the analysis of which revealed the presence of highly superconducting Nb in the mineral. The addition of rare earth elements to superconducting materials can make the superconducting phenomenon easier to realize.

© Science Press 2023
C. Wang, *Theory and Application of Rare Earth Materials*,
https://doi.org/10.1007/978-981-19-4178-8_14

## 14.2 The Main Properties of Superconductors

1. Critical temperature

The critical temperature, i.e., the temperature, at which the electrical resistivity of metal drops to zero, is an important characteristic of superconductor. When the superconductor temperature is close to the critical temperature, its electrical conductivity can be regarded as infinite; therefore, it can carry a large current. As long as this current does not exceed the critical current $(I_C)$, the current flow in the superconducting body can be regarded as unobstructed, and the heat loss is negligible. If a closed loop is composed of superconductors, upon excitation, the current in the loop will be maintained for a long time. Because the resistance of a superconductor is equal to zero, there is no potential difference between any two points in the conductor when the current flows in the superconducting body, and whole superconductor is an isopotential body. Thus, the critical temperature of superconducting materials has become an important research focus (Ma et al. 2007).

2. Critical magnetic field

The superconducting state is not only related to the temperature of the conductor, but also to the strength of the external magnetic field. Even if the superconductor temperature $T$ is smaller than $T_C$, in the presence of a strong external magnetic field, the superconducting state can still be destroyed. This is referred to as the critical value of the external magnetic field, i.e., the critical magnetic field, which is expressed as $H_C$ according to the following equation:

$$H_C = H_0 \left[ 1 - \left( \frac{T}{T_C} \right)^2 \right],$$

*where $H_0$ is the critical magnetic field intensity at $T = 0$ K*, that is, the maximum of the critical magnetic field. When $T = T_C, H_C = 0$.

3. Meissner effect

In a superconducting state, the potential difference between any two points in the conductor is zero; thus, there is no electric field in the superconducting body. This can be represented according to the law of electromagnetic induction as follows:

$$\oint E \cdot dl = -\frac{d\emptyset}{dt} = -\frac{d(B \cdot S)}{dt}$$

The cross-sectional area of a superconductor is invariant because the superconducting body $E = 0$ and $dB/dt = 0$. This means that when the superconductor is in the superconducting state, the internal magnetic field does not change with time. Notably, when the superconductor in the superconducting state is placed in an external magnetic field, as long as its magnetic field intensity $(H)$ is less than the critical magnetic field intensity ($H_C$), the magnetic field intensity ($H$) in the superconducting body remains zero. This is illustrated in a diagram.

The Meissner effect indicates that a superconductor in the superconducting state is a diamagnetic body displaying full diamagnetism. It was discovered by Meissner W in 1933. The diamagnetism of superconductors can be reflected by the experiment illustrated in another diagram. The ring in the figure has a long-lasting steady current passing through, and the ball is made of a superconducting material in the superconducting state. Because of the material diamagnetism, the ball is suspended in air, which is referred to as magnetic levitation.

## 14.3 Bardeen-Cooper-Schrieffer (BCS) Theory of Superconductivity

Since discovering superconductivity in 1911, the microscopic theory of superconductivity has been intensively explored. Following the establishment of quantum mechanics, in 1957, a modern superconducting microscopic theory, namely the superconducting Bardeen-Cooper-Schrieffer (BCS) theory, was founded by John Bardeen, Leon Cooper, and Robert Schrieffer. When the temperature of the metal is less than the critical temperature, the conductor exhibits superconductivity. The BCS theory states that when free electrons move in the crystal lattice, the vibration of the lattice is affected by the attraction between different charges. This makes the local region of the crystal distorted, and the electrons travel from one place to another like a wave.

From a quantum point of view, in the propagation of light waves, a photon is an energy quantum. The quantum energy of a lattice wave propagating from the distortion of a lattice vibration in a crystal is called a phonon, i.e., a "quasiparticle", which is similar to a photon. Phonons can be absorbed by free electrons in a crystal. Moreover, two free electrons can be coupled by exchanging phonons absorbed by another electron. This results in mutual attraction between electrons, which form so-called Cooper pairs. In a Cooper pair, two electrons have equal momentum but opposite direction and spin; therefore, the momentum sum of each Cooper pair is zero. When the metal temperature $T$ is smaller than $T_C$, Cooper pairs begin to form in the metal. All of the pairs are the same size and move in the same direction, resulting in superconducting properties of metal conductors. For direct current (DC) resistance, a large number of Cooper pairs move in the same direction and superconducting currents with little resistance are formed. In contrast, when alternating current (AC) passes through a superconductor, there is still a certain resistance. When the temperature of the conductor $T$ is bigger than $T_C$, the thermal motion causes Cooper pairs to split into single electrons. Consequently, the attraction between the electrons no longer exists, and the conductor loses its superconductivity and becomes a normal conductor.

The distortion of the local region of a crystal is schematically illustrated in a diagaram, and using ○ to indicate normal position of lattice, • to indicate lattice distortion position.

## 14.4 Types of Superconductors

From the material point of view, the numerous known superconductors can be roughly divided into elements, alloys, and compounds. Superconductors with critical temperatures below approximately 25 K are called low temperature superconductors, while those with critical temperatures above 25 K are denoted high temperature superconductors.

1. Superconductors composed of metallic elements

So far, 28 elements have been identified as superconductors at atmospheric pressure. Among them, niobium displays the highest $T_C$. Some transition metals (IIIB–VIII) and non-transition metals only exhibit superconductivity in the form of films and when they are subjected to high pressure and irradiation.

2. Compound superconductors

The structural types and $T_C$ values of compound superconductors are summarized in a table, for example, when structure and type is NaCl, face center cube, typical compound is MoC, $T_C$/K is 14.3, and number is 26. In January 2001, it was found that the superconducting material magnesium diboride ($MgB_2$) exhibiting a hexagonal crystal structure contained three atoms in each cell, which is composed of Mg and B in a ratio of 1:2. The $T_C$ of $MgB_2$ is determined at 39 K, which is considerably higher than those of other cryogenic superconductors.

3. Alloy superconductors

Binary irregular alloys, in which two transition elements can be mixed in any proportion to form solid solutions and display $T_C$ values that are higher, lower, or in between those of the individual elements. The majority of superconducting alloys have at least one component that is an elemental superconductor, and occasionally two elements show superconductivity. Superconducting alloy materials have the advantages of high mechanical strength, low stress and strain, easy production, and low cost. They have been developed and widely applied for many years. Superconducting alloys composed of niobium and titanium display good processing performance and are the most extensively utilized superconducting alloys.

4. Compound superconductors with NaCl structures

Among superconducting compounds, carbides and nitrides have particularly high transition temperatures $(T_C)$. Between 1932 and 1953, NbC $(T_C = 11$ K) and $NbN_{0.7}C_{0.3}$ $(T_C = 17.8$ K) were reported as compound superconductors with NaCl structures and high $T_C$. Notably, NbN shows the highest $T_C$ among binary nitrides (17.3 K). Compounds of particular practical significance include NbN and $NbN_{0.7}C_{0.3}$.

14.5 Oxide Superconductors

5. Compound superconductors with $Cr_3Si$ ($A_3B$) cubic structures

Based on evaluation of $V_3Si$, it was discovered that compounds with $Cr_3Si(A_3B)$ cubic structures are superconducting ($T_C = 17.1$ K). In years 1953–1973, superconductors, such as $Nb_3Sn(T_C = 18$ K) and $Nb_3Ge(T_C = 23.2$ K), were reported and applied in various fields. $V_3Ga$ shows a high critical current density in the high magnetic field region and is often used to fabricate superconducting magnets with magnetic fields exceeding 15 T (T $= 1$ kg/(A•s$^2$)). Numerous other materials, including oxides, chalcogenides, and boron carbides, are classed as oxide superconductors.

## 14.5 Oxide Superconductors

Oxide superconductors exhibiting high $T_C$ values can be composed of various systems, including the most common yttrium and bismuth systems. However, all materials of this type display perovskite structures and contain a Cu–O ion layer. The two-dimensional atomic surface of copper in this structure is directly related to the high temperature superconductivity. Each series begins with a parent phase, which is an insulator, and most of them are antiferromagnetic insulating phases. The insulator becomes a conductor with a low carrier concentration by replacing the constituent elements or changing the chemical ratio, which results in superconductivity. $T_C$ is a function of the carrier concentration. When the carrier concentration is higher, the $T_C$ decreases until the material becomes non-superconducting. A series of superconducting oxide materials with high $T_C$ values are summarized in a table, the items include parent phase, high $T_C$ phase, conduction type, and highest $T_C$/K (approximately). For example, when parent phase is $R_2CuO_4$, high $T_C$ phase is $R_{2-x}M_xCuO_{4-y}$ (M is Th, Ce), conduction type is n, highest $T_C$/K (approximately) is 25, and in it, the R is Pr, Nd, Sm, Eu, etc.; when parent phase is $RBa_2Cu_3O_{<6.4}$, high $T_C$ phase is $RBa_2Cu_3O_{>6.5}$, conduction type is p, highest $T_C$ / K (approximately) is 95, and in it, the R is Y, La, Nd, Sm, Eu, Gd, Ho, Er, etc.

The $YBa_2Cu_3O_{7-\delta}$ superconductor was first discovered in 1987 by the Zhu JW group at the University of Houston as well as by the Zhao ZX team at the Chinese Academy of Sciences. The superconducting transition temperature of this superconductor is 92 K. Thus, the transition temperature is comparable to that of liquid nitrogen, which is a major discovery in the fields of physics and material sciences. The first superconductor with a liquid nitrogen temperature was denoted as Y-123. The 123 oxide superconductors predominantly consist of 12 kinds of $RE_1Ba_2Cu_3O_{7-\delta}$ oxide superconductors (abbreviated as REBaCuO, REBCO, or RE123) and RE, the elements of which include Y, La, Nd, Sm, Eu, Gd, Dy, Ho, Er, Tm, Yb, and Lu. Similarly to previously reported yttrium barium copper oxides (YBCO), the RE123 oxide superconductors display the same atomic ratios and distorted, hypoxic perovskite structures. When the oxygen content of an YBCO decreases, its $T_C$ decreases until superconductivity is lost, the RE123 series oxides can be applied in many kinds of superconductor devices.

The most important characteristic of rare earth superconducting materials is the coexistence of magnetism and superconductivity. It is noteworthy that general conductive metals and semiconductors do not have this feature. Early studies of elements, alloys, and compounds suggested that magnetism and superconductivity could not coexist in the same material due to the interactions between magnetic ions and the spin of conducting electrons, which would destroy the superconducting state. Following the discovery of superconducting ternary compounds containing magnetic rare earth atoms, the related research was further developed.

In the crystal structures of $(RE)Mo_6S_8$, $(RE)MoSe_8$, and ternary rhodium boride $(RE)Rh_4B_4$, the part that provides superconducting electrons is confined to a certain cluster, which is obviously separated from the rare earth 4f electron energy that contributes magnetism. Despite the presence of some interactions between them, very weak, magnetic, and superconducting properties coexist in a certain range.

Following the discovery of YBCO-layered oxide superconductors, researchers have conducted studies on replacing Y with lanthanides. Consequently, $RE_1Ba_2Cu_3O_{7-\delta}$ oxide superconductors with atomic ratios of 1:2:3 have been discovered. The superconducting transition temperatures of these materials are approximately 90 K. Hence, rare earth ions have little effect on the $T_C$ of $RE_1Ba_2Cu_3O_{7-\delta}$. Pm, which is an artificial radioisotope, cannot be used to replace Y. Additionally, Ce, Pr, and Tb is tetravalent ions. After substitution, $PrBa_2Cu_3O_{7-\delta}$ becomes a semiconductor and does not superconduct at low temperature. Ce and Tb form $BaCeO_3$, $BaTbO_3$, $BaO$, and $BaCuO_2$ mixtures under solid-state chemical synthesis conditions. However, the formation of $RE_1Ba_2Cu_3O_{7-\delta}$ does not occur. $RE_1Ba_2Cu_3O_{7-\delta}$ can be prepared by solid-state approaches, while the synthesis of $La_1Ba_2Cu_3O_{7-\delta}$ requires special processes.

YBCO is a typical oxygen deficient compound, the oxygen content of which varies with the $p_{O2}$ value. RE123 oxides are superconductors with high $T_C$. Compared to conventional low temperature superconductors, they exhibit three distinct characteristics, i.e., a very high superconducting transition temperature, short coherence length (0.15–2.5 nm), and highly anisotropy.

Investigations concerning the properties of RE123 oxides predominantly refer to the critical current density ($J_C$), captured magnetic field intensity, and magnetic suspension density. The main factors affecting the performance of these materials are microcracks, weak connections, and magnetic flux pinning. YBCO microcracks are formed for two reasons. The first is the shrinkage of the cell in the tetragonal-orthogonal phase transition following the $c$ axis shortening, which results in tensile stress of the $c$ axis, causing cracks (e.g., intrinsic cracks). The second possible reason is that the YBCO and $Y_2BaCuO_5$ phases display different thermal expansion coefficients. This leads to the formation of cracks at the two-phase interface, which are caused by the shear stress along the $c$ axis during the cooling process. Intrinsic cracks are difficult to eliminate during oxygenation treatment; however, the spacing between intrinsic cracks can be modified according to the size of 211 different particles. For instance, there is a critical $Y_2BaCuO_5$ (211 particle) size. Microcracks form when the actual particle size is smaller than the critical size (0.24 $\mu$m).

Further improvement of the bulk properties involves enhancing the flux pinning effects by doping, preferential oxygen defects, improving twin density, etc. Considering the YBCO blocks, the magnetic suspension performance is an indicator of magnetic levitation in superconducting samples. The theoretical value of the magnetic suspension density is 30 N/cm$^2$, while the current is in the range of 10–15 N/cm$^2$.

## 14.6 Preparation of RE123 Oxide Superconductors

Research on the RE123 preparation process includes analysis of the growth mechanism, technology, single domain, and oxygen treatment process of the superconducting phase transition. RE123 oxide superconductors are crystals, the growth mechanisms of which include crystallization and growth processes. The YBCO growth process can be performed using several approaches, including ① solid-state sintering (1987–1989), which results in the preparation of polycrystalline samples with a grain size of approximately 2 mm$^2$, ② the melt texture method (1989–1993), in which the resulting sample is multi-domain and exhibits a size of approximately 5mm$^2$, and ③ directional growth of the melt texture using large single casting (1993–2002), which leads to the formation of larger grains.

The amount of oxygen atoms entering the lattice and the method of generating the best oxygen content in YBCO superconductors are the key to improving the critical temperature $T_C$ and application performance of the materials. At different temperatures, the oxygen atom occupancy rate inside the YBCO crystals and the absorption rate vary. Nonetheless, they can absorb oxygen at any temperature. To enhance the application performance of bulk YBCO materials, it is necessary to optimize their oxygen content. For example, in the case of $YBa_2Cu_3O_y$, when the oxygen content is $y \geq 6.7$ the $T_C$ value of the sample is above 90 K.

Several following preparation methods have been reported:

1. Melt texture of the top seed crystal

$REBa_2Cu_3O_{7-\delta}$ blocks can be obtained by using a top seed crystal guide and employing a melt texture method. In this approach, a variety of materials can be used as seed crystals, e.g., $MgO$, $CaNdAlO_4$, $SrLaGaO_4$, $SmBa_2Cu_3O_x$, and $NdBa_2CuO_x$. The basic requirements for seed crystal materials include: ① chemical compatibility with the YBCO system, i.e., they should not react with the YBCO system to produce heterophases, which are not conducive to the growth of texture blocks; ② the crystal structure and lattice constants must be similar, leading to effective YBCO growth; ③ the melting point should be higher than the peritectic temperature of YBCO (1015 °C).

The crystal structures of $SmBa_2Cu_3O_x$ and $NdBa_2Cu_3O_x$ are nearly consistent with that of YBCO, and the melting point ratio is 1060 °C and 1080 °C, respectively. Generally, the seed crystals are placed in the middle of the sample surface. During the melting process, YBCO nucleates and grows in an epitaxial manner on the seed crystal surface. The growth of the $ab$ surface is considerably larger than the direction of the $c$ axis due to the anisotropy of the growth rate. Consequently, the liquid phase Y along the $c$ axis is rapidly supersaturated. New RE123 crystal nuclei soon form and grow on the surface. This process is repeated until the growth of large YBCO single domains. The melting processing temperature should be as close to the peritectic temperature of the seed crystal as possible. If $SmBa_2Cu_3O_x$ is used as the seed crystal, the temperature is typically 1050 °C.

The hot seed crystal method involves complete decomposition of the 123 phase in the furnace, and then opening the furnace door to place the seed crystal. Compared with the cold seed crystal method, the time that the seed crystal is subjected to a high temperature is significantly shortened during processing to avoid melting and decomposition of the seed crystal. In addition to the cold and hot seed crystal methods, a compromise approach has also been proposed. In this method, the sample is heated to about 1100 °C, initially without the seed crystal. After the 123 phase is completely decomposed into a 211+ liquid phase, it is quickly cooled to room temperature. For melting processing, the seed crystal is subsequently placed in the furnace and rapidly heated to above the peritectic temperature. This ensures that the 123 phase is completely decomposed and prevents the initial melting of the seed crystal. Appropriate radial temperature gradients and a suitable cooling rate are important for the preparation of large single domains of YBCO.

2. Quenching and melting growth method

The quenching melting growth method and the melting powder growth approach are the key processes adopted by the Japanese Murakami team. They can be employed to improve the performance of YBCO by controlling the distribution of 211 grains above the peritectic reaction temperature in the liquid phase. The approaches initially involve the decomposition of $YBa_2Cu_3O_{7-\delta}$ into an $Y_2O_3$ + liquid phase. Subsequently, a 211+ liquid phase region is obtained by the reaction of $Y_2O_3$ with the liquid phase.

3. Powder melting method

The powder melting method was invented by Zhou L et al. It utilizes 211, $BaCuO_2$, and CuO to obtain the liquid phase region of 211+, which is different from those, achieved using other approaches. Rapid heating of a mixed powder directly into the 211+ liquid phase region, and then slow cooling or moving samples at a certain temperature results in the growth of the texture of YBCO samples. This method has two notable features: ① The size and distribution of the 211 particles can be easily controlled; ② low melting processing temperature, which leads to even distribution of 211 by ball milling or other methods. In addition to the convenience of this approach, the low processing temperature is more conducive to introducing fine 211 particles into the texture YBCO.

## 14.7 Second Generation (2G) High Temperature Superconducting (HTS) Wires

4. Liquid phase elimination method

Preventing the formation of excess liquid phase during peritectic synthesis is important for purifying grain boundaries. There are two ways to avoid excess liquid phase. The first and most commonly used approach involves the addition of excess 211 to the pioneer powder with a nominal composition of 123, so that the liquid phase is completely reacted1. The second method consists of using the substrate to absorb the excess liquid phase. The latter approach is employed in the liquid phase elimination method. Typically, the powder pressing pressure of 211 is used. To control the proportion of 211 captured in texture YBCO better, the porosity of the 211 powder pressing blocks must be considered. In addition, the contact area and interactions between the pressed blocks and 123 phases must be assessed.

5. Solid–liquid phase melting growth

The solid–liquid phase melting growth method involves direct application of the $Y_2O_3$+ liquid phase as the precursor to heat the pioneer blank to approximately 1050 °C. The melting processing temperature is reduced, and the 211+ liquid phases is formed. YBCO blocks prepared by this approach display the similar microstructures to those prepared by the melting powder method.

## 14.7 Second Generation (2G) High Temperature Superconducting (HTS) Wires

In 2015, a Rare Metal Express paper stated that first generation HTS strip (also called the bismuth band) made of superconducting bismuth strontium calcium copper oxide (BSCCO) powder in a metal pipeline moved towards commercialization. Nevertheless, the cost of this material is 100 times higher than that of a copper wire, which limits its widespread application on transmission line market. Hence, to the reduce cost and improve the performance, the development of a second generation HTS strip (also known as an yttrium strip) coated with a superconducting layer of yttrium barium copper oxide (YBCO) on a metal ribbon substrate has attracted considerable attention worldwide.

Power devices made of HTS wires have the advantages of high efficiency, compactness, and low environmental pollution. Consequently, they have become the basis for technological innovation in the areas of power grids, transportation, and material processing, among others. The application of HTS $YBa_2Cu_3O_x$ strip-coated conductors is particularly extensive. The costly technology for the preparation of high quality coating conductors restricts the industrial production and large-scale application of the materials. The most commonly applied superconducting coating is $YBa_2Cu_3O_7$. Superconducting layers are deposited or grown, and the crystal lattice YBCO in the final product is highly aligned to form a single crystalline coating. In this "coated conductor" wire structure, the thickness of the superconducting coating is in the micron order. During production, the higher the crystal arrangement, the higher the current carrying capacity and density is.

At the end of 2001, the realization of nano-scale polishing technology made it possible to directly grow the transition layer and YBaCuO superconducting layer in an epitaxial manner. Large scale continuous ion-assisted deposition (IAD) equipment was developed by Fujikura in 2002. It enabled the industrialization of the baseband buffer layer. In recent years, many companies and institutions in Europe, the United States, and Japan have conducted research on the development of second-generation HTS strips. The Thirteenth National Superconducting Academic Conference in 2015 reported that China produced a second-generation HTS strip of one-thousand-meter grade and with an annual output of 600 km. China is also the world's leader in the mass production of the second-generation HTS materials. The generation of high quality and inexpensive coating superconducting systems has become a hot topic in the research concerning the application of HTS strips.

Two kinds of widely used YBaCuO HTS strips currently exist. The first is a Ni alloy $\rightarrow$ ion beam assisted-deposition (IBAD)/or MgO $\rightarrow$ pulse laser deposition $CeO_2$ eYBaCuO, which was adopted by Fujikura and ISTEC. The second is a mechanism-assisted biaxial texture method used by the American Superconductor Corp. (AMSC), Ni alloy $\rightarrow$ pulsed laser deposition/$Y_2O_3$/YSZ/$CeO_2$ $\rightarrow$ metal organic matter deposition/YBaCuO. The former employs the IBAD technology and uses the orientation of the ion beam to prepare a textured buffer layer. The buffer layers as well as the superconducting layers are all formed by gas phase epitaxy. The latter approach utilizes the deformation texture to form a biaxially textured Ni alloy baseband. The buffer and superconducting layers are then formed by vapor phase epitaxy or solid phase backcasting, which adopts a non-vacuum preparation technology to reduce costs and expand application.

Further research and application of HTS strips are discussed below (Nekrasov et al. 2008; Zurek and Bi 2019).

1. Development profile

For a wide range of practical applications, such as superconducting strips and superconducting magnets, brittle YBaCuO high temperature superconducting oxide materials must first be coated on metal substrates with excellent mechanical properties (i.e., strength, toughness) to reduce or prevent mechanical damage during use. Moreover, the metal substrates must also exhibit sufficient oxidation resistance during the phase formation of YBaCuO superconducting coatings. It is also necessary for them to display good electrical and thermal conductivity to prevent system failure and collapse caused by local overshoot during use.

The layered structure of $YBa_2Cu_3O_x$, a HTS material, leads to extremely strong anisotropy. The current carrying capacity on the $ab$ surface is significantly higher than that in the $c$ axis direction. The material is also remarkably sensitive to lattice mismatch in the $a$, $b$ direction. The current carrying capacity exponentially decreases with the increase of the supporting lattice loss in the $a$, $b$ direction. Epitaxial weaving

# 14.7 Second Generation (2G) High Temperature Superconducting (HTS) Wires

is an indispensable process during the preparation of $a$, $b$ to reduce the supporting role of the lattice loss as well as to decrease the weak connection effect and current carrying capacity. At present, the most recognized substrate materials are Ni-based alloys. There is a certain lattice mismatch on the $ab$ surface of Ni-based alloys and YBaCuO HTS materials. The direct epitaxial growth on the baseband of Ni-based alloys is nearly impossible.

Moreover, during the phase forming heat treatment of YBaCuO, the occurrence of strong mutual diffusion and chemical reaction between Ni and YBaCuO seriously affects the superconducting properties of YBaCuO. Thus, a buffer layer material must be added between the Ni alloy substrate. The existing YBaCuO HTS-coated conductors all exhibit three-layered structures composed of the substrate, buffer layer (at least one layer), and YBaCuO superconducting coating.

A novel HTS wire was reported in 1991 by Fujikura (Japan). The HTS wire displayed a thin layer of a HTS YBCO material deposited on a soft-banded metal matrix. It showed good performance and exhibited a highly crystalline texture. Notably, it could be epitaxially grown to overcome the dislocation of adjacent grains of HTS and hinder the superconducting current. The better the texture or grain arrangement, the lower the grain dislocation and the greater the current flowing through. To obtain high current density, Fujikura adopted a strong long-base band with a highly textured interface and replicated the interface texture through epitaxial YBCO coating. It is the basic principle of coating conductors and second-generation HTS wires. During the process, the insulating material, namely yttrium-stabilized zirconium is deposited in a soft non-textured metal (usually a nickel alloy) belt under vacuum. The performance of Fujikura's initial experiments was not very good; however, Los Alamos subsequently optimized the process. In 1995, it was reported that the critical current density of superconducting samples reached 1 $MA/cm^2$ at 77 K.

There are several reasons for using this approach to develop second-generation conductors.

(1) The critical current density reaches up to 1 $MA/cm^2$. There is a 1 $\mu$m-thick YBCO layer on a 1 cm-wide narrow band, which can pass 100 A at 77 K, and the current is proportional to the bandwidth. Industrial standard for the characteristic performance of second-generation wires is equal to the current divided by the bandwidth, i.e., 100 A/cm. If the layer thickness is 3 $\mu$m and the current density is 3 $MA/cm^2$, the performance is enhanced.

(2) In a magnetic field, second-generation wires can achieve commercial electrical properties at higher temperatures.

(3) The AC loss of second-generation wires is lower. Moreover, it is easier to limit the fault current by certain forms.

2. Selection of baseband

At the beginning of the 1990s, Japan Sumitomo Electric invented the oriented baseband deposition process. It was found that at an angle to the metal surface, YSZ, a simple deposition material, could also produce sufficient texture without the assistance of any additional particles. Although the process is still conducted under vacuum, faster deposition rate significantly reduces the cost. A textured all-silver metal baseband was introduced by Japan's Hitachi and Toshiba. The employed method involved deformation processes, including rolling and recrystallization heat treatment, as well as deposition of YBCO directly on the top of the textured silver band to obtain a moderate current density.

Furthermore, in 1996, the Norton of America's Oak Ridge National Laboratory introduced an outer anti-flushing layer between the metal baseband and YBCO to enable the use of strong and inexpensive metals, such as nickel and nickel alloys. It was determined that following a large amount of deformation of cold rolling and appropriate heat treatment, the Ni metal formed a strong cubic texture. This approach is called the calendaring-assisted biaxial texture technology. The oxide transition layer, which matched the lattice well, was quickly epitaxially formed on the nickel baseband. The YBCO thin film was epitaxially grown on the metal ductile baseband containing a transition layer by vacuum sputtering, affording $J_C$ values of up to 3 × $10^5$ A/cm$^2$. This technique involving mechanical deformation and heat treatment can be applied on a large scale to meet the requirement of the ductile and long baseband for coated superconductors.

Although a highly pure Ni baseband can easily form a strong cubic texture using the above process, it is not the most suitable choice for coating superconducting basebands due to its intrinsic disadvantages, including ferromagnetism, low yield strength, and deep grain boundary after annealing. Consequently, to solve these challenges, various researchers proposed to fix a certain amount of alloy elements in Ni. Currently, the industrial production of Ni-5% (atomic fraction) W alloys has attracted significant attention worldwide. Based on the initial ingot preparation method, the processes predominantly include smelting and powder metallurgy. A study demonstrated that a uniform composition, single orientation, and flat surface cubic texture of the Ni-5% (atomic fraction) W alloy baseband could be obtained by controlling the melting process, particle size of powder metallurgy, cold rolling deformation as well as annealing temperature, time, and atmosphere. Alternatively, composite basebands have also attracted considerable interest. In general, the composite basebands are composed of Ni alloys, which easily form strong cubic textures, while the core layer is a strongly non-magnetic baseband. This method solves the composite problem of the two alloys and limits the influence of the interfacial reaction layer and its strength on the formation of strong recrystallized biaxial texture in the outer layer.

Several research groups in China have produced pure Ni and Ni-5% (atomic fraction) W alloy basebands. However, due to the limitation of the rolling mill, rolling environment, polishing technology, etc., the surface quality of the baseband as well as the uniformity of the cubic texture are significantly different from those of materials commercialized abroad. Hence, most research groups abroad buy Ni-5% W alloys with a short baseband. In China, Ni-5% W with short basebands, strong cubic textures, and high surface finish have been prepared by melting methods.

## 3. Buffer layer

In general, to achieve a good biaxial texture orientation, the method of depositing an YBCO superconducting film directly on a metal substrate is inadequate. It has been found that a buffer layer can be deposited on a metal substrate. The role of the buffer layer is to prevent the mutual diffusion of YBCO and the metal substrate as well as the occurrence of undesired chemical reactions, which are unfavorable to the superconducting properties. Additionally, a metal substrate with a lattice mismatch can be transformed to a lattice-matched substrate through the buffer layer. These roles are important to reduce the weak connections between the grains of YBCO superconducting films and to obtain superconducting films with a high transition temperature and critical current.

There are usually three routes to achieve the desired texture characteristics of the transition layer, namely ion beam-assisted deposition, inclined substrate deposition, and rolling-assisted biaxial texture process, which is a unique rolling annealing process used to obtain better texture features of the baseband itself. American Superconductor has adopted an improved technology of nano surface treatment to produce a buffer layer during continuous strip production. The above processes have enabled effective and continuous preparation of long strips outside of China.

## 4. Selection of the superconducting layer

Deposition of inexpensive superconducting layers can be performed by several relatively complex approaches. In the chemical sol–gel method, the precursor is dissolved in water or solvent to form a uniform solution. The hydrolysis or alcoholysis reaction then results in the formation of particles aggregated to approximately 1 nm. Subsequently, the sol isgenerated and the solved are evaporated to form a gel. Under a certain atmosphere, the gel can form the desired film following heat treatment. The metal–organic deposition (MOD) technology, application of the liquid phase technology, simplicity of the deposition process, full utilization of the precursor materials, and the relatively low cost of the furnace used to treat the precursor into a superconducting state make the coating of the second-generation wire considerably less costly than that obtained by a gas phase method. Ultrasonic atomization is also a cheap chemical method to prepare thin YBCO films.

Application of second-generation wires involves the use of the entire length of the materials, which exhibit strong superconducting properties and uniform mechanical characteristics. Most studies describe the multiplication of the current density of the second-generation wire by the deposition on both sides of the baseband. In the "central axis (NA)" approach, the superconductor coating is placed in the most favorable place to withstand bending deformation. Moreover, a thin layer of silver is deposited on the superconductor coating, forming a complex with the YBCO layer.

Some studies double the current using two layers of YBCO, which is a face-to-face double-sided structure. The middle of the solder displays a lower melting point than Cu–Ag welding and the thin copper surface of the two wires welded together. An intermediate stable layer can be used as a shunt between the YBCO layers when one of the layers is defective. In first-generation devices, the "form fit function" encouraged the user to accept the redesign of second-generation wires. The current focus is on expanding the scale of long-belt manufacturing, improving the production capacity and meeting the needs of commercialization. All HTS applications (e.g., coils and magnets) exhibit magnetic fields, the presence of which makes the performance of HTS wires more complex. The second-generation wires can increase their current density under magnetic field. This can be achieved by the superconducting flux wire pinning process. The diffusion distribution of very fine particles called nanodots can increase the current density under magnetic field. Nanodots consist of nano-scale $Y_2O_3$ or YBCO systems containing non-superconducting components.

Research concerning HTS coating conductors has also been conducted in China. Institutions, such as GRINM, Northwest Institute for Non-ferrous Metal Research, Institute of Physics of the Chinese Academy of Sciences, Tsinghua University, and others have made important progress by adopting different technological routes.

## 14.8 Application of Superconducting Materials

Superconducting materials can be widely used in the fields of energy, transportation, medicine, electronic communication, scientific instruments, mechanical processing, technological engineering, and national defense. HTS materials eliminate expensive liquid helium and can be used for the development of highly reliable and efficient commercial refrigeration systems. Other applications of HTS materials include transmission cables, high current leads, motors, current fault limiters, transformers, flywheel energy storage devices, magnetic levitation trains, etc. The area of HTS electronics is also developing rapidly. Studies on the utilization of HTS wires in domestic projects, such as transmission cables, high current leads, current fault limiters, magnetic resonance imaging, and magnetic levitation trains, have also been conducted.

The utilization of superconductivity can be divided into two categories, i.e., large-scale electric power (strong electricity) and small-scale electronic (weak electricity) applications. Some superconductivity applications are listed in a table. According to an analytical report on the development prospects and strategic investment planning, by 2020, the superconducting industry output is expected to reach $240 billion worldwide. Notably, HTS accounts for 60–70% of this value.

Some existing applications include research on ultra-high field magnets, chemical applications in medical biology, nuclear magnetic resonance spectrometer for chemistry, medical SQUID devices, superconducting magnet for thermonuclear fusion, superconducting magnets for particle accelerators, superconducting synchronous x-ray source, super magnetic separation, high-frequency resonator, superconducting transport system, superconducting magnetic bearings, and magnetic shielding.

The power applications include synchronous machine, synchronous generator, synchronous machine, dc machines, voltage transformer, power transmission, fault limiter, stored energy, a small fast motion smes system, and superconducting magnetic energy storage.

The weak current applications include superconducting quantum interference device, biological applications, nondestructive examination, geographical applications, gravity detection, single flux quantum, josephson voltage standard, information processing, logical processing of initial device, analog–digital converter, electromagnetic wave reception, thermal detection systems and antennas, and superconducting heterodyne receiver.

The following applications have different material requirements:

- Transmission and transmission cables
- HTS motors for ship propulsion
- Fault limiters
- Medical magnetic resonance imaging and nuclear magnetic resonance (NMR)
- Superconducting energy storage devices
- Superconducting magnetic levitation trains.

## 14.9    Research and Application of Rare Earth HTS

Chen et al. (2011, 2012) used the first principles approach combined with the evolutionary algorithm to predict the crystal structure of yttrium when the pressure was higher than 100 GPa. They found that when the superconducting turn temperature ($T_C$) decreased with the increase of pressure, yttrium underwent multiple structural phase transitions. The determination of the crystal structure of yttrium at higher pressure as well as evaluation of the ability to obtain a higher superconducting transition temperature by increasing the pressure remain the key research problems in the area of superconducting and high voltage materials (Freeman et al. 1987; Blackstead 1998; Blackstead and Dow 1998; Liu 2007, 2009; Mumtaz et al. 2021).

Japanese NEDO and Sumitomo Electric conducted an empirical test using a superconducting cable and a strong power supply in the presence of gadolinium, barium, and yttrium materials to produce superconducting cables. The overall energy consumption was only 1/3 of that of an ordinary metal cable. At present, there are plans to gradually replace aging metal cables with the newly developed superconducting ones. German researchers generated a superconductor-efficient wind turbine cooling technology and harnessed the potential of superconductors to produce wind energy, which was one of the research goals of EU-funded projects. A generator of this type can increase power to 10 mW, and reduce volume and mass. In addition, compared to the current permanent magnet wind motors, superconductor motors require less than 1% of its rare earth elements. Consequently, the use of superconductors makes fan equipment efficient and stable. Moreover, application of superconducting technologies saves raw materials, reduces construction, operation, and maintenance costs, and improves the motor service life.

A research team at the Japan Atomic Energy Agency (JAEA) found that yttrium and actinium compounds exhibited superconducting and magnetic properties. JAEA, the University of Tokyo, Kyoto Sangyo University, and University of Osaka used the large "Spring-8" radioluminescence facilities to detect rare earth metals. It was directly observed that rare earth metals and actinium compounds contain heavy electrons and displayed unique properties, including superconductivity. Furthermore, it was determined that compounds possessing heavy electrons have different types of superconducting states compared to metal superconductors. The above researchers also discovered a new HTS substance, an iron-containing compound LaOFeAs, the resistance of which was equal to zero at $-241$ °C.

# References

Blackstead HA (1998) Dependence of high-temperature superconductivity on rare-earth ions. J Appl Phys 83:1536

Blackstead HA, Dow JD (1998) Rare-earths as probes of high-temperature superconductivity. Springer

Chen Y, Hu QM, Yang R (2011) Phys Rev Lett 84:132101

Chen Y, Hu QM, Yang R (2012) Phys Rev Lett 109:157004

Freeman AJ, Evanston IL, Yu J (1987) High temperature superconductivity in transition metal oxides: electronic structure and charge transfer excitations. Mater Sci Forum 37:267–275

German researchers have developed cooling technology for efficient wind turbines with superconductors. http://www.cre.net/show.php?contentid=106829 [2013-02-04]

Japan has completed an empirical test of strong current supply for superconducting cables. http://www.cre.net/show.php?contentid=108968 [2013-06-08]

Kiyohiro K, Semiconductor ceramic and semiconductor ceramic element. https://patents.google.com/patent/US20140159191

Liu GH (2007) Rare earth materials. Chemical Industry Press, Beijing, pp 351–378

Liu XZ (2009) Rare earth fine chemical chemistry. Chemical Industry Press, Beijing, pp 153–166

Ma WW, Xie XS, Zhou YQ (2007) Physics. Version 5, Next volume. Higher Education Press, Beijing, pp 373–378

Mumtaz F, Nasir S, Jaffari GH, Shah SI (2021) Chemical pressure exerted by rare earth substitution in $BiFeO_3$: effect on crystal symmetry, band structure and magnetism. J. Alloys Compd

Nekrasov AI, Pchelkina ZV, Sadovskii MV (2008) High temperature superconductivity in transition metal oxypnictides: a rare-earth puzzle? JETP Lett

The presence of superconductivity and magnetism in rare earth metals was found in Japan. http://www.cre.net/show.php?contentid=256 [2007-08-14]

Zurek E, Bi T (2019) High-temperature superconductivity in alkaline and rare earth polyhydrides at high pressure: a theoretical perspective. J Chem Phys 150(5):050901

# Chapter 15
# Rare Earth Magnetic Materials

## 15.1 Brief Description

As early as the third century BC, the Song Dynasty of China used natural magnetite to make Si Nan (ancient compasses). When magnetized with steel needles, these became a key technology for navigation development.

The first generation of rare earth permanent magnets represented by $SmCo_5$ appeared in the 1960s. The second generation of 2:17-type samarium-cobalt rare earth permanent magnets represented by $Sm(Co, Cu, Fe, Zr)_z$ ($z = 6$–8) appeared in the 1970s. The third generation of rare earth permanent magnets represented by neodymium-iron-boron (Nd-Fe-B) was developed in the early 1980s. These magnetic materials are now broadly applied around the world.

At present, sintered Nd-Fe-B materials occupy the main market of Nd-Fe-B, accounting for ~ 94%. By 2016, the Nd-Fe-B production capacity in China accounted for 89% of global production and the product performance is stable.

In the twentieth century, at least 24 Nobel Prizes were awarded to scientists who made outstanding contributions to the field of magnetism. The study of modern magnetism in China dates back to 1924 when Yeh CS returned home from Harvard University. Under the guidance of Yeh CS, Shi R went to Yale University to study magnetism. In the 1950s, Pan XS, Guo YC and Dai LZ, who returned to China from abroad, organized the first batch of teams engaged in magnetic research within institutions, the education sector and businesses in China.

## 15.2 Magnetic Properties of Substances

Magnetism is one of the basic properties of matter. Magnetic phenomena are associated with various forms of charge motion. Because electron motion and spin inside a material produce a certain size of magnetic field, thereby inducing magnetic behavior, all materials are magnetic. According to the magnetic susceptibility of

© Science Press 2023

C. Wang, *Theory and Application of Rare Earth Materials*,
https://doi.org/10.1007/978-981-19-4178-8_15

matter, magnetic materials can be roughly divided into five categories: paramagnetic, diamagnetic, ferromagnetic, antiferromagnetic and ferrous magnetic materials. Among them, ferromagnetic and ferrimagnetic materials are classed as strong magnetic materials.

The technical magnetic parameters of permanent magnet materials are as follows:

- Magnetic field: a field that acts on moving charged particles.
- Flux density (B): a vector that gives a magnetic field at any point. The force $\vec{F}$ of the charge $Q$ moving at any velocity $\vec{V}$ at that point is equal to the vector product of velocity and flux density multiplied by the amount of charge.

$$\vec{F} = Q\vec{V} \times \vec{B}$$

- Flux ($\varphi$): area fraction of density.
- Magnetic moment ($m$): axial vector associated with a magnetic dipole.

$$\vec{m} = i\vec{A}$$

- Magnetization ($M$): axial vector associated with material volume and equal total magnetic moment of that volume divided by that volume. Saturation magnetization is the maximum magnetization.

$$\vec{M} = \frac{\sum \vec{m}}{V}$$

- Magnetic field intensity ($H$): axial vector related to the flux density at any point in a magnetic field.
- Magnetic constant ($\mu$): a constant with a value of:

$$4\pi \times 10^{-7} \, H/m$$

- Magnetoresistance ($R_\mathrm{m}$): ratio of magneto-magnetic potential to its associated magnetic flux.

$$R_\mathrm{m} = \frac{F_\mathrm{m}}{\phi}$$

- Magnetic susceptibility ($x$): a quantity equal to magnetic intensity multiplied by magnetic field intensity.

$$M = xH$$

- Magnetic domains: in magnetic materials, very small areas with spontaneous magnetization usually form that are substantially uniform in intensity and direction.

## 15.2 Magnetic Properties of Substances

- Curie point/temperature: a temperature value below which the material is ferromagnetic or ferromagnetic and above which the material is paramagnetic.
- Magnetostriction: a phenomenon of elastic deformation that changes with the state of magnetization in a material or object.
- Paramagnetism: in the absence of an external magnetic field, the magnetic moment is disordered and completely disoriented. When an external magnetic field is added, the magnetic moment acquires or tends to obtain the same direction as the external magnetic field.
- Paramagnetic materials: materials whose magnetic phenomena are predominantly paramagnetic.
- Ferromagnetic materials: materials whose magnetic phenomena are dominated by ferromagnetism.
- Magnetic anisotropy: a phenomenon in which the magnetic properties of a substance vary with respect to different directions of a given reference system.
- Magnet anisotropic or isotropic substances: materials that have or present no apparent magnetic anisotropy.
- Permanent magnet materials: magnetic materials with high coerciveness.
- Soft magnetic materials: materials with a coerciveness of less than 0.8 kA/m.
- Hysteresis: irreversible change in magnetic flux density or magnetization associated with a change in magnetic field intensity, independent of a change in velocity.
- Magnetization curve: a curve that is accompanied by a change in magnetic field, indicating the flux density, magnetic polarization and magnetization of the material.
- Static magnetization curves: magnetization curves that change at a slow rate with a minimal effect on the curve.
- Initial magnetization curve: a thermomagnetic neutral state of the material begins to be obtained by a monotone increasing magnetic field intensity from zero.
- Coerciveness: magnetic field strength with zero magnetic flux density.
- Saturation coerciveness ($H_{cb}$, $H_{ct}$ and $H_{cm}$): coerciveness obtained by monotonically varying the magnetic field from the magnetic saturation state in a coerciveness.
- Magnets: objects with external magnetic fields.
- Permanent magnets: magnets that do not require any power to maintain their magnetic field.
- Magnetization: induction of magnetization by an object.
- Magnetic circuit: a combination of media consisting mainly of magnetic materials to form a closed loop through which magnetic flux can pass.
- Dynamic magnetic energy product $(BH)_m$ (or magnetic energy product): useful magnetic energy that attracts heavy objects and is equal to the product of $B_2$ and $H_1$.
- Recovery permeability $\mu_{rec}$: permanent magnets at the working point. When subjected to an additional periodic negative magnetic field $H_a$, the magnet working point will change back and forth between point 1 and 2, and a small return line

called the return permeability is formed, which is expressed as $\mu_{rec} = \frac{B_a}{H_a}$. $\mu_{rec}$ is one of the important performance indexes of permanent magnet materials.

Characteristics of a magnetic field

a. A magnetic field generated around a cut-off direct wire. When a magnetic pole $m$ that is 1 W·m moves around a straight wire, the work done by the magnetic force is given by A:

$$A = \phi H_S ds = ni$$

where $H_S$ represents the magnetic field intensity on the path S, $ds$ is a small path on S, $\phi H_S ds$ represents the ring integral of the amount $\vec{H}$ of the magnetic field intensity along the path, n is the number of the wire and i is the current intensity of each wire.

b. A magnetic field is a continuous field of magnetic flux. Magnetic flux is a function of magnetic field intensity:

$$\phi = \sum Bn\Delta S$$

$$\phi = \iint Bn\Delta S$$

The normal component of magnetic induction intensity (projection of the magnetic induction intensity vector on the normal line at the two media interfaces) is equal, whether it is a homogeneous magnetic medium or an inhomogeneous one. This is determined by the fact that the magnetic inductance line is always a closed curve, which is called as continuity of magnetic flux.

## 15.3 Strong Magnetism and Ferromagnetism

The microscopic source of strong magnetism is the strong interaction between magnetic moments, which makes the atomic magnetic moments arrange in order. The main macroscopic characteristic of a strong magnetic material is that it does not exhibit strong magnetic properties without an external magnetic field. The atomic magnetic moments in a ferromagnetic material are arranged in parallel by the strong interaction inside the material, known as spontaneous magnetization, which can show strong magnetism.

In the demagnetization or neutral state, the atomic magnetic moments in a ferromagnetic material cannot be arranged in parallel in the same direction, the total free energy inside the material is kept at the lowest stable state and many tiny permanent magnets are formed that cancel each other out without showing macroscopic strong

## 15.3 Strong Magnetism and Ferromagnetism

magnetism outside the material. Atomic magnetic moments are arranged in parallel in small regions similar to small permanent magnets, i.e., magnetic domains.

As early as 1928, the theory of ferromagnetic quantum mechanics was put forward by Heisenberg. According to this theory, the electrostatic interaction energy between two electrons in A and B is:

$$E = K_e \pm J_{ex}$$

where:

$$K_e = \iint |\varphi_A(r_1)|^2 |\varphi_B(r_1)|^2 \frac{e^2}{r_{12}} dv_1 dv_2$$

$$J_{ex} = \iint \varphi_A^*(r_1)\varphi_B^*(r_2) \frac{e^2}{r_{12}} \varphi_A(r_2)\varphi_B(r_1) dv_1 dv_2$$

where

$K_e$—Coulomb integral energy;

$J_{ex}$—exchange integral energy;

$\varphi_A(r)$ and $\varphi_B(r)$—ground-state wavefunction of A and B atoms, respectively;

$r$—electronic coordinates;

$dv$—volume element near $r$ point;

$\varphi_A^*(r)$—conjugate function of $\varphi_A$.

The exchange integral energy is obtained from the position coordinates of the exchange electron, which can only be obtained from the theory of quantum mechanics. To satisfy the stability condition with minimum energy, if the exchange integral energy $J_{ex}$ is positive, the electron spin in A and B must be parallel, i.e., ferromagnetism is obtained. If the $J_{ex}$ is negative, the spin in A and B must be antiparallel, with antiferromagnetism obtained. In contrast, when the electron spin in A and B is not equal, but antiparallel, subferromagnetism is obtained. Therefore, from a microscopic point of view, subferromagnetism is a special case of antiferromagnetism. Heisenberg's spontaneous magnetization theory is to extend the hydrogen molecular model to a system of N atoms. The exchange energy is given by:

$$E_{ex} = -2\sum_{ij=1} A_{ij}\vec{\sigma}_i \cdot \vec{\sigma}_j \tag{15.1}$$

The exchange integrals of the electrons of A between adjacent atoms are assumed to equal in N atomic systems, i.e.:

$$A_{ij} = A$$

get,

$$E_{ex} = -2A \sum_{adjacent} \vec{\sigma}_i \cdot \vec{\sigma}_j \tag{15.2}$$

Because the exchange action only exists between neighbors, the exchange action of each atom can only be calculated once. Equation (15.2) sums the number of terms only to have 1/2NZ, where $Z$ is the nearest neighbor number of atoms, which can be written as:

$$E_{ex} = -2A \cdot \frac{1}{2} NZ |\vec{\sigma}_i \cdot \vec{\sigma}_j|_{average} \tag{15.3}$$

The electron spin angular momentum square of the total vector of the $N$ atomic system is given by:

$$\left(\sum_{i=1}^{N} \vec{\sigma}_i\right)^2 = \sum_{i=1}^{N} \vec{\sigma}_i^2 + \sum_{ij=1}^{N} \vec{\sigma}_i \vec{\sigma}_j \tag{15.4}$$

where $\left(\sum_{i=1}^{N} \vec{\sigma}_i\right)^2$ is the square of the total spin angular momentum of the $N$ atomic system.

If the total spin quantum number of this system is $S$, then:

$$\left(\sum_{i=1}^{N} \vec{\sigma}_i\right)^2 = S(S+1) \tag{15.5}$$

If the square of the electron spin angular momentum is $(\vec{\sigma}_i)^2 = S(S+1)$, then the square sum of the $N$ electron system spin angular momentum is:

$$\sum_{i=1}^{N} \sigma_i^2 = NS(S+1) \tag{15.6}$$

We can substitute Eqs. (15.5) and (15.6) into (15.4):

$$S(S+1) = NS(S+1) + \sum_{i \neq j=1}^{N} \vec{\sigma}_i \cdot \vec{\sigma}_j$$

so:

$$\sum_{i \neq j=1}^{N} \vec{\sigma}_i \cdot \vec{\sigma}_j = S(S+1) - NS(S+1) \tag{15.7}$$

In Eq. (15.7), each value may be different and we replace it with an average value:

15.4 Magnetic Origin of Rare Earth Elements

$$\sum_{i\neq j=1}^{N} \vec{\sigma}_i \cdot \vec{\sigma}_j = N(N-1)|\vec{\sigma}_i\vec{\sigma}_j|_{\text{average}}$$

That is:

$$\frac{|\vec{\sigma}_i\vec{\sigma}_j|}{\text{average}} = \frac{1}{N(N-1)}[S(S+1) - NS(S+1)] \qquad (15.8)$$

We then substitute Eq. (15.8) into (15.3):

$$E_{\text{ex}} = -2A\frac{1}{2}NZ\frac{1}{N(N-1)}[S(S+1) + NS(S+1)] \qquad (15.9)$$

The spin quantum number of an electron in Eq. (15.9) is $S = 1/2$:

$$E_{\text{ex}} = -\frac{ZA}{N-1}\left[S(S+1) - N\frac{3}{4}\right]$$

where
 $A$—exchange integral constant;
 $Z$—nearest neighbor atom number;
 $S$—total spin quantum number of $N$.

## 15.4 Magnetic Origin of Rare Earth Elements

The electronic effects of atoms, ions or molecules come from electron orbital and spin motion, so their magnetism is a combination of orbital and spin magnetism. Orbital magnetism is determined by orbital angular momentum and spin magnetism is produced by spin angular momentum. The magnetism of atoms or ions depends on their total orbital angular momentum $L$, total spin angular momentum $S$ and the total angular momentum $J$. Their magnetic moment is given by:
$$\mu = g\sqrt{J(J+1)}(B.M.)$$
where $g$ is the Lande factor, which has a value of:

$$g = 1 + \frac{J(J+1) + S(S+1) - L(L+1)}{2J(J+1)}$$

Expressed by the Bohr magneton B.M., $h$ is the Planck constant, $e$ is electron charge, $m$ is electron mass and $c$ as speed of light:

$$\text{B.M.} = \frac{eh}{4\pi mc} = 9.273 \times 10^{-21} \text{ erg/G} = 9.273 \times 10^{-24} \text{ Am}^2/\text{mol}$$

In some atoms or ions, when electron orbital-spin coupling is basically negligible, the effective magnetic moment of atoms or ions is expressed as:

$$\mu_{L+S} = \sqrt{L(L+1) + 4S(S+1)}(\text{B.M.})$$

For some transition elements in the d region of the periodic table, the effective magnetic moment of atoms or ions is more consistent with the pure spin magnetic moment:

$$\mu_S = \sqrt{4S(S+1)}(\text{B.M.})$$

The orbital-spin coupling of the atoms or ions of rare earth elements is larger. The calculated magnetic moments of rare earth ions and experimentally obtained magnetic moments are shown in a table. At room temperature, all atoms or ions are actually on the basis of multiple states, so their effective magnetic moments are given by:

$$\mu_{\text{eff}} = g\sqrt{J(J+1)}(\text{B.M.})$$

With the exception of $Sm^{3+}$ and $Eu^{3+}$, the measured data for rare earth elements are similar to those from the above formula.

1. Magnetic properties of rare earth ions

The electron configuration of rare earth ions determines their magnetic properties.

(1) With the exception of La, Lu, Sc and Y, rare earth ions contain single electrons, so they all have paramagnetism and most trivalent ions have larger magnetic moments than d transition elements ions.

(2) Unlike d transition element ions, $RE^{3+}$ magnetic moments depend on the magnitude of ground state J. In lanthanides, the magnetic moment varies with ground state J.

(3) The magnetic moments of trivalent rare earth ions in compounds are less affected by their environment and are basically close to the theoretical magnetic moments of trivalent ions. Since the single electron of rare earth ions is in the 4f shell of the inner layer of the ion, it is shielded by the $5s^2$ and $5p^6$ shell from the environment and is therefore less affected by it, so the magnetic moment of its compounds is consistent with the theoretical magnetic moment of trivalent ions.

The magnetic moment of non-trivalent rare earth ions are basically the same or close to that of isoelectronic trivalent ions. The magnetic moment of $Pr^{4+}$ is 2.48 B.M., similar to that of isoelectronic $Ce^{3+}$ (2.56 B.M.), but the magnetic moment of $Ce^{4+}$ is not zero as for $La^{3+}$, which is an exception. The magnetic moment or magnetic susceptibility of bivalent $Sm^{2+}$, $Eu^{2+}$ and $Yb^{2+}$ is similar to that of isoelectric $Eu^{3+}$, $Ga^{3+}$ and $Lu^{3+}$. The magnetic moment or magnetic susceptibility of divalent

$Sm^{2+}$, $Eu^{2+}$ and $Yb^{2+}$ is basically similar to those of ions with equal electrons. For example, under the SI international system of units, for $Eu^{3+}$ at 20 °C, the molar magnetization ($25,800 \times 4\pi \times 10^{-12}\chi m$) is close to the molar susceptibility ($25,700 \times 4\pi \times 10^{-12}\chi m$) of $Gd^{3+}$. The magnetic moment of $Yb^{2+}$ is similar to that of $Lu^{3+}$, approaching zero.

2. Magnetic properties of rare earth metals

Rare earth metals are paramagnetic at room temperature, where the La, Yb and Lu magnetic moments are less than one. As the temperature decreases, they change from paramagnetic to ferromagnetic or antiferromagnetic.

The magnetism of rare earth metals is mainly related to its unfilled 4f shell. In addition to the lanthanide 4f electrons in the inner layer, the $5d^1$ and $6s^2$ electrons in its metal state are conduction electrons, so the effective magnetic moment of most rare earth metals (except Sm, Eu and Yb) is almost the same as the trivalent ion magnetic moment that loses $5d^1$ and $6s^2$ electrons. Eu and Yb metals provide only two conducting electrons per atom to maintain a stable state of the 4f shell when semi or fully filled. A number of heavy rare earth metals (Tb, Dy, Ho, Er and Tm) change from diamagnetic to ferromagnetic at lower temperatures, while Gd changes from paramagnetic to ferromagnetic directly. The magnetic moments of adjacent metals are similar.

## 15.5 Magnetic Properties of Rare Earth Metals and 3d Transition Metal Compounds

Buschow (1977) did a survey of the physical properties, composition and crystal structure of intermetallic compounds formed between rare earth elements and 3d transition elements.

The magnetic moment of rare earth metal atoms is larger than that of 3d transition metal atoms, and has strong magnetocrystalline anisotropy and magnetostriction. However, the disadvantage of rare earth metals is that Curie temperature is too low, and chemical nature is very lively, so they cannot be directly used as magnetic materials. The researchers want rare earths to form stable metal compounds with other elements. Mössbauer spectroscopy was applied to investigate the relationship between $Y_2(Co, Fe)_{17}$ compound and iron atom lattice position (Wu and Zhuang 1990).

Rare earths and other metals can form various intermetallic compounds. Only rare earth and 3d metal (Mn, Fe, Co and Ni) with non-zero magnetic moments compounds have important magnetic properties, like $RECo_5$, which have been applied to a class of permanent magnetic materials in industry. They can form intermetallic compounds composed of $RE_mB_n$. The Sm-Co system forms seven compounds of different compositions and their magnetism varies according to the composition. The two series of compounds with $REB_5$ and $RE_2B_{17}$ are the most important and their magnetic characteristics are as follows (Li et al. 2009).

Shen (2017) emphasizes that it is of great significance to strengthen study of structure, phase transition, magnetism and magneto-thermal effects of rare earth-transition metal compounds, and there is also huge challenge.

## 15.6 Types of Rare Earth Permanent Magnetic Materials

Rare earth permanent magnet materials were discovered in the 1960s (Strnat 1970). So far, two kinds and three generations of rare earth permanent magnet materials with large-scale production and practical value have been established. The first category is Sm-Co permanent magnets, which also include two generations, namely, the first generation of rare earth permanent magnets 1:5 type $SmCo_5$ permanent magnets and the second-generation 2:17 type Sm-Co alloys, all of which are based on metal cobalt. The second category is RE-Fe-B or iron-based rare earth permanent magnets and the third generation is Fe-based rare earth alloys represented by NdFeB alloys (Knoch 1996; Liu 2007).

1. First-generation are earth permanent magnet

The first-generation rare earth permanent magnet $SmCo_5$ was first reported in 1967 and is a binary intermetallic compound. It consists of a RE-Co permanent magnet composed of RE atoms and other metal atoms in a ratio of 1:5. The melting point of $SmCo_5$ is 1350 °C and it can be defined as single or multiphase. Single phase refers to a $RECo_5$ permanent magnet with a single compound in principle from a magnetic phase. A multiphase 1:5 type Sm-Co permanent magnet material refers to the 1:5 phase as the matrix. The permanent magnet material usually contains a small amount of 2:17 type and has Cu included.

The first-generation rare earth permanent magnet alloy $SmCo_5$ possesses a $CaCu_5$-type structure, which is a hexagonal crystal system, as shown in a diagram, that is, $CaCu_5$ crystal structure ($RECo_5$), including (a) spatial diagram of crystal structure, (b) $CaCu_5$ structure cellular diagram, and (c) $CaCu_5$ structure atom projection. The rare earth occupies a crystal sites and cobalt occupies c and g crystal sites. This structure can be seen as an alternating stacking of two atomic layers, where one layer consists of rare earth and cobalt atoms (A layer) and the other consists of cobalt atoms (B layer), with ABAB stacking. There are $RECo_5$ phases in the $CaCu_5$ structure of RE-3d series compounds. The$REFe_5$ phase does not exist.

2. Second-generation rare earth permanent magnet

Second-generation rare earth permanent magnet alloy 2:17-type magnetic phase compounds have the $Th_2Ni_{17}$ crystal structure at high temperature and transform into the isomeric $Th_2Zn_{17}$ structure at low temperature. The spatial stereogram of the $Th_2Ni_{17}$ structure belongs to a hexagonal crystal system shown in a diagram ($Th_2Ni_{17}$ hexagonal crystal structure), where a single cell is composed of two $Th_2Ni_{17}$ molecules, in which the rare earth occupies b and d crystal sites, cobalt or iron occupies g, k, f and j crystal sites and 4f is a dumbbell crystal sites, which is the replacement

## 15.6 Types of Rare Earth Permanent Magnetic Materials

result of rare earth atoms on the $CaCu_5$-type structure c crystal sites by cobalt atom pairs.

The 2:17 series compounds of RE-cobalt and RE-iron have the $Th_2Zn_{17}$ structure at low temperature and its spatial stereogram is shown in another diagram ($Th_2Ni_{17}$ rhombohedral crystal structure), which belongs to the rhombohedral system (or triangular system). One unit cell contains three $Th_2Zn_{17}$ molecules, in which rare earths occupy c crystal sites and cobalt or iron occupy the d, f, h and c crystal sites.

3. Third-generation rare earth permanent magnet

The spatial structure of the third-generation rare earth permanent magnet alloy magnetic phase $RE_2Fe_{14}B$ compound belongs to a tetragonal crystal system. It is made up of four $RE_2Fe_{14}B$ molecules, with 68 atoms in a cell (eight rare earth, 56 iron and four boron atoms). Rare earth atoms occupy the 4f and 4g crystal sites, boron is at the 4g crystal sites and iron occupies six different crystal sites at 4c, 4e, $8j_1$, $8j_2$, $16k_1$ and $16k_2$. All rare earth elements with iron and boron can form $RE_2Fe_{14}B$ compounds, in which $RE_2Fe_{14}B$, $Pr_2Fe_{14}B$, $(Pr_1Nd)_2Fe_{14}B$ or $(Ca, Pr, Nd)_2Fe_{14}B$ can be made into practical permanent magnet materials. The addition of metal boron and other elements plays a decisive role in the formation of the tetragonal phase $Nd_2Fe_{14}B$.

Experimental results show that boron-free Nd-Fe alloys consist of $\alpha$-Fe and $Nd_2Fe_{17}$ phases. When the boron mole fraction increases to 4%, the $Nd_2Fe_{17}$ phase disappears and the $Nd_2Fe_{14}B$ phase begins to appear. When this value increases to 7%, the $\alpha$-Fe phase disappears (cooling sample) and the alloy consists of $Nd_2Fe_{14}B$ and neodymium- and boron-rich phases. Carbon substituted boron atoms can form $Nd_2Fe_{14}C$ compounds and their crystal structure is the same as that of $Nd_2Fe_{14}B$, but it has high intrinsic magnetic properties, form and decompositions at high temperature are difficult to work with, so it cannot be used to prepare significant permanent magnetic materials. Although silicon and carbon have the same atomic valence, the atomic radius of silicon (0.1316 nm) is much larger than that of boron (0.098 nm) and can only replace the crystal position of iron.

4. $RE_2Fe_{17}N_x$ and $REFe_{12-x}M_xN_y$ interstitial compounds

$RE_2Fe_{17}N_x$ and $REFe_{12-x}M_xN_y$ interstitial compounds are part of a new family of magnetic materials. $RE_2Fe_{17}$ and $REFe_{12}$ absorb nitrogen to form a class of interstitial permanent magnetic compounds containing nitrogen. The crystal structure of compounds such as $RE_2Fe_{17}$ can be regarded as derived from the $CaCu_5$-type structure. The common characteristic of crystal structure derivation is that some calcium atoms in the $CaCu_5$-type structure are replaced by iron atom pairs parallel to the c axis (dumbbell atom pairs) and their replacement relations can be expressed as:

$$RE_{1-\delta}M_{2\delta+5} = REM_Z \quad Z = (2\delta + 5)/(1 - \delta)$$

When $\delta = 0$, $Z = 5$, i.e., the 1:5 structure, when $\delta = 1/3$, $Z = 8.5$, 1/3 of the calcium atoms in the $CaCu_5$ structure is replaced by atomic pairs of transition metal elements M, which forms the 2:17 type structure. Others such as $\delta = 1/2$, $Z = 12:00$, form the 1:12 structure; $\delta = 2/5$, $Z = 9.6$, form the 3:29 structure. $RE_2Fe_{17}$ forms the $RE_2Fe_{17}N_x$ series of compounds after nitrogen absorption. Nitrogen exists in the $RE_2Fe_{17}$ lattice as an interstitial atom and the cell volume expansion is 6–7%, but the $RE_2Fe_{17}N_x$ crystal structure is still the same as that of $RE_2Fe_{17}$, which is rhombohedral and hexagonal.

In the $RE_2Fe_{17}$ phase, adding M plays the role of stabilizing the phase structure. $REFe_{12-x}M_xN_y$ interstitial compounds are formed after nitrogen absorption and their structure is of the $ThMn_{12}$ type. There is a diagram to show $YFe_{11}TiN_{0.51}$ compound unit cell structure, in the $YFe_{11}TiN_y$ structure, the rare earth occupies 2a crystal sites, iron and titanium occupy 8f, 8i and 8j crystal sites and nitrogen occupies 2b crystal sites. The 3:29 type compound also requires the addition of a third component on the basis of the RE-Fe binary system and forms the $RE(Fe_{1-x}M_x)_{29}N_y$ interstitial compound with a monoclinic crystal structure after nitrogen absorption.

Yang YC of Peking University began to study the magnetic properties of rare earth alloys in the 1970s. He explored the physical connotation of the Sm-Co alloy and extended the research to 3d (transition metal)-4f (rare earth metal) intermetallic compounds, confirming that iron is the best carrier for 3d electrons. A new rare earth iron phase with the $ThMn_{12}$ structure was synthesized in 1980, which has become an important series of rare earth permanent magnet materials that has also been extended to develop new stealth absorbing materials.

In 1990, his research group discovered the interstitial atomic effect of nitrogen in rare earth ferroalloys and new materials, such as neodymium-iron–nitrogen and praseodymium-iron–nitrogen. The team calculated the crystal field action of nitride rare earth 4f electrons and iron 3d electron band structure based on crystal diffraction and data. The physical roots of interstitial atomic effects of nitrogen in these alloys are theoretically elucidated and thus a new family was found in rare earth magnetic materials, i.e., rare earth interstitial compounds.

## 15.7 Phase Diagram of Rare Earth Permanent Magnet Materials

Phase diagrams are the basis for understanding phase relationships, changes of tissue structure and material compositions, and are also an important basis for improving the properties of materials and exploring the best preparation process of materials. There are a total of more than 1000 binary phase diagrams between rare earth metals (except Pm) and ~ 60 other elements in the periodic table (except for the super-uranium elements, inert gases and other extremely rare elements) (Department of Metals, Sun Yat-sen University 1978). More than 600 binary phase diagrams have been determined experimentally.

The binary phase diagrams between rare earth metals and other elements can be divided into four categories.

15.7 Phase Diagram of Rare Earth Permanent Magnet Materials 261

- Infinite solid solutions: including La-Ce, Pr-Nd, Gd-Tb, Gd-Er, Gd-Y, Eu-Gd, Ce-Tb and Sc-Hf.
- The binary phase diagram of Y, Gd, Er, Ti, Zr, Y, Er and H. It is the diagram of IV and VI elements and rare earth metals.
- Phase diagram of N, Ta and Nb elements and rare earth metals, partial crystalline rare earth metals. There is little solubility in both the liquid and solid states.
- A rare earth metal phase diagram for the formation of compounds. For example, Cu, Ag and Au in IB; Zn and Cd in IIB; Si, Sn, Pb and Ge in IVA; As, Sb and Bi in VA; O, S, Se and Te in VIA; Mn and Re in VIIB.

The further away from rare earth elements in the periodic table, the more compounds are formed with rare earth metals, the easier to form intermetallic compounds and the better their stability. Elements adjacent to rare earths do not form compounds with rare earths.

Rare earth elements with the IVA family of metals can form six to seven compounds. Be and Mg in the IIA family can form up to five compounds. Mn in the VIIB group and rare earth elements can form up to three compounds. In the same family, the number of compounds formed with rare earth elements decreases with increasing atomic number. In the same cycle, the number shows an increased trend.

1. Rare earth-cobalt binary system phase diagram and compounds

Several binary phase diagrams of rare earths and Co, taking La-Co, Ce-Co, Pr-Co, Nd-Co and Sm-Co binary phase diagrams as examples, are shown in a series of diagrams, and illustrated in a table that items include rare earth element, number of compounds formed, and compound formula and the ratio of rare earth metal atom to Co atom (3:1, x:y, 2:1.7, 2:3, 1:2, 1:3, 2:7, 5:19, 1:5, and 2:17), with the Gd-Co binary phase diagram as a comparison.

The result shows:

- Most compounds in the RE-Co system have the same molecular formula;
- Compound numbers formed by RE and Co were seven to 10, except for six in the La-Co series.

The crystal structure parameters of RE and Co compounds is summarized in a table, which shows that compounds with the same molecular formula have almost the same crystal structure. The items include intermetallic compounds, crystal symmetry, structure type, space group, and lattice constant/Å. For instance, to $RE_3Co$ compounds, taking its intermetallic compounds $La_3Co$ as an example, the crystal symmetry is orthogonal, structure type $Fe_3C$, space group is *Pnma*, and lattice constant/Å is $a = 7.277$, $b = 10.020$, $c = 6.575$.

For example, Pan (2011) and his group showed that when a $SmCo_5$ sample was heated to 400–750 °C in the JEM-1000 ultra-high pressure electron microscope, it precipitated $Sm_2Co_{17}$ and $SmCo_7$ and eutectoid decomposition occurs 750 °C. The new phase indices are:

$$3SmCo_5 - Co + Sm \rightarrow 2Sm_2Co_7$$

$$3RECo_5 - RE + 2Co \rightarrow RE_2Co_{17}$$

$$3SmCo_5 - Sm + 2Co \rightarrow Sm_2Co_{17}$$

2. Phase diagram of Sm-Co–Cu system

To study the Sm-Co–Cu ternary system phase diagram, we briefly introduce the Sm-Cu and Co–Cu binary system phase diagrams. Another one shows the Co–Cu binary system phase diagram. $SmCo_5$ and $SmCu_5$ are fully fused within 800–1200 °C. The study shows that there are 2:17 and 1:5 phases inside the as-cast $Sm(Co_{0.84}Cu_{0.16})_{6.9}$ alloy, the matrix is 2:17 RE phase and 1:5 is the precipitated phase. The as-cast $Sm(Co_{0.87}Cu_{0.13})_{7.8}$ alloy has the characteristics of a soft magnetic field, where the coerciveness reached a maximum after solution treatment at 800 °C with tempering for 4 h.

3. Phases of Sm-Co–Cu-Fe system

In the Sm-Co–Cu system, Fe can be used to replace the expensive and low-volume Co, the compound composition is $Sm(Co_{1-x-y}Cu_xFe_y)_z$, but the quantity of Fe replacing Co cannot be too much. If the amount of Fe substituting Co is greater than 10%, there is microstructure in the alloy and an Fe-Co soft magnetic phase occurs, which makes the coercivity decrease sharply. In addition, replacing part of the Sm with a small amount of Ce can play the role of thinning the cellular tissue.

4. Sm-Co–Cu-Fe system 2:17-type analysis

The Sm-Co–Cu-Fe-M permanent magnet, where M is Zr, Ti or Ni, with the best alloy properties is the Sm-Co–Cu-Fe-Zr 2:17 alloy. These alloys are now commercialized. The alloy composition determines the phase transition of 2:17-type permanent magnets at various temperatures.

- High coercivity can be obtained when Sm is at 25.5% (mass fraction).
- In the $Sm(Co_{0.675-x}Cu_{0.078}Fe_{0.22+x}Zr_{0.027})_{8.22}$ alloy, the magnetic properties increase with increasing Fe content, which should also increase the Zr content.
- The development direction of the 2:17 type is high Fe and low Cu.
- There is diagram to show the Sm-Co-7.0%Cu-22.0%Fe (Cu < Fe) longitudinal section phase diagram.
- There is diagram to show the Sm-Co-7.0%Cu-22.0%Fe-2.0%Zr phase diagram of longitudinal section (Cu < Fe, containing Zr).
- Phase diagram and compounds of RE-Fe binary system

Understanding the RE-Fe binary phase diagram and compounds is good for understanding the relationship between properties of iron-containing permanent magnets and the content of each element, as shown in a series of diagrams, including La-Fe, Sm-Fe, Nd-Fe, Ce-Fe, and Fe-B binary phase diagram.

15.8 Application of Alloy Phase Transition

As can be seen from the figures, La-Fe system does not form compounds; Sm-Fe forms three compounds, namely $SmFe_2$, $SmFe_3$, $Sm_2Fe_{17}$; Nd-Fe forms $Fe_{17}Nb_2$ and $Fe_2Nd$ compounds; Ce-Fe system forms $CeFe_2$ and $Ce_2Fe_{17}$ compounds; B-Fe system forms $Fe_2B$ and FeB compounds.

6. Phase diagram and compounds of Nd-Fe-B ternary system

The ternary system phase diagram concisely describes the alloy state under equilibrium conditions (Sagawa et al. 1987). The diagram of Nd-Fe-Bernary system (B $\leq$ 50%) at room temperature section shows a room temperature cross section of Nd-Fe-B (B < 50%, atomic fraction) by Che GC team's research, indicating that there are three ternary compounds in Nd-Fe-B: $Nd_2Fe_{14}B$, $Nd_8Fe_{27}B_{24}$ and $Nd_2FeB_3$. However, these two compounds are different from the literature results, which may be related to different temperatures and experimental conditions. There are ten parts (from I to X) in the diagram, that is:

I.     $\alpha$-Fe + $Nd_2Fe_{17}$ + $Nd_2Fe_{14}B$;
II.    $Nd_2Fe_{14}B$ + $Nd_2Fe_{17}$ + Nd;
III.   $Nd_2Fe_{14}B$ + $Nd_8Fe_{27}B_{24}$ + Nd;
IV.   $Nd_2Fe_{14}B$ + $Nd_8Fe_{27}B_{24}$ + $\alpha$-Fe;
V.     $Nd_8Fe_{27}B_{24}$ + $Fe_2B$ + $\alpha$-Fe;
VI.   $Fe_2B$ + FeB + $NdB_4$;
VII.   $Fe_2B$ + $Nd_8Fe_{27}B_{24}$ + $NdB_4$;
VIII. $Nd_2FeB_3$ + $NdB_4$ + $Nd_8Fe_{27}B_{24}$;
IX.   $Nd_8Fe_{27}B_{24}$ + $Nd_2FeB_3$ + Nd;
X.     Nd + $Nd_2FeB_3$ + $Nd_2B_5$.

There is diagram to show the Nb-Fe-B system stereo phase diagram, which is a rich-Fe ternary system; another one shows the projection of the liquid phase line of the Nb-Fe-B ternary system phase diagram with a B atomic fraction less than 35%.

To obtain high performance RE-Fe-B permanent magnets, the Fe content should be more than 80%, B less than 6% and Nd should be 13%.

## 15.8 Application of Alloy Phase Transition

In the physicochemical concept, the phase is any homogeneous part of the system with an interface between phases. The difference between alloy phases is that their crystal structures are different, i.e., the arrangement of atoms is different. The microstructure contains phase and interphase interfaces. The phase change is from the parent phase to the new phase. The first-order phase transition, such as the transition between the $\alpha$ and $\beta$ phases, in which the chemical phase is equal but the first-order partial derivative is not equal, i.e.:

$$\mu^\alpha = \mu^\beta$$

$$\left(\frac{\partial \mu^{\alpha}}{\partial T}\right)_{P} \neq \left(\frac{\partial \mu^{\beta}}{\partial T}\right)_{P}$$

$$\left(\frac{\partial \mu^{\alpha}}{\partial P}\right)_{T} \neq \left(\frac{\partial \mu^{\beta}}{\partial P}\right)_{T}$$

Because of the thermodynamic function relation:

$$\left(\frac{\partial \mu}{\partial P}\right)_{T} = V$$

$$\left(\frac{\partial \mu}{\partial T}\right)_{P} = -S$$

then:

$$V^{\alpha} \neq V^{\beta}$$

$$S^{\alpha} \neq S^{\beta}$$

Most common phase transitions, such as the desolvation of a solid solution, the martensite phase transition, and the crystallization of metals and alloys, all belong to the first-order phase transition. The Clapeyron equation describes the relationship between temperature and pressure at such phase change points, as shown in a diagram (free enthalpy, entropy and volume changes during first order phase transition).

The second-order phase transition is characterized by no entropy and volume change at constant temperature and constant pressure, but the free enthalpy second derivative has discontinuous change. In a magnetic field is order–disorder phase transition and ferromagnetic–paramagnetic phase transition (free enthalpy, entropy and volume changes during second order phase transition).

The third phase transition is a phase transition with an unequal partial derivative of the third order.

The application of rare earth permanent magnet alloys of high temperature phase change:

Pan SM studied systematically the phase transition process of rare earth permanent magnet alloy from room temperature to 960 °C using a JEM1000 kV ultra-high pressure electron microscope. The result showed that alloys are multiple element recombination (a certain range of components). At high temperature, the atomic arrangement is disordered, while at low temperature, it presents an ordered state. This disorder-ordered transition is reversible and can improve magnet performance by technics.

Different rare earth permanent magnet alloys have different aging processes. $Nd_{15}Fe_{77}B_8$, for example, obtains permanent magnets with $(BH)_{max} > 286.6$ kJ/m$^3$ by sintering for 1–3 h at 1080–1100 °C, then 1–2 at 600 °C; $Nd_{14}DyFe_{74}AlNb_2B_8$

needs 1100 °C and sintering for 1–3 h, after 900–950 °C, 1–1.5 h and then 550–650 °C, 1–1.5 h to room temperature, the more coercive rare earth permanent magnet alloy can be obtained. The Sm $(Co, Cu, Fe, Zr)_{7.4}$ permanent magnet alloy phase transition belongs to diffusion phase transition and its manufacture process is based on the abovementioned high temperature phase change.

## 15.9 Preparation of Rare Earth Permanent Magnet Materials

Sintered or bonded rare earth permanent magnet materials are prepared by powder metallurgy. First, it is necessary to melt the parent or single alloy with a certain chemical composition. There are different processes for the preparation of rare earth permanent magnet alloys, including vacuum induction melting, vacuum arc melting, vacuum thermal reduction diffusion and so on. The vacuum induction melting method is the most widely used and the vacuum thermal reduction diffusion method has been developed in recent years. Vacuum arc melting method is mostly used in laboratory research. The alloy obtained by smelting can be directly cast or alloy ingot with certain crystalline structure or it can be alloyed or controlled by a mechanical alloying method, melt fast quenching method or gas atomization method (Shi 2007; Hajime et al. 2016).

The main preparation methods include vacuum induction melting, vacuum arc melting process, vacuum thermal reduction diffusion, mechanical alloying, rapid melt quenching, gas atomization, and characteristics (Nakamura et al. 2012). Their corresponding characteristics of alloy microstructure are macro-segregation in slow cooling; approach to balance phase, less segregation; when heat diffusion is insufficient, the composition is not uniform; nanocrystalline structure, amorphous phase, nonequilibrium phase; amorphous phase, nonequilibrium phase, microcrystalline grain, less segregation; and amorphous phase, nonequilibrium phase, microcrystalline grain, less segregation. The features low price, suitable for mass production, suitable for experiment and small production, no crushing, low cost, powder raw material, low temperature treatment, easily crushed, and spherical powder, not crushed.

The melting of rare earth permanent magnet alloys means that the rare earth metal, iron, cobalt, aluminum, copper, zirconium, boron and so on, which are well weighed according to the formula, are placed in the crucible of intermediate frequency induction furnace, vacuumed (up to $10^{-3}$–$10^{-2}$ Pa), filled with argon to protect, avoid oxidation and nitriding, and then high power transmission, so that the furnace material can melt quickly to reduce the volatilization of volatile metals. After all furnace materials are melted, high power electromagnetic stirring is used for a little time to ensure uniform composition of alloy, according to formula composition, decide whether to use the refining process.

The rare earth Fe-based permanent magnet alloy (Pr, Nb)-Fe(M)-B system, rapid solidification technology has been developed since 1998 to make the thick band or squama ingot, so as to achieve the purpose of manufacturing high performance sintered rare earth iron-based permanent magnet alloys. The technology and principle of manufacturing scale ingot are shown in a diagram of rapid solidification of thick band or scale, in it, 1-medium frequency power supply, 2-vacuum unit, 3-nozzle, 4-tundish, 5-crucible, 6-cooling roller wheel, 7-infrared camera, 8-scale thick belt, 9-thick belt cooling slot. Its main equipment is an intermediate frequency induction power supply, tundish, cooling roller and vacuum unit.

## 15.10 Application of Rare Earth Permanent Magnet Materials

Half of the production of rare earth permanent magnet materials is used to make various permanent magnet motors. According to AC and DC classification, permanent magnet motors can be divided into more than 10 kinds of motor names. Permanent magnet AC motors include synchronous generators, AC speed measuring motors, induction generators, ignition magnet motors, synchronous motors and so on. Permanent magnet DC motors includes DC servo motors, DC speed measuring motors, DC torque motors, stepping motors, DC brushless motors, piezoelectric motors, Hall motors, limited angle motors, sound coil motors and so on.

The emergence of the third-generation rare earth permanent magnet-rare earth iron-based permanent magnet materials promotes the development of permanent magnet motors. Because of the high magnetic properties of third-generation rare earth permanent magnets, the motor has the advantages of high power, high efficiency, smaller volume and lighter mass. The rare earth iron-based permanent magnet motor has no excitation loss, then the efficiency is higher than that of the general motor, and the magnet itself does not heat. Under the same heating conditions, the output ratio of the permanent magnet motor made of rare earth permanent magnet material is large and the efficiency is high.

At present, most of magnets in high efficiency small motors use rare earth permanent magnets, this kind of motors are widely used in automobile, which promotes the development of rare earth permanent magnet motor, as shown in a diagram of permanent magnet devices for ordinary cars, a common car needs 25 permanent magnet devices. A modern car has at least 80 places that use rare earth permanent magnets.

These components include 1-door lock motor, 2-tailgate motor, 3-fogging motor, 4-car horn, 5-auto temperature control, 6-shade motor, 7-magnetic tape motor, 8-windscreen cleaning pump, 9-speedometer, 10-uhf control, 11-level indicating controller, 12-saving and pollution control, 13-front light door motor, 14-ignition system, 15-coolant fan motor, 16-starting dynamo, 17-windscreen wiper motor, 18-wind door and crankshaft position sensor, 19-heating regulator, 20-antenna lift motor,

21-seat adjustment motor, 22-debris collect, 23-oil pump motor, 24-windows motor, and 25-lubrication device.

In addition, computers and computer peripherals, magnetic separation equipment, magnetic cooler, earphone, hearing aid, voice coil motor, magnetic vibrator, vibration motor, magnetic control tube of radar, communication equipment, magnetic levitation, nuclear magnetic resonance device, magnetic clutch, magnetic control tube, electronic gun control, magnetic therapy device, and all fields of defense technology need permanent magnetic materials (Velicescu 1999; Zepf 2013; Zhao and Guo 2013).

## 15.11 Rare Earth Super Magnetostrictive Materials

Ferromagnetic materials, due to the change of magnetization state, its length and volume will change slightly, this phenomenon is called magnetostriction, which refers to the conversion effect of magnetic energy and mechanical energy existing between magnetic state and mechanical form. The magnetostrictive effect was discovered by Joule in 1842, called Joule effect. Rare earth super magnetostrictive materials mainly refer to RE-Fe intermetallic compounds. This kind of material has fast mechanical response, and the power density is high, so it can efficiently convert electric energy into mechanical energy, or conversely (Liu 2009).

1. Magnetostriction actions of RE-iron system

   - When the lattice of the material is distorted, the exchange energy also changes, which can be isotropic or anisotropic.
   - Interaction between magnetic dipole moments of atoms can also cause magnetostriction.
   - Magnetostriction caused by the interaction of atomic orbitals and crystal fields and spin–orbit interaction, which is anisotropic and the main cause of large magnetostriction. Among rare earth metals, alloys or intermetallic compounds, the large magnetostriction is mainly caused by localized 4f electrons in rare earth ions.

2. Eigenvalues of magnetostrictive materials

   - Saturation magnetostrictive coefficient ($\lambda$s)
   - Sensitivity constant (d)
   - Magnetic-elastic coupling coefficient (k).

3. Some rare earth super magnetostrictive materials

   In the early 1970s, Clark AE of the U.S. Naval Surface Warfare Center began looking for materials that still have large magnetostriction at room temperature. He systematically studied the magnetostrictive properties of RE-Co, RE-Ni, RE-Fe series of compounds and alloys, and found that $TbFe_2$ is the material with maximum magnetostrictive at room temperature. Magnetic crystal anisotropy of important binary REFe$_2$ compounds is too large, which makes the practical application of materials difficult. Clark et al. used materials with the same magnetostrictive symbols and opposite anisotropic constant symbols to form a compensated binary REFe$_2$ system.

4. Preparation of RE ultra magnetostrictive materials

The preparation of polycrystalline supermagnetic telescopic materials is generally arc furnace melting, casting and then at high temperature (800–1000 °C) for a long period of about one week annealing.

5. Application of rare earth super magnetostrictive materials

Rare earth ultra magnetostrictive materials are widely used in sonar systems, high-power ultrasonic devices, precision control systems, various valves, drives and robots (Oettinger 2015), for example:

- Rare earth ultra magnetostrictive material brakes: The brake is a stepping motor for micrometers or finer positioning; realizable resolution with constant temperature control, get displacement resolution of 10 nm, high energy density; low voltage, fast response, wide use temperature range, extremely stable.
- Sound signal generator: Sound signal is the main means of underwater communication, detection, reconnaissance and remote control. The core components of sonar device are piezoelectric material and magnetostrictive material. Magnetic restrictive materials Tb$_{0.3}$dy$_{0.7}$Fe$_2$ are widely used in sonar system; the super magnetostrictive material is made into a high power ultrasonic transducer for ultrasonic cleaning, processing and dispersion.
- Electrical–mechanical transducers: widely used in automobile, aircraft, spacecraft, robot, precision machine tool, precision instrument, computer, optical communication.
- Physical sensors and electronic devices: rare earth supermagnetostrictive materials can make pocket magnetometer; physical sensors for detecting displacement, force and acceleration, as well as adjustable surface acoustic wave components, can be designed and manufactured; as well tunable acoustic devices, sonar phase sensors, oscillators.

## 15.12 Magnetic Refrigeration and Magnetic Storage Materials

Magnetic refrigeration system is by adding magnetic field to magnetic material, through heat exchange to achieve refrigeration purpose. Compared with gas freezing, magnetic refrigeration has the advantages of low noise and low vibration because it does not need compressor and can make the refrigeration system miniaturized.

1. Rare earth magnetic refrigeration

In 1881, Warburg first observed a thermal effect of metallic iron in an external magnetic field. In 1895, Langeriz discovered a magnetocaloric effect. In 1926, Debye theoretically deduced adiabatic demagnetization cooling technology, made magnetic refrigeration technology develop. In 1976, Brown in NASA first studied ferromagnets as refrigerants. Currently, $Gd_3Ga_5O_{12}$ and $Dy_3Al_5O_{12}$ are the magnetic cooling materials used in low temperature regions.

2. Rare earth magnetic cold storage materials

In order to improve the characteristics of commonly used small chillers, a cooler must be used. Most of the cold storage materials used in recent studies at low temperature range are rare earth metal compounds, such as $(ErDy)Ni_2$, $Er(NiCo)_2$, $GdErRh$, $RE_3T(RE = Er, Nd, T = Ni, Co)$, $Er_{1-x}Dy_xNi_2$, $Er(Ni_{1-x}Co_x)_2$, $GdRh$ and $Gd_{0.5}Er_{0.5}Rh$.

3. Rare earth magnetic refrigeration and magnetic storage materials

Rare earth magnetic refrigeration and magnetic storage materials are mainly used in helium freezer, which is mainly used in magnetic levitation trains, nuclear magnetic resonance imaging tomography, superconducting magnets and so on.

In addition to the abovementioned, rare earth magnetic materials also have rare earth ferrite, magnetic semiconductor, superconducting magnetic ordering, nanomagnetic powder materials, nanocomposite magnetic materials and so on.

# References

Adamas Intelligence (2020) Comprehensive industry review and market outlook, Rare Earth Magnet Market Outlook to 2030 (Part)

Buschow KHJ (1977) Intermetallic compounds of rare-earth and 3d transition metals. Rep Prog Phys 40(10):1179

Department of Metals, Sun Yat-sen University (1978) Rare earth physicochemical constant. Metallurgical Industry Press, Beijing, pp 9–79, 181–217

Hajime N, Takehisa M, Koichi H (2016) Method for preparing rare earth permanent magnet material. FreePatentsOnline

Knoch KG (1996) Rare earth permanent magnet materials and their properties. In: IEE colloquium on conference: impact of new materials on design, January

Li M, Liu ZG, Wu JX et al (2009) Rare earth elements and analytical chemistry. Chemical Industry Press, Beijing, pp 49–56

Liu GH (2007) Rare earth materials. Chemical Industry Press, Beijing, pp 150–193

Liu XZ (2009) Rare earth fine chemicals. Chemical Industry Press, Beijing, pp 134–151

Nakamura H, Hirota K Minowa T (2012) Preparation of rare earth permanent magnet material. FreePatentsOnline

Oettinger M (2015) High performance, rare earth-free supermagnetostrictive structures and materials. ibridgenetwork.org

Pan SM (2011) Principle manufacturing and application of strong magnets-REE permanent magnet materials. Chemical Industry Press, Beijing, pp 1–11, 14–108

Sagawa M, Hirosawa S et al (1987) Nd–Fe–B permanent magnet materials. Jpn J Appl Phys, Jpn Soc Appl Phys 26:785

Shen BG (2017) Rare-earth magnetic materials. Sci Focus 12(4):27–30

Shi F (2007) Preparation of rare earth permanent magnet materials. Metallurgical Industry Press, Beijing, pp 1–52

Strnat K (1970) The recent development of permanent magnet materials containing rare earth metals. IEEE Trans Magn

Velicescu M (1999) Progresses in the development of rare earth permanent magnet materials. Rom Rep Phys

Wu CH, Zhuang YZ (1990) Introduction to structure and magnetism of rare earth-transition metal compounds. Physics 19(2):83–84

Zepf V (2013) Rare earth elements: a new approach to the nexus of supply demand and use: exemplified along the use of neodymium in permanent, magnets, Springer

Zhao WJ, Guo YM (2013) Synthesis report of the fourth international rare earth summit

# Chapter 16
# Rare Earth Luminescent Materials and Laser Materials

## 16.1 Brief Description

In 1906, Becquerel observed a particularly sharp spectral line from an ore containing rare earth and transition-metal elements. At that time, rare earths were merely regarded as impurities in minerals, and received scant attention from scientists. After Bohr's atomic theory in 1913, quantum mechanics in 1926, Bethe crystal field theory in 1929, and the subsequent atomic spectrum theory, scientists finally confirmed that the aforementioned sharp spectral line was due to a transition in the 4f shell of a rare earth ion.

After the Second World War, the study of rare earth spectra attracted more attention, and the spectral theory and spectroscopy of rare earth elements developed during the 1950s. In recent decades, with the application of synchrotrons, high-energy particle colliders, and laser technology in spectroscopy, new methods for studying the energy levels and fine spectral behavior of high excited states of rare earth ions, $4f^n n'l'$, have come to the fore. Consequently, the theory of rare earth spectroscopy and the application of rare earth luminescent materials have developed greatly (MPL 2014).

Since the 1970s, China's many research institutions and universities have been at the forefront of studies on the spectral and luminescence properties of rare earths, contributing to different aspects (Hong 2011).

## 16.2 On Luminescent Materials

1. Luminescence

The essence of light is electromagnetic waves. Purple light has the shortest wavelength, while that of red light is the longest. Light consisting of a single wavelength is referred to as monochromatic. However, strict monochromatic light is almost nonexistent, as all light sources produce light within a wavelength b and, although this

© Science Press 2023
C. Wang, *Theory and Application of Rare Earth Materials*,
https://doi.org/10.1007/978-981-19-4178-8_16

can be quite narrow. A laser is considered as coming closest to producing ideal monochromatic light (Hao et al. 2010; CRE 2010).

The electromagnetic radiations at the purple and red ends of the spectrum, bookending the visible range, are referred to as ultraviolet light and infrared light, respectively. Light radiation can be divided into two categories: balanced radiation and unbalanced radiation. Balanced radiation is also called thermal radiation. Non-equilibrium radiation is the radiation produced by an object deviating from its original thermal equilibrium state under the excitation of some external action. If the excess energy of the object is emitted in the form of light radiation during the recovery to the equilibrium state, it is called luminescence.

The characteristics of optical radiation are generally described by five macro-optical parameters, namely brightness, spectrum, coherence, polarization, and radiation time.

2. Classification and application

Various luminescent materials are made into different light-emitting devices according to certain technical requirements, and such devices emit light under external excitation. The use of light as a source to make various lights is a revolution in lighting technology.

The luminescence types include photo-, radio-, X-ray, electro-, cathodic ray, thermo-, acoustic, chemi-, bio-, and frictional luminescence.

3. Spectrum and energy levels of luminescent materials

Spectral measurement is an important method for obtaining relevant information on luminescent materials.

(1) Absorption spectra

When a continuous beam of light passes through a transparent medium, if the energy of the light wave and the energy interval from the ground state to the excited state in the medium are equal, the matter in the medium will be excited from its ground state to its excited state, and the light will be attenuated by such absorption. The absorption ability is different at different excited states. The pattern of spectral lines obtained when light is passed through a transparent medium is called the absorption spectrum.

Light absorption by luminescent materials follows the Beer–Lambert law:

$$I(\lambda) = I_0(\lambda)e^{-k_\lambda x}$$

or

$$\lg \frac{I_0}{I} = k_\lambda x$$

Here,

## 16.2 On Luminescent Materials

$I_0 (\lambda)$—light intensity of wavelength $\lambda$ used to irradiate the substance;

$I (\lambda)$—light intensity passing through a luminescent material of thickness $x$;

$k_\lambda$—absorption coefficient (a function independent of the intensity of the light, but dependent on its wavelength).

### (2) Diffuse-reflectance spectra

Most practical luminescent materials are not single crystals, for which absorption spectra may be easily measured. Rather, they are powders, which present great difficulties for the accurate measurement of absorption spectra, such that light absorption can only be estimated by measuring the reflectance spectrum of the material. When light is projected onto a rough surface, it is scattered and reflected in all directions, which is known as diffuse reflection.

### (3) Excitation spectra

Excitation spectrum refers to the intensity change of a certain spectral line or band of luminescence with excitation wavelength or frequency, which reflects the effect of different wavelengths of light on excited materials (shown in a diagram of $Y_3Al_5O_{12}$ and $Y_3Al_5O_{12}$:Ce diffuse reflectance spectra (dashed lines) and $Y_3Al_5O_{12}$:Ce excitation spectra (solid lines)). The excitation spectrum represents the wavelength range of the excitation light that elicits luminescence.

### (4) Emission spectra

The emission spectrum of a luminescent material, or luminescence spectrum, is characterized by the distribution of luminescence energy according to wavelength or frequency. Some are continuous bands, as shown in ZnS:Cu emission spectra.

Some materials have relatively narrow luminescence bands and exhibit split structures at low temperatures (liquid nitrogen or liquid helium temperature), whereas for other materials the emission spectrum at room temperature is a spectral line. The spectral structure of the trivalent rare earth ion in $Y_2O_2S$:$Eu^{3+}$ is very similar to that of the free trivalent rare earth ion, so the source of each line can be determined according to its position, as shown in a diagram of $Y_2O_2S$:$Eu^{3+}$ luminescence spectra.

### (5) Spectral line shape

Generally, spectrum shape can be represented by a Gaussian function, that is:

$$E_V = E_{V_0} \exp\left[-\alpha(v - v_0)^2\right]$$

Here,

$v$—frequency;

$E_V$—relative value of luminous energy density near frequency $v$;

$E_{V_0}$—relative energy at peak frequency $v_0$;

$\alpha$—a positive constant.

The wavelength difference between two half-maximum intensity points on the distribution curve of the radiation spectrum is the spectral line width $\Delta v$, as shown in a diagram of spectral line shape; $AB$ is the spectral line width.

(6)  Stokes law and anti-Stokes luminescence

The luminescence wavelength of a luminescent substance is generally greater than the excitation wavelength, in accordance with Stokes law. The difference between the excitation energy and the emission energy is called the Stokes shift. The energy-level structure of a luminescent center includes some levels, such as $E_{01}$, $E_{02}$, $E_{03}$, ... at ground state, and $E_{11}$, $E_{12}$, and $E_{13}$, ... at excited state.

In the late 1960s, various up-conversion materials were discovered. Upon excitation in the near-infrared region (ca. 1000 nm), they emit red, green, or blue luminescence. Such materials can convert low-energy photons into high-energy photons. The absorption of two excited photons to emit a high-energy photon is referred to as anti-Stokes luminescence.

(7)  Configuration coordinates

The experimental spectrum of a luminescent material can be rationalized in terms of the configuration coordinate, which describes the relationship between the energy of a system and the position of the lattice ions.

The masses of ions are much larger than that of electrons, and hence their motion is much slower. Therefore, in the process of rapid transition of electrons, the relative position and velocity of motion between ions in crystals can be considered to remain approximately the same, which is known as the Franck–Condon principle. Positional coordinates can be used to represent a system: the vertical coordinate represents the potential energy of the luminous center in the crystal, and the transverse coordinate represents the "configuration coordinate" of the central ion and the surrounding ions. After the electronic transition, the ion may be at a very high vibrational energy level, so it can release a large number of phonons in the process of returning to the equilibrium position. This is referred to as a lattice relaxation process, as shown in a configuration coordinate diagram.

Applying the configurational coordinate to crystal luminescence, the following features can be rationalized:

- the Stokes shift between the excitation and emission energies;
- broadening of the absorption spectrum at high temperatures;
- thermal quenching of crystal luminescence, as shown in a diagram of temperature quenching bitmap of luminescence;
- material without glow, as shown in a diagram of bitmap of non-luminous materials.

(8)  Intensity and efficiency of luminescence

Luminescence is a process in which energy is absorbed in some way and converted into light radiation. It can be categorized as thermodynamic non-equilibrium radiation, and one of its important features is brightness, the light energy emitted by a

## 16.2 On Luminescent Materials

light source in unit time. Light intensity refers to the physical quantity of strong or weak luminescence in a certain direction, generally expressed as $I_V$.

The unit of luminous intensity is the candela (cd), a basic unit in photo metry, and the other units in photometry are derived from it. In the international system of units, brightness is expressed in $cd/m^2$. For a light source, the ratio of luminous flux $\phi$ emitted to the electrical power consumed is referred to as its luminous efficiency, expressed in units of lm/W.

The light efficiency of a common incandescent lamp is 10–15 lm/W, that of a high-temperature incandescent tungsten lamp is 26–28 lm/W, that of a metal halide lamp is 70–90 lm/W, and that of a rare earth tri-color lamp is 80–100 lm/W.

For any lighting source, the aim is to maximize the light efficiency. An important way to improve light efficiency is to select an appropriate luminescent material such that more radiation is emitted in the visible region.

### 4. Luminescence lifetime

A fundamental difference between luminescence and other light-emission phenomena is its duration. Assuming that some ions in a luminescent material are excited and that no energy transfer occurs, the electrons excited to the higher energy state will eventually return to the ground state. The number of electrons $dn$ that fall to the ground state in interval $dt$ if $n$ electrons are at a certain excitation energy level at time $t$ should be proportional to $ndt$; that is:

$$dn = -\alpha n dt$$

The proportionality constant $\alpha$ represents the probability of electronic transition to the ground state.

$$n = n_0 e^{-\alpha t}$$

Here, $n_0$ is the number of electrons excited at the initial moment ($t = 0$). The average lifetime of an electron in the excited state is $1/\alpha = \tau$, and the luminescence intensity $I$ is proportional to the rate at which electrons revert to the ground state, that is:

$$I \propto dn/dt$$

Therefore,

$$I = I_0 e^{-\alpha t}$$

The attenuation of many luminescent materials has exponential form. While exponential decay is a basic form of the attenuation process that of some luminescent materials can be decomposed into a superposition of $n$ exponentials. The change of luminous intensity over time can be written as:

$$I = I_{01} e^{-\alpha t_1} + I_{02} e^{-\alpha t_2}$$

The decay of long-afterglow materials does not conform to an exponential form, that is:

$$I = At^{-\alpha}$$

This formula does not apply to the initial process, but becomes applicable to the middle and late stages of attenuation. Indeed, many long-afterglow materials conform to this formula. Sometimes, the decay process can be expressed as:

$$I = I_0(1 + \alpha t)^{-\alpha}$$

If $t$ is large, equation $I = I_0(1 + \alpha t)^{-\alpha}$ becomes equation $I = At^{-\alpha}$.

There are free positive charges and negative charges in a luminescent material, and each charge can be combined with any one of several different charges, such that:

$$I = -\frac{dn}{dt} = bn^2$$

Here,

$$n = \frac{n_o}{1 + n_0 bt}$$

$$I = \frac{n_0 b}{(1 + n_0 bt)^2}$$

The fluorescence intensity decreases exponentially with time:

$$I_t = I_0 e^{-t/\tau}$$

where,

$I_0$—initial fluorescence intensity when excitation is removed;

$t$—time;

$\tau$—average time for which an atom or molecule is in its excited state.

It should be noted that the attenuation of luminescence is not equal to the fluorescence lifetime. At this time, the luminescence is also called the afterglow (afterglow time). Currently, the longest afterglow times of rare earth long-afterglow materials are in excess of 10 h.

5. Energy transfer and transport

The luminescence process is often divided into three stages: excitation, transmission, and emission. Here, we mainly discuss the energy transmission (transfer and transport) phenomenon. The energy absorbed by one part of a crystal excited by

## 16.2 On Luminescent Materials

external influences is often transferred to another part of the crystal in some way. This illustrates that there is sure to be energy transmission between the two parts.

### (1) Ways of transferring and transporting energy

"Energy transfer" refers to the process by which the center of an excitation transfers all or part of the excitation energy to another center. "Energy transport" refers to the process of transporting excitation energy from one part of the crystal to another by means of electrons, holes, excitons, and so on. "Energy transmission" refers to both of these processes.

Through the energy transport of carriers, hole diffusion can mediate the intrinsic absorption energy of crystals and their glow, for example, Cu center excitation due to hole migration after intrinsic absorption.

### (2) Quenching role

If a very small amount of a quencher is introduced into a luminescent material such as ZnS:Cu, the luminescence efficiency will be greatly reduced. For example, Ni is such a quencher, as shown in a diagram of energy transfer model between Cu and Ni centers.

### (3) Energy transmission of excitons

When an excited-state molecule binds to a corresponding ground-state molecule, the transient excited-state dimer formed is known as an exciton. Exciton excitation spectra and temperature quenching of rare earth Eu-activated potassium iodide and strontium fluoride show that the excited-state energy of the free exciton is higher.

### 6. Sensitized luminescence

Sensitized luminescence is a manifestation of energy transmission in crystals. Many rare earth elements have been found to display characteristic sensitized luminescence phenomena. These elements as matrices, activators, or sensitizers are used in lamps, displays, and up-conversion devices. The phenomenon of mutual sensitization by rare earths is also very common. In the mutual sensitization of rare earth elements, there are two items to be considered, that is, sensitizer/activator and ground substance, for example, when sensitizer/activator is Er $\rightarrow$ Tm, the ground substance is $CaMoO_4$, $Y_3Al_5O_{12}$; when sensitizer/activator is Dy $\rightarrow$ Tb, the ground substance is silicate glass.

### 7. Light and color

(1) Color generation: the relationship between the wavelength at which a material absorbs light and the color presented covers both the light-absorbed (wavelength / Å, frequency /cm$^{-1}$, and color) and color observed. For example, when wavelength / Å is 4000, frequency /cm$^{-1}$ is 25,000, and color is purple, then color observed is green yellow.

The colors of rare earth ions in crystals or aqueous solution cover four aspects, that is, atomic number, ion, 4f electron number, and color. For example, when atomic number is 61, ion is $Pm^{3+}$, 4f electron number is 4, and color is pink; when atomic number is 58, ion is $Ce^{3+}$, 4f electron number is 1, and color is None.

(2) Principle of tri-base color and chromaticity diagram: almost all colors in Nature can be made up of three primary colors. Using this principle, fluorescent lamps with various color requirements can be made. Colorimetry has solved the problem of the quantitative description of color through the CIE (Commission Internationale de l'Éclairage) chromaticity diagram. Most widely used is the 1931-CIE standard chroma system.

## 16.3 Spectral Properties of Rare Earth Ions

1. Electronic configurations of rare earth atoms

The rare earths are composed of a total of 17 elements, namely scandium (21), yttrium (39), and the lanthanides from lanthanum (57) to lutetium (71). The electronic configurations of their atoms are as follows.

For the scandium atom:

$$1s^2 2s^2 2p^6 3s^2 3p^6 3d^1 4s^2 \left(\text{or } [Ar]3d^1 4s^2\right)$$

For the yttrium atom:

$$1s^2 2s^2 2p^6 3s^2 3p^6 3d^{10} 4s^2 4p^6 4d^1 5s^2 \left(\text{or } [Kr]4d^1 5s^2\right)$$

For lanthanide atoms:

$$1s^2 2s^2 2p^6 3s^2 3p^6 3d^{10} 4s^2 4p^6 4d^{10} 4f^n 5s^2 5p^6 5d^m 6s^2 \left(\text{or } [Xe]4f^n 5d^m 6s^2\right)$$

Among the trivalent rare earth ions, $Sc^{3+}$ and $Y^{3+}$, without 4f electrons, and $Lu^{3+}$, with a fully occupied $(4f^{14})$ layer, are colorless ions and are optically inert, and as such are suitable as matrices for optical materials. On incrementally filling the f layer from $Ce^{3+}$ $(4f^1)$ to $Yb^{3+}$ $(4f^{13})$, a multitude of f-f transitions become available. These 4f electronic transitions are responsible for the luminescence and lasing properties (Zhang 2008).

The ground-state spectral quantum numbers $L$, $S$, and $J$ of trivalent lanthanide ions, the spin–orbit coupling coefficients, and the energy differences between the ground state and other multistate closest to it, are listed in a table (omitted here). The items include $RE^{3+}$, f electron number, $L, S, J, \Delta / cm^{-1}$, and $\xi_{4f} / cm^{-1}$. Taking $Ce^{3+}$ and $Yb^{3+}$ as examples, these items are 1, 3, 1/2, 5/2, 2200, and 640; 13, 3, 1/2, 7/2, 10,300, and 2880, respectively.

16.4   f-f and f-d Transitions of Rare Earth Ions                                    279

The relationship between $L$, $S$, and $J$ and atomic number is shown in a diagram (omitted here).

2.   Spectral terms and energy levels of lanthanide ions

Generally, lanthanide ions are trivalent in their compounds. The transitions observed in the visible or infrared regions are those occurring in their $4f^n$ configurations, that is, f-f transitions; Transition between $4f^n$ configuration and other configurations is generally in the UV region.

The energy-level numbers of several lowest-excited-state configurations, such as $4f^{n-1}5d$, $4f^{n-1}6s$, and $4f^{n-1}6p$, of lanthanide ions is shown in a table of energy level numbers for each configuration of rare earth ions. For example, when $RE^{2+}$ is Yb, $RE^{3+}$ is Lu, ground state is $^1S_0$, the energy level numbers $4f^n$ is 1, $4f^{n-1}5d$ is 20, $4f^{n-1}6s$ is 4, and $4f^{n-1}6p$ is 12, then the total is 37; when $RE^{2+}$ is Gd, $RE^{3+}$ is Tb, ground state is $^7F_6$, the energy level numbers $4f^n$ is 295, $4f^{n-1}5d$ is 3006, $4f^{n-1}6s$ is 654, and $4f^{n-1}6p$ is 1928, then the total is 5883.

Dieke and Crosswhite (1963) at Johns Hopkins University analyzed the spectra of all of the rare earth ions, giving their energy-level distributions, as shown in a energy level diagram of trivalent rare earth ions observed. Electron mutual exclusion, spin–orbit coupling, and crystal and magnetic field effects influence the position and splitting of lanthanide free-ion levels (energy difference / $cm^{-1}$), as shown in a schematic diagram of splitting caused by $4f^n$ configuration perturbation. The levels are Unperturbated $f^n$ configuration, mutural exlusion between electrons ($\sim 10^4$), spin–orbit coupling ($\sim 10^3$), effect of crystal field ($\sim 10^2$), and effect of magnetic fiels. It can be seen that the magnitudes of the effects decrease in the order: electron mutual exclusion > spin–orbit coupling > crystal field > magnetic field.

The absorption spectra of rare earth pentaphosphate ($LnP_5O_{14}$) crystals at room temperature are shown in a diagram, taking $CeP_5O_{14}$, $PrP_5O_{14}$, $NdP_5O_{14}$, and $SmP_5O_{14}$ as representative examples.

As mentioned in Chap. 1, in total, 1639 energy levels are available in the $4f^n$ configurations of trivalent rare earth ions, and the number of possible transitions between them is as high as 192,177. Thus, many new luminescent materials may still be discovered.

## 16.4   f-f and f-d Transitions of Rare Earth Ions

1.   f-f transitions

Rare earth luminescence can be divided into two categories, namely $f^n$ configurational transitions in linear spectra, referred to as f-f transitions, and f-d transitions in broadband spectra.

Related theoretical work was published simultaneously by Judd and Ofelt in 1962. The luminescence characteristics of f-f transitions include a linear spectrum that is little affected by temperature, emission wavelength, or concentration; the effect of temperature quenching is little, still glow even at 400–500 °C; and the spectral line extends from the ultraviolet to the infrared.

### 2. f-d transitions

#### (1) Luminescence characteristics

Some trivalent rare earth ions, such as $Ce^{3+}$, $Pr^{3+}$, and $Tb^{3+}$, have low energy level ($< 50 \times 10^3$ cm$^{-1}$), and their 4f-5d transitions can be observed in the visible region. The most useful of these is $Ce^{3+}$, the absorption and emission of which can be observed in both the ultraviolet and visible regions. The 4f-5d transitions of some stable divalent rare earth ions, such as $Eu^{2+}$, $Sm^{2+}$, $Yb^{2+}$, $Tm^{2+}$, $Dy^{2+}$, and $Nd^{2+}$, can also be observed. The most useful of these is $Ce^{3+}$ and $Eu^{2+}$.

#### (2) $Ce^{3+}$ f-d transition luminescence

The band position of the $Ce^{3+}$ emission varies greatly in different matrices, ranging from the UV region to the infrared region. Its wide variation range of $> 20,000$ cm$^{-1}$ exceeds those of other trivalent rare earth ions.

Certain laws concerning $Ce^{3+}$ luminescence can be summarized as follows.

#### (1) Matrix effects

Energy level splitting of the 5d configuration of $Ce^{3+}$ by spin–orbit coupling amounts to $10^3$ cm$^{-1}$, while the total amount of energy level splitting induced by the crystal field can reach $10^4$ cm$^{-1}$ or more. The properties of the emitted light are shown in a diagram of emission spectra of $Ce^{3+}$ in some matrices, there are six curves, that is, 1-YPO$_4$:Ce; 2-YAl$_3$B$_4$O$_{12}$:Ce; 3-YOCl:Ce; 4-YBO$_3$:Ce; 5-Ca$_2$Al$_2$SiO$_7$:Ce; 6-Y$_3$Al$_5$O$_{12}$:Ce. The unit $\lambda$/nm is from 300 to 600.

There is corresponding table is $Ce^{3+}$ luminescence properties in some matrices, the main items include matrix (composition, crystal, space group), energy conversion efficiency $\eta$ / %, quantum efficiency $q$ / %, emission peaks / nm, excitation peaks / nm, Stokes displacement / $10^3$ cm$^{-1}$, fluorescence lifetime $\tau$/ns (cathode rays (CR), burst of ultraviolel). For example, when matrix composition is YAl$_3$B$_4$O$_{12}$:Ce, crystal is tetragonal, space group is $R$32, energy conversion efficiency $\eta$ / % is 2, quantum efficiency $q$ / % is 40, emission peaks / nm is 368, 344, and 305, excitation peaks / nm is 290, 240, 205, and 195, Stokes displacement / $10^3$ cm$^{-1}$ is 1.9, fluorescence lifetime $\tau$/ns at cathode rays (CR) is 30, and at burst of ultraviolel is 25.

#### (2) Effects of cations

When it is necessary to vary the emission peak position in a small range, the method of changing the cation in the matrix can be adopted. Another role of cations in the matrix is as diluents ions.

#### (3) Effect of partial substitution of anionic groups in the matrix

16.5 $Eu^{2+}$ Spectrum

The anionic groups in a matrix sometimes fall into two categories, namely those directly coordinated to the cation and those indirectly coordinated. The position of the $Ce^{3+}$ emission peak can be varied within a certain range by partially replacing the indirectly coordinated anions.

(4) $Ce^{3+}$ energy transmission and sensitization

$Ce^{3+}$ has a strong and broad 4f-5d absorption band, such that it can effectively absorb energy. This can make $Ce^{3+}$ itself glow, or the energy may be transferred to other ions to play a sensitization role. The absorption bands of $Ce^{3+}$ in most matrices are in the purple region, and its emission peaks are in the purple and blue regions. Therefore, it is most suitable as a sensitizing ion in lamp luminescent materials.

# 16.5  $Eu^{2+}$ Spectrum

Among 4f-5d transitions of divalent rare earth ions, the spectrum of $Eu^{2+}$ has attracted more attention in recent years. Because $Eu^{2+}$ exhibits a broadband fluorescence spectrum (d-f transitions), emitting blue light in many matrices, $BaMgAl_{10}O_{17}:Eu^{2+}$ has been used as an important blue light material for lamps, and $BaFCl:Eu^{2+}$ has been used as an X-ray sensing screen material. The fluorescence spectrum of $Eu^{2+}$ in alkaline-earth metal fluoroaluminates features a peak emission due to the f-f transition of $^6P_{7/2} \rightarrow {}^8S_{7/2}$ in its 4f configuration.

The electronic configuration of $Eu^{2+}$ is $[Xe]4f^7 5s^2 5p^6$. The seven electrons in its ground state self-arrange into a $4f^7$ configuration. The spectral term of the ground state is $^8S_{7/2}$, and the lowest excited state may be composed of the $4f^7$ configuration inner layer or the $4f^6 5d^1$ configuration. Hence, $Eu^{2+}$ cations are located in different crystal-field environments, and their electronic transition forms are also different. Electron transitions in $Eu^{2+}$ occur in two main ways:

(1)  f-d transitions: allowed transitions from the $4f^6 5d^1$ configuration to the ground state.
(2)  f-f transitions: forbidden transitions within the same configuration.

Generally, the $4f^6 5d^1$ configuration of $Eu^{2+}$ is lower in energy than the $4f^7$ configuration at room temperature; that of the free $Eu^{2+}$ ion is 50803 $cm^{-1}$. Most activated $Eu^{2+}$ materials do not display f-f transitions.

The luminescence behavior of rare earth ions mainly depends on their properties in the lattice, and the environment plays only an interference role. For $Eu^{2+}$, as for $Ce^{3+}$, the 5d electrons are exposed to the outer layer without shielding, so they are more affected by the crystal field. The spectrum of $Eu^{2+}$ depends not only on the crystal structure of the matrix lattice, but also on the other cations present.

## 16.6 Charge-Transfer Band of Rare Earth Ions

A charge-transfer band corresponds to the energy absorbed or emitted by an electron when it is transferred from one ion to another. For rare earth luminescent materials, electrons migrate from the filled molecular orbitals of ligands (oxygen, halogen, etc.) to the partially filled 4f shells of the rare earth ions, resulting in a wide charge-migration band in the spectrum ($3000$–$4000$ $cm^{-1}$). This is seen for trivalent $Sm^{3+}$, $Eu^{3+}$, $Tm^{3+}$, and $Yb^{3+}$ and tetravalent $Ce^{4+}$, $Pr^{4+}$, $Tb^{4+}$, $Dy^{4+}$, and $Nd^{4+}$.

The f-f transitions of rare earth ions belong to the band of low-intensity forbidden transitions in their excitation spectra, which are not conducive to the absorption of light. This is one of the reasons for the low luminescence efficiency of rare earth ions. Therefore, one of the ways to improve the luminescence efficiency of rare earth ions is to study and utilize the absorption of light at the charge-transport band and energy transfer within the activated ions.

As an activator of red phosphors, $Eu^{3+}$ is widely used in luminescent materials for color TV sets and trichromatic lamps. Hence, the study of its charge-transfer bands is important. Among trivalent rare earth ions, the charge-migration band of $Eu^{3+}$ is the lowest in energy. Since the $4f^6$ configuration of $Eu^{3+}$ mostreadily accepts electrons from ligands to form a stable half-filled $4f^7$ configuration, compared to $Sm^{3+}$, $Tm^{3+}$, and $Yb^{3+}$, its charge-transfer band has the lowest energy and it is most easily reduced to the divalent cation.

Less electronegative elements and d-block element M ($Ti^{4+}$, $Zr^{4+}$, $Nb^{5+}$) induce a larger red-shift in the charge-migration band $E_{ct}$ of $Eu^{3+}$ than more electronegative elements and p-block element M ($Si^{4+}$, $Sn^{4+}$, $Sb^{5+}$).

In oxygenated compounds, when $Eu^{3+}$ and $Sm^{3+}$ are added to orthorhombic systems $A_3RE_2(BO_3)_4$ ($A^{2+} = Ca^{2+}$, $Sr^{2+}$, $Ba^{2+}$; $RE^{3+} = La^{3+}$, $Gd^{3+}$, $Y^{3+}$), the charge-transport results show that $E_{ct}$ of $Eu^{3+}$ is lower than that of $Sm^{3+}$, and that $E_{ct}$ decreases in the order $Ca^{2+}$, $Sr^{2+}$, $Ba^{2+}$, corresponding to decreasing electronegativity (Ca1.0, Sr1.0, Ba0.9) and increasing ionic radius.

## 16.7 Energy Transfer Between Rare Earth Ions in Crystals

Rare earth ions have abundant energy levels, especially in crystals due to the action of the crystal field (FJIRSM 2012). These energy levels produce splitting, and with increasing numbers they become more dense. Hence, there is increased likelihood of the energy difference between two levels of a rare earth ion matching that of another ion. Emissive or radiation less processes can then occur between the respective energy-level pairs to realize energy transfer between ions.

1. Energy-transfer mode

Energy-transfer modes can generally be divided into radiative and radiationless transfer processes. In a radiative transfer process, radiation is emitted by an ion.

# 16.7 Energy Transfer Between Rare Earth Ions in Crystals

If the radiation coincides with the energy of the absorption spectrum of another ion present, the radiation will be absorbed by this ion and energy transfer between ions will occur. The f-f transitions of rare earth ions in crystals give rise toil near spectral patterns, and the emission and absorption intensities are weak.

In a radiationless transfer process, the energy-transfer effect is high, and indeed this is the main route for energy transfer. Radiationless transfer processes can be divided into three forms, namely resonance, cross-relaxation, and phonon-assisted transfer, as illustrated in a schematic diagram of energy transfer without radiation (a)–(c), respectively. The (a) is resonance transfer; (b) is cross relaxation transfer; (c) is phonon-assisted transfer.

2. Concentration quenching of rare earth ions

The concentration quenching of the same rare earth ion varies in different crystals due to many factors. These include variations in crystal matrix compositions, absorption spectra, and the environment of the rare earth ion inducing different energy-level splittings. Differences in energy-level matching will affect the concentration quenching.

Rare earth ions intrinsically have many energy levels, and hence there are many opportunities for pair wise energy-level matching. Concentration quenching occurs when the ion concentration reaches a certain level, as illustrated for $Pr^{3+}$, $Nd^{3+}$, and $Sm^{3+}$ self-quenching cross-relaxation process, and cross-relaxation process of $Tb^{3+}$ and $Dy^{3+}$ self-quenching, respectively.

3. Energy transfer of different rare earth ions

The energy matching between respective rare earth ions can be very good. Energy transfer occurs between them, so that the fluorescence of one ion is enhanced while that of the other is weakened.

Energy transfer between some rare earth ions is illustrated as follows:

- Schematic diagram of energy transfer between $Sm^{3+}$ and $Yb^{3+}$;
- Schematic diagram of energy transfer between $Nd^{3+}$ and $Yb^{3+}$;
- Schematic diagram of energy transfer between $Pr^{3+}$ and $Er^{3+}$;
- Schematic diagram of energy transfer between $Pr^{3+}$ and $Gd^{3+}$;
- Schematic diagram of energy transfer between $Tb^{3+}$ and $Dy^{3+}$, $Tm^{3+}$.

Energy transfer between matrix and rare earth ions:
Strong characteristic luminescence can be obtained from rare earth ions, which shows the phenomenon of energy transfer from the matrix to the rare earth ions. For example, $Eu^{3+}:YVO_4$ crystals constitute a good luminescent material.

Energy transfer between ions or between a matrix and ions can be summarized as follows: the absorption energy of the energy level pair of the sensitizing ion or matrix group must be strong enough; it must match the energy of the energy level pair of the excited ion; radiation life of sensitized ion and matrix group is short enough than that of activated ion the radiation lifetime of the sensitized ion or matrix group should be shorter than fluorescence lifetime of the activating ion. Energy transfer can only occur effectively if the above three conditions are satisfied.

## 16.8 Application of Rare Earth Luminescent Materials

Luminescent materials with a rare earth as activator or a rare earth compound as matrix are called rare earth luminescent materials (Liu 2007; Li et al. 2009; Liu 2009). They can be classified in different ways (China Electronics Technology Group and ACCU 2013). In luminescent materials with rare earth ions as activators, besides doping with a rare earth ion, co-activators or sensitizers are also used. Rare earth compounds, alone or in combination with transition metals, are used as matrix materials (Wang et al. 2015).

1. Classification of rare earth luminescent materials by excitation mode

- photo luminescent materials (UV, visible, and infrared excitation);
- electroluminescent materials (DC or AC excitation);
- cathode-ray luminescent material (electron beam excitation);
- high-energy photonic excitation (X- or γ-rays);
- photo excited luminescent materials (excited by light);
- thermoluminescent materials (thermally excited).

Rare earth luminescent materials are widely used in many fields, such as new light sources, displays, imaging, optoelectronics, nuclear physics, and radiation field detection and recording (Yin et al. 2011; Ju 2014).

2. Application of rare earth phosphors

The excellent performance the red-based phosphor historically used in color TV picture tubes is due to the use of an yttrium compound, $Y_2O_2S$ or $Y_2O_3$, as the matrix and europium ($Eu^{3+}$) as the activator. Various rare earth phosphors have been used in black-and-white TV picture tubes, X-ray sensing screens, radar picture tubes, fluorescent lamps, and high-pressure mercury vapor lamps.

Rare earth activators include Ce, Pr, Nd, Sm, Eu, Tb, Dy, Ho, Er, and Tm, which are deployed in multiple bases. Some luminescent materials also contain co-activators to play a synergistic activating role. The fluorescent screen of a color TV picture tube is composed of red, green, and blue trichromatic phosphors, which not only require high efficiency and good color saturation, but also high brightness.

(1) Rare earth red phosphors

In the mid-1960s, yttrium vanadate ($YVO_4$:$Eu^{3+}$) red phosphor activated by europium was developed, followed by $Y_2O_3$:$Eu^{3+}$ and $Y_2O_2S$:$Eu^{3+}$ high-efficiency red phosphors, facilitating a more than twofold increase in the image brightness. $Y_2O_2S$:$Eu^{3+}$ is widely used because of its excellent comprehensive performance, with an annual market value of hundreds of millions of dollars.

Trace $Tb^{3+}$ and $Pr^{3+}$ have strong fluorescence enhancement effects on luminescent materials with $Y_2O_2S$ as the matrix, such that their luminescence efficiencies increase exponentially. Nowadays, color TV is the largest user of $Y_2O_3$ and $Eu_2O_3$.

## 16.8 Application of Rare Earth Luminescent Materials

### (2) Rare earth green phosphors

In all-color video displays, the green brightness needs to be the largest, with a proportion of about 60%. Therefore, the choice of green powder is particularly important. Only the yttrium aluminum garnet ($Y_3Al_5O_{12}$) system shows sufficient performance in color TV electron-beam tubes and projection tubes to be of practical application value.

Replacing Al in yttrium aluminum garnet with gallium (Ga) affords a new type of green luminescent material that is resistant to high energy density.

Other rare earth green phosphors include $Tb^{3+}$-activated yttrium orthosilicate ($Y_2SiO_5:Tb^{3+}$), lanthanum bromide ($LaOBr:Tb^{3+}$), lanthanum chloride ($LaOCl:Tb^{3+}$), and indium borate ($InBO_3:Tb^{3+}$). Some other novel cathode-ray green luminescent materials are under development.

### (3) Rare earth blue phosphors

Rare earth luminescent materials mostly provide narrow-band emissions and serve as blue fluorescent bodies suitable for projection TV. $Tm^{3+}$ is the most ideal activator of blue fluorescence among trivalent rare earth ions, it almost rarely reaching current saturation under high-current–density excitation, and its temperature quenching characteristics are good. However, it is not stable during long-term electron bombardment, and the particles show flaking. $LaOBr:Ce^{3+}$, $(La, Y)OBr:Ce^{3+}$, and $(La, Gd)OBr:Ce^{3+}$ are highly efficient blue luminescent materials, but they are chemically unstable and decompose in water.

### 3. Rare earth photo luminescent materials

The luminescence phenomenon caused by the excitation of luminescent materials with ultraviolet, visible, or infrared light is called photoluminescence. It can be divided into the emissions from fluorescent materials, long-afterglow materials, and up-conversion luminescent materials. In 1974, Philips Corporation of the Netherlands first developed the rare earth aluminate system tri-color phosphors, also known as rare earth narrow-band emission phosphors, which realized the combination of high light efficiency (100 lm/W) and high color rendering. Subsequently, other researchers have developed rare earth-activated phosphates and borate system phosphors.

Currently, the main components of rare earth tri-color phosphors for lamps are barium magnesium aluminate ($BaMg_2Al_{16}O_{27}:Eu^{2+}$) activated by europium with blue light (peak 450 nm), magnesium polyaluminate ($MgAl_{11}O_{15}:Ce^{3+}, Tb^{3+}$) activated by cerium and terbium with green light (peak 543 nm), and yttrium oxide ($Y_2O_3:Eu^{3+}$) activated by europium with red light (peak 611 nm).

The rare earth tri-chromatic phosphor has the following excellent properties:

- strong ability to withstand short-wavelength (185 nm) ultraviolet radiation;
- a powder layer surface that resists the formation of a mercury atomic layer and reduces light decay;
- high temperature resistance; quenching temperature above 800 °C; high brightness maintained when working at 120 °C;

- a quantum efficiency of over 80%;
- narrow emission kurtosis and high color purity;
- the three emission spectra are relatively concentrated in the sensitive region of the human eye, so that visual function values are high;
- rare earth ions have abundant spectral transition energy levels and so can emit light of different colors under ultraviolet radiation at 254 nm.

Rare earth tri-chromatic phosphors have become the only phosphors used in compact fluorescent lamps. The luminous efficiency of a 9 W rare earth three-color fluorescent lamp is equivalent to that of a 60 W incandescent lamp. However, its high cost is its main shortcoming.

## 16.9 Rare Earth Long-Afterglow Luminescent Materials

Long-afterglow luminous materials are also called night-light materials. They are a class of materials that absorb and store light energy (sunlight or artificial light), then slowly release it in the form of light after the excitation source is removed. They do not consume electricity, but can store the absorbed natural light, showing bright and discernible visible light in a dark environment. As such, they play a role in indicating and decorative lighting.

Dating from the Tang Dynasty (AD 618–907), one of the sentences in Wang H poem was: "Grape fine wine night-light-cup, ...". This night-light-cup was made of jade, and could shine at night. Now, this night-light material can be made manually.

1. Rare earth long-afterglow luminescent materials

The traditional long-afterglow materials are mainly zinc sulfide (such as ZnS:Cu, producing yellowish-green light) and calcium sulfide (such as CaS:Bi and SrS:Bi) fluorophores. In recent years, rare earth-activated sulfides and aluminates have become the main body of long-afterglow materials, and their industrialization has developed rapidly.

Rare earth aluminate long-afterglow luminescent materials are mainly composed of two kinds of compounds: alkali orthoaluminates [$MAl_2O_4$:($Eu^{2+}$, $RE^{3+}$) (M = Ca, Sr, Ba)] and alkali polyaluminates [$Sr_4Al_{14}O_{25}$:($Eu^{2+}$, $Dy^{3+}$)]. These materials are centered on rare earth-ion activated $CaO$-$Al_2O_3$ and $SrO$-$Al_2O_3$, with activators such as $Eu_2O_3$, $Dy_2O_3$, and $Nd_2O_3$ in a co-solvent of $B_2O_3$. They are mainly used as lamp powders, and are formed into luminescent coatings, inks, plastic, paper, fibers, ceramics, enamel, and glass. They can also be used in building decoration, traffic signs, military facilities, fire emergency signage, instrumentation, electrical switches, information storage, high-energy-ray detection, and consumer goods.

Long-afterglow luminous markings play an important role in evacuation from buildings. For example, in 911, people in the World Trade Center saw the long-afterglow tape on stairs, guiding the escape direction.

16.10 White Light-Emitting Diodes

2. Mechanism of rare earth long-afterglow luminescence

The luminescence mechanism of photo luminescent materials with long afterglow can be summarized as follows. Upon excitation with UV light, electrons transition to the excited state, and the trap energy level with a certain energy depth captures a sufficient number of these electrons and stores them. When the UV source is removed, the released electrons revert to the ground state and produce the characteristic luminescence. One of the mechanisms is that a new energy level is generated in the matrix by the introduction of $Dy^{3+}$.

3. Preparation of alkaline aluminate long-afterglow material

High-temperature solid-state reaction is the earliest and most widely used method for producing long-afterglow materials, and it is also the main method for industrial production. For example, $SrAl_2O_4$:($Eu^{2+}$, $Dy^{3+}$) can be obtained from processing $SrCO_3$, $Eu_2O_3$, $Dy_2O_3$, $Al_2O_3$, and, for example, $B_2O_3$. Long-afterglow luminescent materials can also be prepared by a sol–gel method.

4. New rare earth long-afterglow materials

Long-afterglow materials based on silicate are new materials developed in recent years. They rely on rare earth ions as activators, and usually also incorporate a certain amount of a boron or phosphorus compound to improve their long-afterglow properties. Luminescent materials with different light colors can be obtained by doping with appropriate ions.

A great deal of research has been directed towards the long afterglow of various materials, such as $SrAl_2SiO_6$:$Eu^{2+}$, emission wavelength 510 nm, afterglow time 24 h; and (Sr, Ca)$MgSi_2O_7$:$Eu^{2+}$, co-doped with $Dy^{3+}$ ions, emission wavelength 490 nm, afterglow time 20 h.

Lei BF et al. reported that $Sm^{3+}$ and $Mn^{2+}$ induce red long afterglow in $CdSiO_3$. The emission spectrum of $CdSiO_3$:$Sm^{3+}$ consists of a broadband emission with a peak at 400 nm due to the self-activated luminescence of the $CdSiO_3$ matrix, and three sharp emissions located at 566, 603, and 650 nm, respectively, due to the transition emissions of $Sm^{3+}$. The origin of the glow is unequal charge substitution of $Cd^{2+}$ by $Sm^{3+}$, forming a certain number of partially charged defect centers. These hole or electron defects can trap and store energy, facilitating long-afterglow luminescence when the excitation source is removed.

# 16.10 White Light-Emitting Diodes

White light-emitting diodes (LEDs) are composed of an LED chip and phosphors that can be effectively excited, and represent a new light source for the lighting market. Their development depends on diode development (CIAC 2013).

1. Characteristics of white LED lighting (Wu et al. 2013, 2014a, 2014b, 2014c)

(1)  An LED light source has the longest life of all light sources, up to 100,000 h;
(2)  Efficiency is 2–3 times that of ordinary incandescent lamps;
(3)  Impact resistance and anti-seismic performance are much better than those of other traditional light sources;
(4)  Wide spectral range, covering the entire visible region;
(5)  Long visual distance due to the narrow half-width;
(6)  No release of environmentally harmful substances during production and use;
(7)  Energy-saving: light efficiency can reach 300 Im/W;
(8)  Safety: low voltage, low temperature;
(9)  Good color rendering, the index reaching 80%;
(10)  Short response time, only $10^{-3}$ for incandescent lamps;
(11)  No stroboscopic, infrared, or ultraviolet radiation;
(12)  Small size, easy to shape and design.

2. Basic principle and structure

LEDs not only have a diode rectifier function, but also display light-emitting characteristics. Basic structural diagram of the LED shows the basic structure, and LED luminescence schematic diagram shows the principle of luminescence.

The chip structures of LEDs made of different materials can vary, and the luminescence situation may be different. However, the basic principle remains the same. According to the materials used, various energy levels are available. By modulating the photon energy generated by the composite, various spectra and colors can be obtained.

3. Synthetic white LEDs from red, green, and blue

Through the combination of red, green, and blue LEDs, the color purity of white light is very high. Typical products are white LEDs, combined with blue InGaN LED chips and Ce-activated garnet $(Y \cdot Gd)_3 (Al \cdot Ge)_5 O_{12}:Ce$, namely YAG:Ce phosphors capable of effective excitation.

## 16.11  Rare Earth Laser Materials

Laser materials convert light, electricity, and ray energy into coherent light (Ikesue et al. 2007; CRE 2008, SICCAS 2012). Such light is produced when a laser working material is excited. When the number of particles in this state exceeds that in the lower energy state, the substance can amplify the light radiation at a certain wavelength because of stimulated radiation. A laser is a good monochromatic, directional, and coherent light source, with high energy and brightness (Xie et al. 2015). For the 14 rare earth ions that can produce lasers, the matrix crystals involved include fluoride, oxide, compound fluoride, compound oxide, and more than 170 other crystals and

## 16.11 Rare Earth Laser Materials

other morphological substances. More than 90% of the existing laser materials are doped with rare earth ions as activators. American and German scientists (2014) use the laser technoloy to study life science, and won the Nobel Prize.

1. Rare earth laser principle

There is a diagram to show the lasers generated by three- or four-level systems of some rare earth ions; and a schematic diagram of rare earth solid or liquid lasers.

In a four-level system, in order to reduce the threshold, the rare earth laser material must meet the following requirements:

- Have broad or numerous absorption bands and match the optical pump emission wavelength to maximize energy absorption.
- Relax rapidly from energy level 4 through a radiation-free process to meta stable state of laser generation 3.
- The relaxation process from energy level 2 to ground state 1 must be rapid, otherwise the particles will accumulate and can not be excreted at energy level 2, thus affecting to realize the particle number inversion at energy level 3.
- No absorption or scattering by the matrix or internal impurities.
- Matrix materials require good optical uniformity, good thermal conductivity, and a low thermal expansion coefficient.
- The solid matrix must have good mechanical properties, with minimal nonlinear refractive index so as to avoid self-focusing and impairment of performance.

The active ion commonly used in rare earth laser materials, meeting the above requirements, is $Nd^{3+}$, its $^4F_{3/2} \rightarrow {}^4I_{11/2}$ transition providing a 1.08 $\mu$m laser output through a four-level system.

Of the 14 rare earth ions that have achieved laser output, the shortest laser emission wavelength is that of $Gd^{3+}$, and the longest is that of $Dy^{3+}$. Those of $Pr^{3+}$, $Tb^{3+}$, $Ho^{3+}$, $Eu^{3+}$, and $Sm^{3+}$ appear in the visible region; those of $Nd^{3+}$, $Yb^{3+}$, $Er^{3+}$, $Tm^{3+}$, $Tm^{2+}$, and $Dy^{3+}$ appear in the infrared region.

2. Rare earth solid laser materials

Solid-state laser materials are mostly electrolyte materials, with the laser center using rare earth ions in optical pump mode. These materials can be divided into crystal laser materials, glass laser materials, and fiber laser materials.

(1) Rare earth crystals as laser materials

The main difference between laser crystals and laser glass is that the activating ion in the former is in an ordered crystal structure, whereas in the latter it is in a disordered environment. The physicochemical properties of an activating ion in a matrix material are modulated by its environment, as are its spectral and lasing properties.

There are only a few laser crystals in widespread current use, all of which are doped with $Nd^{3+}$, such as $YAG:Nd^{3+}$, $LYF:Nd^{3+}$, $YAP:Nd^{3+}$, and $Y_3Al_5O_{12}:Nd^{3+}$. $YAG:Nd^{3+}$ is one of the most widely used rare earth solid materials, present in more than 90% of the devices in current use. Its laser crystals have been widely used in guidance, target indication, ranging, cutting, drilling, medical treatment, spectrometry, and micro-area applications.

(2)   Rare earth glass laser materials

Research on rare earth glass laser materials have focused on silicate, borate, and phosphate glasses, as well as on fluorophosphate glass, fluorozirconate glass, and germanate glass.

The advantages of rare earth glass laser materials include high output power, good optical uniformity, ease of preparation, and low cost. They can be formed into different sizes and shapes by hot-forming and cold-processing techniques. The components of a glass can be greatly varied, thus changing its refractive index to the laser wavelength, and adjusting its optical properties, such as temperature coefficient, stress optical coefficient, thermal light constant, and nonlinear refractive index. Such glasses have been widely used in laser drilling, welding, ranging, medical treatment, and instrument manufacturing.

3.   Rare earth fiber laser materials

With progress in the technology of integrated optics and optical fiber communication, micro-lasers and amplifiers are needed. Thus, rare earth fiber laser materials have been developed in recent years. The fiber can be glass fiber, fused quartz fiber, or fluoride glass fiber. $Er^{3+}$-based fused quartz fiber amplifiers (FFAs) have achieved practical results in the optical fiber communications field, and researchers regard the development of such fiber amplifiers as being of great importance. Erbium-doped fiber amplifiers have been used in new code-speed submarine communications and long-distance experiments.

4.   Rare earth liquid laser materials

Rare earth ions such as $Eu^{3+}$, $Tb^{3+}$, $Ga^{3+}$, and $Nd^{3+}$ can produce lasers in some liquids. Their spectral characteristics include wide absorption bands and linear emission bands, similar to those in a glass. High-frequency oscillation of the solvent causes nonradiative relaxation of excited electronic states. These materials can be divided into two categories, namely organic liquid laser materials using rare earth chelates, and rare earth inorganic liquid laser materials using aprotic solvents.

# References

American and German scientists won the 2014 Nobel Prize in Chemistry. http://news.xinhuanet. com/2014-10/08/c_1112740145.htm.2014-10-8
Dieke GH, Crosswhite HM (1963) The spectra of the doubly and triply ionized rare earth. Appl Optics 2(7)
Domestic holmium laser minimally invasive treatment machine has been successfully developed. http://www.cre.net/show.php?contentid=108092.2013-05-02

# References

Fujian Institute of Research on the Structure of Matter (FJIRSM) has made progress in the photopophysics of rare earth doped disordered structure crystal materials. http://www.cre.net/show.php?contentid=105887.2012-12-10

German scientists have made progress in quantum memory research. http://www.cre.net/show.php?contentid=113969.2014-06-03

Hao Y, Qin X, Wang S, Zhang Q (2010) Fabrication and optical properties of a highly transparent Nd:YAG ceramic. Rare Met 5:88–91

Hong GY (2011) Rare earth luminescent materials-basics and applications. Science Press, Beijing, pp 1–308, 344–370, 372–406, 443–487, 492–536

Ikesue A, Yan LA, Yoda T, Nakayama S, Kamimura T (2007) Fabrication and laser performance of polycrystal and single crystal Nd:YAG by advanced ceramic processing. Opt Mater 29(10):1289–1294

Ju SG (2014) To improve the luminescence properties of rare earth complexes. Univ Phys Exp 7(1):9–12

Li M, Liu ZG, Wu JX et al (2009) Rare earth elements and their analytical chemistry. Chemical Industry Press, Beijing, pp 56–79

Liu GH (2007) Rare earth materials. Chemical Industry Press, Beijing, pp 194–244

Liu XZ (2009) Rare earth fine chemical chemistry. Chemical Industry Press, Beijing, pp 44–67, 123–132

Shanghai Institute of Ceramics (SICCAS) has made important progress in large-size and low-scattering-loss laser ceramics study. http://www.cre.net/show.php?contentid=104613.2012-09-24

Space building power stations catch the sunlight. http://www.cre.net/show.php?contentid=372. 2008-11-03

The United States LED lighting penetration rate is expected to reach 46 per cent by 2020. http://www.cre.net/show.php?contentid=90073.html. 11 May 2010

The Changchun Institute of Applied Chemistry (CIAC) and other institutions develop new generation of AC-LED lighting technology. http://www.cre.net/show.php?contentid=107308.html. 13 Mar 2013

Wang ML, Li YZ, Huang ZH, et al (2015) Advances in organic, inorganic and nanocomposite fluorescent materials. 27(6):788–795

Wu YF, Wang BL, Zhang QJ et al (2014a) Recovery of rare earth elements from waste fluorescent phosphors: nao molten salt decomposition. J Mater Cycles Waste Manage 16(4):635–641

Wu YF, Yin XF, Zhang QJ et al (2014b) The recycling of rare earths from waste tricolor phosphors in fluorescent lamps: a review of processes and technologies. Resour Conserv Recycl 88:21–23

Wu YF, Zhang QJ, Yin XF et al (2013) Template-free synthesis of mesoporous anatase yttrium-doped $TiO_2$ nanosheet-array films from waste tricolor fluorescent powder with high photocatalytic activity. RSC Adv 3(25):9670–9967

Xie XQ, Yang XY, Gao J et al (2015) Near infrared fluorescent probe and its application in immunoassay. J Food Biotechnol 34(3):225–231

Yin M, Wen J, Duan CK (2011) Transition selection rules of rare-earth in optical materials. Chin J Lumin 32(7):643–649

Zhang SY (2008), Spectroscopic properties and spectral theory of rare earth ions. Science Press, Beijing, pp 1–48, 102–126, 131–156, 182–209

# Chapter 17
# Rare Earth Functional Ceramics

## 17.1 Brief Description

The chemical and physical properties of a substance or material are here collectively referred to as physical properties. Chemical bonds, structures, and states also belong to physical properties, which include mechanical, thermal, magnetic, optical, and electrical properties. In addition, the various functions shown by physical effects also belong to physical properties. The properties of the material are inherent in the material itself.

A functional physical property, that is, the physical effect of a physical property, refers to a property in which the material can convert this action into another function when it is applied to a material under certain conditions and within a certain limit (Balaram 2019; Fuertes et al. 2019). Therefore, a physical property is not a simple physical property but can exhibit a certain function of the property. The expression "functional ceramics" refers to materials that mainly utilize their nonmechanical properties (Jiang et al. 1999; Li et al. 2008; Ward et al. 2019). They have one or more functions, such as electrical, magnetic, light, and heat; some also have coupling functions, such as piezoelectric, electrooptic, piezomagnetic, magneto-optic, and acousto-optic.

Rare earths have been widely applied in functional ceramics materials (Zhang 1985), such as dielectric ceramics, piezoelectric ceramics, sensitive ceramics, magnetic (ferrite) ceramics, yttrium-stabilized zirconia ceramics, and superconducting ceramics (Shiokawa 1993; Ma and Sun 2006; Liu 2007; Mancheri et al. 2013; Pan et al. 2016; IEEE IRDS).

## 17.2 Piezoelectric Ceramics

Some crystals, when they are subjected to mechanical action, such as pressure, strain, polarization phenomenon, show positive and negative charges on their two ends.

© Science Press 2023
C. Wang, *Theory and Application of Rare Earth Materials*,
https://doi.org/10.1007/978-981-19-4178-8_17

This phenomenon is called the positive piezoelectric effect. In contrast, if a voltage is applied to this crystal, it will cause polarization, and the crystal will deform or produce mechanical stress, this is called the inverse piezoelectric effect. The amount of charge generated by the positive piezoelectric effect is proportional to stress, and the deformation generated by the inverse piezoelectric effect is proportional to voltage. The positive and inverse effects are collectively called piezoelectric effects (Zhan 2005).

In 1894, Voigt pointed out that of the 32 crystal lattices, only the 20 noncentrosymmetric crystal lattices could show a piezoelectric effect, and there is no piezoelectric effect in isotropic objects. There are still many problems in the study of piezoelectricity, which are not completely consistent with the experimental results.

Materials showing a piezoelectric effect are called piezoelectric materials, through which mechanical energy and electrical energy can be converted into each other. There are three main uses of piezoelectric materials in electronic technology:

- Electromechanical transducers,
- Vibrators,
- Transmission medium for vibrating waves.

There are many types of piezoelectric materials, such as single crystal, ceramic body, and single-crystal film. Compared with crystal piezoelectric materials, piezoelectric ceramic materials have the advantages of easy processing, suitable for mass production, cheap price, and the piezoelectric properties can be controlled and improved by doping.

With the research on piezoelectric materials, the exploration of rare earth piezoelectric materials has also made progress.

After the discovery of barium titanate, which has a perovskite structure, in the 1940s, the ceramics made from it have been polarized to obtain materials with piezoelectric properties. The piezoelectric ceramics have been developed in addition to one-component, two-component, three-component, four-component, and more-component systems, such as the one-component system $BaTiO_3$, two-component system $PbTi_{0.48}Zr_{0.52}O_3$, three-component system $Pb(Mg_{1/3}Nb_{2/3})O_3-PbTiO_3-PbZrO_3$, and four-component system $Pb(Sn_{1/2}Sb_{1/2})O_3-PbTiO_3-PbZrO_3 + MnO_2$.

In the 1950s, lead zirconate titanate as a piezoelectric ceramic with excellent piezoelectric properties appeared. $PbTiO_3$ is the basic composition of lead zirconate titanate. Because pure $PbTiO_3$ ceramics are difficult to sinter and had no practical application for a long time, scholars over the world have extensively studied its preparation methods. At present, dense $PbTiO_3$ ceramics are obtained by doping modification and reducing grain boundary energy. After the 1990s, people began to pay attention to the design of modification schemes from the point of view of improving the properties of materials, in which the modified materials of $Ca^{2+}$ and $Sm^{2+}$ have greater piezoelectric anisotropy and have advantages in sonar application. However, $La^{3+}$–modified materials have greater pyroelectric coefficient and quality ratio, and have a broad application prospect in infrared detection and other fields.

Pyroelectric ceramics made from materials such as $PbTiO_3$ with pyroelectric phenomena can be used as infrared sensors, such as alarms, fire alarms, noncontact

## 17.2 Piezoelectric Ceramics

temperature measurements, nondestructive testing, satellite global pollution detection and resource investigation, and missile detection. Rare earth oxide doping has an important effect on the pyroelectric properties of ceramics.

1. Characteristics of piezoelectric materials

(1) Piezoelectric constant

In materials showing piezoelectricity, the interaction between elastic quantities (such as stress and strain, also called mechanical quantities) and dielectric quantity through the piezoelectric effect is called electromechanical coupling (Liu 2009). It can be expressed in the following formulas:

$$D_i = d_{iR} T_R (i = 1, 2, 3; \quad R = 1, 2, 3 \ldots, 6)$$
$$S_n = d_{in} E_i (i = 1, 2, 3; \quad n = 1, 2, 3 \ldots, 6)$$

Here,

$D$—Displacement, $C/m^2$;

$E$—Electric field, V/m;

$T$—Stress, $N/m^2$;

$S$—Strain, strain is relative deformation, a dimensionless physical quantity;

$d$—Piezoelectric constant, C/N (or m/V);

$i$—Direction of electrical quantities;

$R$—Direction of mechanical quantities.

(2) Electromechanical coupling coefficient

There is a coupling relationship between mechanical energy and electrical energy of piezoelectric materials:

$$K = \frac{U_I}{\sqrt{U_M U_E}}$$

Here,

$K$—Electromechanical coupling coefficient,

$U_I$—Interaction energy density,

$U_M$—Elastic energy density,

$U_E$—Dielectric energy density.

The electromechanical coupling coefficient is an important parameter to reflect the performance of piezoelectric materials comprehensively.

## (3) Dielectric constant

The dielectric constant ($\varepsilon$) reflects the dielectric properties (or polarization properties) of the material, defined by the following formula:

$$\varepsilon = Cd/A$$

.

$C$—Capacitance (F);

$d$—Electrode distance (m);

$A$—Electrode area ($Q_m$).

## (4) Mechanical quality factor, $Q_m$

When an electrical signal is input to a piezoelectric element (ceramic or wafer) with a certain shape and size, if the signal frequency, $f_r$, is consistent with the mechanical resonance frequency of the element, the piezoelectric element will produce mechanical resonance due to the inverse piezoelectric effect. The mechanical quality factor, $Q_m$, reflects the loss of piezoelectric elements in resonance, its approximate calculation formula is:

$$Q_m = \frac{1}{4\pi (C_0 + C_1) R_1 \Delta f}$$

Here,

$C_0$—Electrostatic capacitance of components,

$C_1$—Resonant equivalent capacitance,

$R_1$—Equivalent resistance of the element at resonance,

$\Delta f$—Difference between the resonant frequency and the antiresonant frequency, Hz;

$Q_m$—Mechanical quality factor.

## (5) Ferroelectricity of crystals

Ferroelectricity refers to spontaneous polarization within a certain temperature range. Under the action of an external electric field, the spontaneous polarization energy is reoriented, and the relationship between the electric displacement vector and electric field intensity is characterized by the hysteresis loop phenomenon. Crystals with ferroelectricity are called ferroelectric crystals; substances containing ferroelectric crystals are called ferroelectrics. Such materials have correspondence with many physical properties of ferromagnets.

## 17.2 Piezoelectric Ceramics

Ferroelectrics have many domains, and the orientation of permanent dipole moments between different domains is inconsistent. When there is no external electric field, the domains are irregular, the net polarization intensity is zero. The polarization intensity increases with the increase of electric field intensity. The relationship between the polarization intensity $P$ and the applied electric field is shown for hysteresis loop of ferrite.

2. Structural characteristics of $BaTiO_3$ ceramics

Ferroelectric ceramics are ferroelectric ceramic materials. Ferroelectric ceramics with $BaTiO_3$- or $[Pb(ZrTi)O_4]$-based solid solution as the main crystalline phase are mainly used as piezoelectric and other materials.

$BaTiO_3$ ceramics are ceramics with $BaTiO_3$ or its solid solution as the main crystal phase, which can be regarded as a $TiO_6$, $BaO_{12}$ nested structure; oxygens form an octahedral structure, $Ti^{4+}$ is located in the center of the oxygen octahedron, $Ba^{2+}$ in the voids between oxygen octahedral structures. There are hexagonal, cubic, tetragonal, oblique, and tripartite equilateral crystalline phases. In the production of ferroelectric ceramics, the hexagonal crystal phase should be avoided. Cubic, tetragonal, oblique, and tripartite phases all belong to a variant of the perovskite structure (Gao and Huang 2002).

The cubic $BaTiO_3$ structure is an ideal perovskite ($CaTiO_3$)-type structure. Taking $Ba^{2+}$ as the origin, it is as shown for unit cell of $Ba^{2+}$ as origin; if $Ti^{4+}$, it is as shown for unit cell of $Ti^{4+}$ as origin.

Ferroelectrics have the lowest temperature of loss of spontaneous polarization and disappearance of domain structure, that is, Curie temperature ($T_c$). The phase transition temperature of $BaTiO_3$, quaternary phase and cubic phase is 120 °C.

3. Rare earth piezoelectric ceramics

Piezoelectric ceramics are aggregates of many small grains. In general, this type of ceramic does not show piezoelectricity. Only a strong current is applied, the electric domains in the ceramic are aligned along the direction of the electric field, and the ceramics treated by this "polarization" process can show the piezoelectric effect. At present, the most widely used piezoelectric ceramics are $BaTiO_3$, $PbTiO_3$, $Pb(Zr_xTi_{1-x})O_3(PZT)$, and $PbNb_2O_6$, usually with a certain amount of rare earth oxides added, such as $Y_2O_3$, $La_2O_3$, $Ce_2O_3$, $Nd_2O_3$, and $Sm_2O_3$, to $PbTiO_3$, PZT, and $PbNb_2O_6$ ceramics, which can greatly improve the piezoelectric properties of these materials, and some new properties will appear when supplemented with appropriate process conditions (Fu et al. 2008).

The effects of La, Ce, Nd, Sm, Eu, and Gd on lattice parameters, piezoelectricity, and dielectric properties of $PbTiO_3$ ceramics have been systematically studied by the Shanghai Silicate Institute of the Chinese Academy of Sciences. Because the radius of these lanthanide ions is close to the radius of $Pb^{2+}$, these rare earth ions can easily replace $Pb^{2+}$. To maintain the electrical properties, a Pb vacancy will be formed, and the formation of this vacancy will change the lattice constant of $PbTiO_3$ and make the sintering density of $PbTiO_3$ ceramics reach 96% of the theoretical value. Besides, it is also suitable for high-sensitivity and high-resolution ultrasonic transducers.

Because $Pb(Zr_xTi_{1-x})O_3$ has a large electromechanical coupling coefficient, piezoelectric material has a new application because of its large piezoelectric coefficient and high Curie temperature, and can be adjusted in a wide range by changing the composition. Types of rare earths doped have $La_2O_3$, $Sm_2O_3$, $CeO_2$, $Nd_2O_3$, and $Dy_2O_3$.

The paper "Rare earth Modified Conductive Ceramic Materials" by Hao and Zhang (2009) introduced the preparation and electrical properties of Nd, Sm, Gd, and Dy; Er-doped $BaTiO_3$; preparation and electrical properties of rare earth-modified barium zirconate titanate ceramics; preparation and electrical properties of rare earth-modified $PbTiO_3$ ceramics; rare earth modified conductive ceramic powder; preparation and properties of doped-type conductive coatings.

4. Application of piezoelectric ceramics

Piezoelectric ceramics are widely used in the national economy and military, and are mainly used in electronics and sensors. For example, applications in the field of electronics include processing optical signals, storing display information and images, transmitting and receiving ultrasonic waves in water, solids, and underground, automatic control, standard signal sources, long-distance calls, electronic watches, alarms, ship sonar, sea buoys, cleaning, cutting, welding, flaw detection, penetration, remote control, theft prevention, ignition, detonation, high-voltage sources, television picture tubes, and electric field therapeutic apparatus.

5. Application of piezoelectric ceramics

In preparing rare earth piezoelectric ceramics, the main process is: Ingredients $\rightarrow$ presintering (synthesis) $\rightarrow$ ball mill $\rightarrow$ molding $\rightarrow$ sintering upper electrode polarization.

According to the formula, accurately weigh the raw materials, which are generally carbonate or oxide (predrying or dehydration), prefired synthesis. The reaction is carried out in two steps, first at 850–900 °C, and then at 1,200 °C. During sintering, the blank is buried in the clinker of the same composition. The spontaneous polarization in the sintered billet is cluttered oriented and has no piezoelectricity. Polarization treatment should be carried out before use. A silver electrode is generally used, the silver slurry is coated on the surface of the sintered billet after grinding, and silver oxide in the silver slurry is reduced to silver at 750°C for 10–20 min, and then infiltrated into the ceramic surface.

6. High-polymer piezoelectric material

A piezoelectric body must be a dielectric body, and the piezoelectric constant $d$ is a third-order tensor with different component values in different crystal axis directions. On crystalline polymer samples, $d$ is negative when the tensile stress in one direction increases the thickness of the sample and the electrode charge decreases; while the stress decreases the thickness in the other direction of the sample, the electrode charge increases and $d$ is positive; if the state is unchanged before and after the shear, and no piezoelectric effect occurs, $d = 0$.

## 17.2 Piezoelectric Ceramics

Since 1940, many natural polymers, such as ramie and animal bones, tendons, and skin, have been found to be piezoelectric. In 1960, it was found that synthetic polymer materials also have piezoelectric properties. In 1969, it was found that polarized polyvinylidene fluoride (PVDF) had strong piezoelectricity, and piezoelectric polymer gradually became practical. At present, in addition to PVDF and its copolymers, there are poly(vinyl chloride), polycarbonate, and nylon 11.

Piezoelectric polymer materials are flexible and tough, have low mechanical impedance, can be made into a large area of film, are convenient for large-scale integration, and thus have wide application. Polymer materials with practical value can be divided into three categories: natural polymer piezoelectric, synthetic polymer piezoelectric, and composite piezoelectric materials.

(1) Natural polymeric piezoelectric materials

Natural polymers and synthetic polypeptide piezoelectric materials are natural polymers with lattice symmetry. For example, the piezoelectricity of wood is caused by the natural orientation of its basic component cellulose single crystals along a certain direction; bone collagen crystals such as bone tendon and wool have some degree of piezoelectricity. The study of natural polymer piezoelectrics is not only to seek piezoelectric materials as piezoelectric devices but also to seek the mystery of biological growth and promote the development of biomedicine. Piezoelectricity of bone has been used in the treatment of fractures, and can be used to control the function of growth for plastic surgery. Synthetic peptides have both crystalline and amorphous configurations, and only synthetic peptides with high crystallinity and orientation have piezoelectric properties.

(2) Synthesis of polymeric piezoelectric materials

For polyethylene, polypropylene, and other polymer materials, there are no polar groups in the molecule, so there is no polarization due to dipole orientation in the electric field, such materials have no obvious piezoelectricity.

If polar polymers such as PVDF, poly(vinyl chloride), nylon 11, and polycarbonate are softened or melted at high temperature, and they are subjected to high DC voltage (or magnetic field, light, force, heat, microwave, etc.), and through cooling, solidification, and high polarization state, and after cooling and solidification before removing the electric field. It makes the highly polarized state of the polymer freeze, then the electric field is displayed, this semi-permanently polarized material is called a polymer electret. The charge maintained in the electret includes real charge and dielectric polarization charge. The real charge refers to positive and negative charges trapped in the body or on the surface; the dielectric polarization charge refers to aligned and frozen dipoles. If the polymer film is stretched before polarization, strong piezoelectricity can be obtained. Polymer electret is one of the most practical piezoelectric materials because of its highly polarized molecules, which have completely different physical effects at each end.

Among all piezoelectric polymers, PVDF not only has excellent piezoelectricity and thermoelectricity but also has excellent mechanical properties. It is only 1/4 the density of piezoelectric ceramics, the elastic compliance constant is 30 times larger than ceramic, it is soft, tough, and impact resistant; it can be processed into several microns thick film, but also bent into any shape, suitable for curved surfaces, easy to process into a large area or complex shape, and also conducive to device miniaturization. Because of its low sound resistance, it can match well with a liquid level.

The crystallinity of PVDF is 35–40%. When the film is formed by casting or hot pressing, the main component is the $\alpha$ crystalline phase, which is nonpolar, nonpiezoelectric. Only when the film is stretched or rolled by uniaxial tension at about 120°C can a part of it be converted into a piezoelectric $\beta$ phase. After polarization treatment, an electret of permanent polarization intensity, high voltage electric constant, and thermoelectric constant can be obtained.

(3)  Composite piezoelectric materials

Polymer piezoelectric materials have flexure, but their piezoelectric constants are small, and there are limitations in use. However, if piezoelectric ceramic powders with high polarization strength, such as $BaTiO_3$ or $PbTiO_3$, are incorporated into polymer piezoelectric materials, flexible polymer composite piezoelectric materials with strong piezoelectricity can be obtained after polarization.

The properties of composite polymer piezoelectric materials are related to the properties of the polymer matrix and piezoelectric ceramic powder, and appropriate materials should be selected. The mass content of piezoelectric ceramic powder is generally more than 70%. Composites are formed by the roll method and casting method. The roll method is to mix the polymer and ceramic powder near the softening point of the polymer, and then granulate or squeeze into sheets or other shapes. The casting method is to dissolve the polymer matrix in the appropriate solvent (such as ketones), then add ceramic powder, and make it form a mud-like complex by ball milling and so on. The optimal polarization conditions vary according to the types of polymers and ceramics. Polarization voltage, holding time, and polarization temperature are approximately the same as for the piezoelectric ceramics.

7.  Polymeric thermoelectric materials

The reason for the thermoelectricity of polymer materials is the same as that of piezoelectricity, which is mainly caused by the orientation of dipoles in the molecule, change of impurity charge distribution, and the spontaneous polarization of the material caused by the electrode injection effect. PVDF is both an excellent piezoelectric and excellent thermoelectric material.

When the temperature changes, the spontaneous polarization intensity also changes, and the surface charge balance of the material is destroyed, resulting in a hot current. In the composite with PVDF as the matrix, there is almost no hot current when unpolarized, and the maximum value of hot current can be observed at about 80°C after polarization. The properties of polymer thermoelectric composites mainly depend on the polymer matrix materials.

## 17.2 Piezoelectric Ceramics

8. Application of polymer piezoelectric and thermoelectric materials

(1) Electroacoustic transducer

Using the transverse and longitudinal effects of the polymer piezoelectric film, sound equipment such as loudspeakers, headsets, and microphones can be made, and it can also be used to measure a chord vibration. A polymer loudspeaker is not sensitive to mechanical vibration, shock, and electromagnetic interference, and has all the advantages of a capacitive loudspeaker, but the structure is much simpler, the price is low, and it is also suitable for hearing aids, sound level tables, cameras, etc.

(2) Dual-voltage wafers

Two piezoelectric films are bonded in reverse, and the opposite bias is applied. Using the piezoelectric effect, when one side is stretched and the other side is compressed, the film will bend, so it can be made into an electrically controlled displacement element. Compared with electromagnetic displacement components, the energy consumption is low, reliability is good, and the structure is simple. This principle can be used to make optical fiber switches, vibration sensors, pressure detectors, etc. The output voltage at the same stress is about seven times that of a sensor fabricated from lead zirconate titanate piezoelectric material.

(3) Thermoelectric transducer

The PVDF thermoelectric properties are very obvious, and a temperature change of 1 °C can produce about 10 V, so the sensitivity as a temperature-measuring device is very high, and even a small change in temperature of 1/1,000,000 °C can be measured. The heat diffusivity is small and the chemical inertness is high, so it is suitable for secondary information processing; the heat picture is also clear, and it can, therefore, be used as a thermal light conduction camera tube or infrared radiation photodetector.

(4) Ultrasonic and acoustic transducers

PVDF piezoelectric film, similar to water acoustic resistance, is flexible, and can be made into a large-area film and many array sensing points, and made at low cost, so it is an ideal material for making water sounders. The film can also be used to monitor submarines, fish, or underwater geophysical detection, but also can be used for the reception and emission of ultrasonic waves in liquid or solid.

(5) Medical instruments

The acoustic impedance of PVDF matches well with the human body, so it can be used to measure a series of data such as the sound of the human body, heart, heart rhythm, pulse, pH, blood pressure, current, and respiration, and can also be used to simulate human skin.

## 17.3 Rare Earth Transparent Electrooptical Ferroelectric Ceramics

Under the action of an external electric field, the phenomenon in which the refractive index of the medium changes is called the electrooptic effect, a medium with an electrooptic effect is called electrooptic material, and ferroelectric ceramics with an electrooptic effect are called electrooptic ferroelectric ceramics (Zhan 2005).

In the early 1970s, Haerting developed high transparency lead (Plumbum)–Lanthanum–Zirconate–Titanate (PLZT) ferroelectric ceramics, bringing ceramic materials into the functional optics field. After that, a series of ceramics, such as niobate and titanate, which are transparent and have electrooptic properties, were developed. Almost all of these components are doped with lanthanum oxide as one of the important components (Zhang 2006).

(1) Principle of the electrooptic effect

Let the applied bias electric field be $E$, medium refractive index $n$, the relationship between $n$ and $E$ can be expanded in the series form:

$$n - n' = a_1 E + a_2 E^2 + \cdots$$

$a_1, a_2$—Constants,
    $n'$—Refractive index at $E = 0$.

    $a_1 E$ is the primary term, the refractive index change caused by this term is called the primary electrooptic effect, the change of refractive index is proportional to the external field strength, the so-called Pockels effect; and the electrooptic effect caused by the quadratic term, $a_2 E^2$, is called the secondary electrooptic effect, or Kerr effect.

The Pockels effect is expressed as:

$$\Delta n = n_0 - n_e = n_0^3 \gamma E$$

where,

$n_0$—Refractive index of constant light (0 light),

$n_e$—Refractive index of nonconstant light (e light),

$\gamma$—Electrooptic coefficient of the medium,

$E$—External field intensity.

    The essence of the electrooptic effect is that molecules that make up a substance are polarized under the action of an external electric field, which changes the intrinsic electric moment of the molecule, so the refractive index of the medium changes.

    Because $PbZrO_3$ and $PbTiO_3$ can be inter-dissolved in all components, and $La_2O_3$ solubility in $PbZrO_3$–$PbTiO_3$ solid solution is very high, it is possible to prepare various PLZT solid solutions with different chemical compositions within a broad

## 17.3 Rare Earth Transparent Electrooptical Ferroelectric Ceramics

range. $La^{3+}$ enters the solid solution A-position, and displaces $Pb^{2+}$ to maintain electrical neutrality, and a Pb vacancy will form. For example:

$$0.01La_2O_3 + Pb(ZrTi)O_3 \rightarrow Pb_{0.97}La_{0.02}(Pb \text{ defect site})_{0.01}(Zr, Ti)O_3 + 0.03 PbO$$

The chemical formula for A defects $Pb_{1-x}La_x\Box_{x/2}(Zr_{1-y}Ti_y)O_3$:
$La^{3+}$ may result in B ion defects after entering the solid solution.
The chemical formula for B defects: $Pb_{1-x}La_x(Zr_{1-y}Ti_y)_{1-x/a}\Box_{x/4}O_3$
The experiment shows that there are A and B ion defects in PLZT and their ratio is related to the zirconium–titanium ratio. High zirconium levels produce A ion defects, as titanium content in the solid solution increases, B ion defects increase.

The greatest role of lanthanum in PLZT materials is to improve the transparency of ceramics. For ceramics to be transparent, the following conditions must be met:

- High density,
- The chemical composition is uniform,
- Grinding into optical surfaces.

The reasons lanthanum can improve the optical transparency are as follows:

Adding a small amount of lanthanum can reduce the anisotropy of the oxygen octahedral unit cell, thus reducing the light scattering caused by multiple refractions on grain boundaries. With the high solubility of lanthanum in PLZT, and in a large range of mutual dissolution into a uniform composition, the appropriate process to improve the chemical composition of powder uniformity can effectively reduce the second phase caused by light scattering. A certain amount of lanthanum can lead to the formation of a considerable number of lattice defects in PLZT ceramics, which may be beneficial to the material during sintering, and thus promote the densification of ceramics and generate a fairly uniform microstructure, so that the bulk density of PLZT ceramics reaches more than 99% of the theoretical value.

The composition and process factors of the material are closely related to the transparency of the material. Ceramics with lanthanum content (mole fraction) of 8–16% have the greatest transparency. PLZT material has great absorption ability for ultraviolet light, and ultraviolet rays with wavelengths below 370 nm will be fully absorbed while the material remainstransparent to visible and infrared light.

PLZT ceramics have a high electrical constant, and changing its composition can make its dielectric constant change in the 970–5000 range; a dielectric loss of 0.3–6.0%; high breakdown strength of about 3 MV/mm.

Besides the electrooptic ferroelectric materials mentioned above, PLZT ceramics are also photoferroelectric materials, which are both photosensitive and ferroelectric. Photoelectric materials have the effect of photo induced domain steering and being photorefractive. The photo induced domain steering effect is that as photosensitive materials, UV and visible light can excite free electrons in the lattice structure of PLZT materials; while as ferroelectric materials, the polarization intensity of PLZT materials will turn with the external electric field. If the two processes act on a PLZT ceramic at the same time, the photo induced domain steering effect will occur. The photorefractive effect is that PLZT electrons excited by the wafer under UV irradiation can migrate and be "recaptured" in the "low-illumination" region through thermal diffusion or electric field action, thus forming a space charge within the PLZT wafer. This space charge changes the refractive index and birefringence of the material. This phenomenon is called the photorefractive effect.

PLZT ceramics are also pyroelectric ceramics, the necessary condition for the pyroelectric effect is spontaneous polarization, but it is also due to temperature changes that produce strain, the latter stress, through which the piezoelectric effect deals with spontaneous polarization.

(2) Preparation of rare earth transparent electrooptic ferroelectric ceramics

There are several sintering methods for preparing rare earth transparent electrooptic ferroelectric ceramics. Its main processes are: mixing PbO powder, zirconium alcohol salt solution, titanium alcohol salt solution, and lanthanum acetate aqueous solution; stirring; drying; calcination; ball milling; calcination; cold pressing; and atmosphere sintering. Because PbO is volatile, excess PbO can be made in the powder to form a PbO atmosphere. It can not only avoid the volatilization of lead but also, because the excess PbO in the powder is in the liquid phase in sintering and promotes the mass transfer of the sintering, can effectively eliminate the pores and prepare the ideal transparent PLZT ceramics. The drying temperature was 100 °C, the calcination temperature was 500 °C, ball-milling in acetone, and the final sintering was at 1200 °C.

(3) Application of rare earth transparent electrooptic ferroelectric ceramics

Transparent electrooptic ferroelectric ceramics are characterized by their transparency and are used as goggles for military and industrial applications. After artificial polarization, the material also has piezoelectric and optical birefringence characteristics, which have been developed to make optical modulators, optical switches, holographic storage, and the page compose of optical data processing processes and grating.

## 17.4  Rare Earth Dielectric Ceramics

A dielectric is also called a dielectric material. Dielectric ceramics refer to ceramic materials with a resistivity greater than $10^8$ $\Omega \cdot m$, which can withstand a strong electric field without being broken down. According to polarization characteristics in an electric field, the dielectric ceramics can be divided into electrically insulated and capacitor ceramics, and rare earth elements are commonly used in the latter (Zhan 2005).

## 17.4 Rare Earth Dielectric Ceramics

### 1. Eigenvalues of dielectrics

#### (1) Static dielectric constant

The polarization intensity of the dielectric is the vector sum of the dipole moment per unit volume:

$$P = \frac{\left|\sum \gamma\right|}{\Delta V} = n\overline{\mu}$$

where,

$P$—Polarization intensity of dielectric under electric field,

$\gamma$—Dipole moment vector,

$\Delta V$—Volume,

$n$—Number of molecules per unit volume,

$\overline{\mu}$—Average dipole moment.

The relation between static permittivity $\varepsilon$ and polarization $P$ is:

$$\varepsilon = \varepsilon_0 + P/E$$

where,

$\varepsilon_0$—Dielectric constant of vacuum,

$E$—Electric field intensity.

This formula illustrates that the greater the polarization intensity, $P$, of the dielectric, the greater the $\varepsilon$, and the relative static permittivity $\varepsilon_r$ is:

$$\varepsilon_r = \varepsilon/\varepsilon_0$$

$\varepsilon$ is called the absolute dielectric constant.

#### (2) Dynamic permittivity

The polarization of dielectric molecules takes a certain amount of time, and the time to complete the polarization is called the relaxation time ($\eta$) and its reciprocal is the relaxation frequency ($f$). The polarization of molecules includes three parts: electron, atomic (ionic), and orientation polarizations. The $f$ of electron polarization is about $10^{15}$ Hz, atom (ion) is about $10^{12}$ Hz, and oriented polarization is $10^0$–$10^{10}$ Hz.

Under the action of an alternating electric field, the response of polarization to an electric field change is different because of the different frequency of the electric field.

When $f < (10^0–10^{10})$ Hz, all three polarizations can be established. When $10^{13}$ Hz $< f < 10^{15}$ Hz, hysteresis of polarization is observed. In an alternating electric field, the dielectric constant is expressed as a complex number, also known as the dynamic dielectric constant.

$$\varepsilon = \varepsilon' - j\varepsilon''$$

Here,

$\varepsilon'$—Real part,

$\varepsilon''$—Imaginary part.
The $\varepsilon''$ to $\varepsilon'$ ratio is

$$\varepsilon''/\varepsilon' = \tan \delta$$

$\tan \delta$—Loss tangent, also called loss factor,
$\delta$—Phase angle of inductance and electric field.
Because $\varepsilon$ changes with $P$, it also changes with $f$.

(3) Dielectric loss ($W$)

In an alternating electric field, the energy consumed per cubic meter of dielectric per second is called the dielectric loss.

$$W = 2\pi f E_0^2 \varepsilon'' = 2\pi f E_0^2 \varepsilon' \tan \delta$$

Here,

$E_0$—Maximum amplitude of alternating electric field,

$f$—Frequency of alternating electric field,

$\tan \delta$—Loss tangent,

$\varepsilon'$—Real part of the complex permittivity,

$\varepsilon''$—Imaginary part of the complex dielectric constant.

(4) Volume resistivity ($\rho$)

$$R = \rho L / S$$

$R$—Resistance,

$\rho$—Volume resistivity,

$L$—Length of the dielectric,

$S$—Cross-sectional area of the dielectric.

## 17.4 Rare Earth Dielectric Ceramics

### (5) Breakdown voltage ($U$)

When the voltage of the electrolyte exceeds a certain value, it loses its ability to insulate and is broken down. This voltage is called the breakdown voltage.

### 2. Rare earth dielectric ceramics

Rare earth dielectric ceramic materials require the dielectric constant to be as high as possible in their properties to reduce the volume of the capacitor; the dielectric loss tangent to be small to make full use of it in high-frequency circuits; the volume resistivity to be higher than $10^{10}$ $\Omega$·cm, to ensure that it can work at high temperature; and to be able to work reliably and stably in high-frequency, high-temperature, high-pressure, and harsh environments.

$MgO·TiO_2–La_2O_3·TiO_2$ ceramic is a type of high-frequency thermally stable and thermal compensation capacitor material, its dielectric constant is 20–40, insertion loss tan $\delta = 5 \times 10^{-4}$, and its dielectric constant temperature coefficient is (100–650) $\times 10^{-6}$ $°C^{-1}$.

The Shanghai Institute of Ceramics of the Chinese Academy of Sciences (SICCAS) has studied $BaO–Nd_2O_3–TiO_2$ series materials, the results are:

(1) With the increase of $Nd_2O_3$ fraction, the temperature coefficient moves in a positive direction, and the dielectric constant increases, tan $\delta$ decreases.
(2) Adding $2Bi_2O_3·TiO_2$ increases the dielectric constant.
(3) The temperature can be reduced to about 1150 °C by adding an appropriate flux. Rare earth (RE) microwave dielectric ceramic materials include $MgTiO_3–CaTiO_3–RE_2O_3$, $MgTiO_3–CaTiO_3–RE_2Ti_2O_3$, $RE_2TiO_7–(BaPb)TiO_3–TiO_2$, and $BaO–TiO_2–SnO_2–RE_2O_3$.

### 3. Preparation of rare earth dielectric ceramics

Rare earth dielectric ceramic materials can be obtained using a general ceramic preparation process. After proportioning, the mixture is uniform, and high-quality ceramics can be prepared by sintering for 8 h at 1180–1250 °C. In production, attention should be paid to prevent the introduction of impurities and to ensure that there is sufficient density after molding, while ensuring that the sintering is done in a sufficiently oxidizing atmosphere.

### 4. Application of rare earth dielectric ceramics

Ceramic capacitors are a type of very important electronic component, which are small in size, large in capacity, simple in structure, excellent in high-frequency characteristics, low in price, and convenient for mass production. They have been widely used in household appliances, communication equipment, industrial instruments, instrumentation, and other fields. Rare earth dielectric ceramics are used to manufacture microwave devices such as microwave filters, microwave integrated circuit substrates, components, dielectric waveguides, dielectric antennas, output windows, attenuators, matching terminals, and tube clamping rods.

## 17.5 Rare Earth Semiconductor Ceramics

Semiconductor ceramics are polycrystalline ceramics with semiconductor properties. They belong to sensitive ceramics, especially to changes of heat, humidity, light, electricity, and electricity, and can be used to make heat-, gas-, humidity-, and pressure-sensitive components. Of the rare earth semiconductor ceramics, currently, the most widely used are titanate ceramics, $BaTiO_3$ and $SrTiO_3$, doped with rare earth oxides (Tang 2005; Yao and Xu 1995).

1. Band structure of the semiconductor

The intrinsic semiconductor band structure is shown in a diagram, which the bottom is the valence band. The valence band is full of electrons when the atoms of pure semiconductor are at absolute zero, so it is a full valence band. Above is the conduction band, and is empty. The band zone between the full valence band and the empty conduction band is forbidden band. The forbidden band width $E_g$ is relatively narrow, generally around 1 eV. After the electrons in the valence band are excited by energy, if the excitation energy is greater than that of the $E_g$, electrons can transition from the valence band to the conduction band, while leaving a hole in the valence band. The energy of the hole is equal to the electron energy before excitation.

The electrons in the valence band of a semiconductor are excited to the conduction band by means of heat, electricity, and magnetism. Semiconductors that meet intrinsic excitation are called intrinsic semiconductors. By incorporating impurity elements into pure elements, the electrons are excited from the impurity energy band to the conduction band, or the electrons are excited from the valence band to the impurity energy band, the excitation of holes in the valence band is nonintrinsic or impurity excitation. Such a semiconductor is called an impurity semiconductor. An impurity semiconductor itself also has intrinsic excitation, so it has impurity and intrinsic excitation.

The impurity semiconductor can be divided into n-type and p-type semiconductors according to the difference of valence electron of the doped element and pure element. An n-type semiconductor is electron type or donor type; a p-type semiconductor is a hole type or acceptor type.

The energy band structure of an impurity semiconductor is shown in a diagram of n-type, and p-type.

2. Grain boundary effect of rare earth semiconductor ceramics

Ceramics are multiphase systems composed of grains, grain boundaries, and pores. The phenomenon that the composition of grain boundaries and structural changes significantly changes the electrical properties of grain boundaries is called the grain boundary effect. Grains in both rare earth-doped semiconductors $BaTiO_3$ and $SrTiO_3$ ceramics belong to n-type semiconductors. If there is an acceptor state in the grain boundary between the two grains, the electrons of the n-type semiconductor grains at contact with grain boundaries will be captured, which decreases the grains in a

## 17.5 Rare Earth Semiconductor Ceramics

thin layer carrier near the grain boundaries, and changes the electrical properties of grain boundaries, thus significantly changing ceramics' electrical properties.

3. Types of rare earth semiconductor ceramics

(1) Ceramic materials of positive temperature coefficient thermal resistor (PTCR)

The material positive temperature coefficient (PTC) effect means that the resistivity of the material at room temperature is $10^{-3}$–$10^{-2}$ $\Omega\cdot$cm, and when the temperature rises near the phase transition temperature, $T_c$, the resistance rises sharply, having a positive resistance temperature coefficient.

The ceramics introduced micro rare earths (e.g., $La^{3+}$, $Sm^{3+}$, $Y^{3+}$, and semi-conductive $BaTiO_3$) as a class of positive temperature coefficient resistor (PTCR) ceramic materials. Such materials have resistance–temperature, current–time, and current–voltage characteristics.

1) Resistance–temperature characteristics

By changing material formulation and preparation conditions, the $\alpha$ value of the resistance temperature coefficient of such materials can be changed from 5–10% $°C^{-1}$ to 15–50% $°C^{-1}$. According to the resistance–temperature characteristics, materials can be divided into three categories: slow PTCR material, low Curie temperature mutant PTCR material, and self-controlled generator based on their resistance–temperature relation.

2) Current–time characteristics

Current–time characteristics refer to the characteristics of current change with time during the application of voltage to PTCR elements. When you start to increase voltage, the instantaneous current is called the starting current. The current at 1 s or 3 s after electrification is called the damping current, and the current at equilibrium is called the residual current. The exponential decay of PTCR element's current as a function of time can be written as:

$$I = I_0 \exp(-kt)$$

$k$—Decay coefficient,

$t$—Time.

The higher the $k$ value, the faster the current decays with time.

$$k = \frac{V_0 \alpha_T \lg e}{C R_0}$$

Here,

$V_0$—Voltage applied at both ends of the thermistor,

$\alpha_T$—Resistance temperature coefficient of the thermistor,

*C*—Heat capacity,

$R_0$—Resistance,

*e*—Constant.

Using the time characteristics of PTCR components, demagnetizing elements can be made.

3) Current–voltage characteristics (volt–ampere characteristics)

Current–voltage characteristics of PTCR elements generally refer to static current–voltage characteristics, indicating the relationship between the voltage at both ends of the specimen and the steady-state current in static air at room temperature.

(2) Ceramic material for grain boundary layer capacitor

Grain boundary layer capacitor ceramic materials are also characterized by semi-conductive grains and insulated grain boundaries. The whole ceramic should be seen as being made of numerous capacitors and resistors in parallel and in series. The dielectric coefficient of ceramics can be expressed as follows:

$$\varepsilon = \frac{\varepsilon_b\left[d_1 d_2 (\sigma_2 - \sigma_1)^2\right]}{(d_1\sigma_2 + d_2\sigma_1)^2}$$

Here,

$\varepsilon_b$—Dielectric coefficient of grain boundary insulating layer,

$\sigma_1$—Conductivity of grain boundary insulation,

$\sigma_2$—Conductivity of semiconductor grains,

$d_1$—Thickness of grain boundary insulation layer,

$d_2$—Grain size.

It can be seen from this formula that a thin insulation layer with large grain size and high dielectric coefficient is the cause and guarantee of a high effective dielectric coefficient of such ceramics.

Grain boundary layer capacitor ceramic materials can be either semiconductive $BaTiO_3$ ceramics doped with trace rare earth elements or doped $SrTiO_3$-based ceramics. The impurities incorporated are donor impurities such as $Y_2O_3$, $La_2O_3$, $CeO_2$, $Nd_2O_3$, and $Dy_2O_3$. There are two main ways to form the insulating layer of a grain boundary: one is to coat the surface of the sintered semiconductor ceramic matrix with the acceptor impurity, and a second diffusion is carried out at a high temperature so that the acceptor impurity enters the grain boundary to form the insulating layer, a method called secondary sintering. The other is to add the acceptor impurity directly to the raw material, and then the grain boundary is insulated by the segregation of the acceptor impurity on the grain boundary during sintering. This method is called primary sintering.

## 17.5 Rare Earth Semiconductor Ceramics

The $SrTiO_3$ ceramic grain boundary layer capacitor commonly uses $PbO \cdot Bi_2O_3 \cdot B_2O_3$ multiple oxide as the coating material. Grain boundary layer capacitors made of rare earth semiconductor ceramics have the advantages of high specific capacitance, good temperature stability, and good frequency characteristics.

(3) Gas-sensitive ceramic materials

Using the surface properties of gas-sensitive ceramics may produce a gas-sensitive element, which has both sensitivity and stable physical chemistry properties. Rare earth gas-sensitive ceramic materials include the incorporation of rare earth elements in the $SnO_3$ system. The prepared ceramic materials can be used for the detection of $C_2H_5OH$ gas at 250–300 °C; composite oxide systems, such as $La_{1-x}Sr_xCoO_3$ systems, are highly sensitive to ethanol; while $BaTiO_3$ ceramics doped with rare earth oxides and ZnO mixed oxides can be used for the detection of exhaust gas at 100–400 °C.

Gas-sensitive ceramic components have the advantages of simple structure, high sensitivity, convenient use, and low price, and are mainly used in disaster prevention, alarm, and detection.

(4) Wet-sensitive ceramic material

Humidity-sensitive ceramic materials refer to ceramic materials that are sensitive to the moisture content in air or other gases, liquids, and solid substances. Humidity-sensitive resistance can convert humidity changes into electrical signals, with which it is easy to realize the automation of humidity indication, recording, and control. Rare earth humidity-sensitive ceramic materials include $Sr_{1-x}La_xSnO_3$ and $BaTiO_3$-doped $La^{3+}$ series. The sensitivity, reproducibility, and stability of humidity-sensitive ceramic elements need to be further improved.

4. Preparation of rare earth semiconductor ceramics

The doped $BaTiO_3$ or $SrTiO_3$ raw powder can be prepared by oxide synthesis, a sol–gel method, and alcohol–brine method; adding antihybrids such as $1/3Al_2O_3$–$3/4SiO_2$–$1/4TiO_2$, $SiO_2$, $GeO_2$, $B_2O_3$ to prevent the influence of the main impurities ($K^+$, $Na^+$, $Fe^{2+}$, $Fe^{3+}$, $Mg^{2+}$ etc.) during preparation. These additives form a glass phase with acceptor impurities in grain boundaries, add a certain amount of adhesive and lubricant when forming. The PTC effect can only be obtained by sintering in an oxidizing atmosphere or heat treatment in an oxidizing atmosphere above 900 °C, while it does not appear in a reducing atmosphere.

5. Application of rare earth semiconductor ceramics

PTCR ceramic materials containing introduced rare earths can be used for temperature control, measurement, liquid level control, overheating protection, and using its delay characteristics to predetermined delay protection. The material of the grain boundary layer capacitor can be used as a high-performance component for isolation, direct coupling, and bypass filtering in radio and television communication, computers, and all types of household appliances. Gas-sensitive and humidity-sensitive ceramics can be used as gas and humidity sensors, respectively.

## References

Balaram V (2019) Rare earth elements: a review of applications, occurrence, exploration, analysis, recycling, and environmental impact. Geosci Front 10(4July):1285–1303

Fu P, Xu ZJ, Chu RQ et al (2008) Research and prospect of rare earth oxides in ceramic materials. Ceramics 12:7–10

Fuertes V, Fernández JF, Enríquez E (2019) Enhanced luminescence in rare-earth-free fast-sintering glass-ceramic. Optica 6(5):668–679

Gao CH, Huang XY (2002) Application analysis of rare-earth oxides in electric ceramics. Ceramics Sci Art 36(1):14–16

Hao SE, Zhang JS (2009) Rare earth modified conductive ceramic materials. Defense Industry Press, Beijing, pp 28–185

Jiang MT, Wu NP et al (1999) Application of rare earth in high aluminum porcelain. Ceramic Eng 4:9–13

Li FS, Cao WB, Qiu WH et al (2008) Nonmetallic conductive functional materials. Chemical Industry Press, Beijing, pp 315–356

Liu GH (2007) Rare earth materials. Chemical Industry Press, Beijing, pp 275–288

Liu XZ (2009) Rare earth fine chemical chemistry. Chemical Industry Press, Beijing, pp 199–243

Ma WM, Sun XD (2006) $ZrO_2(Y_2O_3)$ /$Al_2O_3$ composites preparation by the vacuum sintering method. Acta Metall Sin 42(4):431–436

Mancheri N, Sundaresan L, Chandrashekar S (2013) Dominating the world, China and the rare earth industry, International Strategy & Security Studies Programme (ISSSP), National Institute of Advanced Studies, Apr 2013

Pan JB, Chen HH, Shi Y (2016) Rare earth ceramics. Metallogical Industry Press

Semiconductor Materials. IEEE IRDS. https://irds.ieee.org/topics/semiconductor-materials

Shiokawa J (1993) The latest application technology of rare earths. Translated by Zhai YZ, Yu ZH. Chemical Industry Press, Beijing, pp 186–192

Tang ZY (2005) Application of rare earth oxides in ceramics. Shandong Ceram 2(28):16–19

Ward RE et al (2019) Using metadynamics to obtain the free energy landscape for cation diffusion in functional ceramics: dopant distribution control in rare earth-doped $BaTiO_3$, Advanced Functional Materials, 11 Dec 2019

Yao L, Xu JW (1995) Application of rare earth oxide in ceramic industry. Bull Chin Ceram Soc 2:41–44

Zhan ZH (2005) Application and market prospect of rare earth in functional ceramics. Rare Earth Inf 3:10–12

Zhang SQ (1985) Introduction to new inorganic materials. Shanghai Science & Technology Press, Shanghai, pp 1–82, 147–175

Zhang ZG (2006) Rare-earth oxides are widely used in transparent ceramics. Bull Chin Ceram Soc 25(1):10

# Chapter 18
# Gemstones, Jade, and Rare Earth Optical Glasses and Ceramics

## 18.1 Brief Description

Jade is a common name of the gemstone and is further distinguished under two forms: jadeite and nephrite, and nephrite are much more abundant type of material compared to jadeite.

The books "Gemstone and Jade Course" edited by Guo SG (Science Press 1998) and "Gemstones and Rare Stones" by Hall C (China Friendship Publishing Company 2005) provide a wealth of information on natural gemstones and jade, reference material for artificial gemstones, as well as theoretical support for the artificial manufacture of rare earth optical glasses and rare earth ceramics.

Gemstones are beautiful, durable, and rare, and harmless decorations are made by the cutting and grinding of single crystal minerals, such as diamond, ruby, and sapphire. Jade is a rock composed of polycrystalline or amorphous minerals. According to its color, size, and shape, it can be transformed into a variety of works of art and decorations, such as jadeite, greenstone.

Recent years, researchers have done extensive studies of rare earth elements in jadeite and nephrite mines, including geochemical characteristics, influencing factors, micromechanism, and geological significance, etc (Gonçalves et al. 2002; Sun et al. 2002; Liu 2007; Yan 2012; Wang 2013).

## 18.2 Origin and Classification of Precious Gemstones/Jade

1. The main gemstones and jade include the following kinds (Zhang 1985; Zhang and Qiu 2007).

(1) Natural gemstones: natural gemstones, jade, and organic gemstones.
(2) Improved gemstones: optimizing-, handling-, blending gemstones, etc.

© Science Press 2023
C. Wang, *Theory and Application of Rare Earth Materials*,
https://doi.org/10.1007/978-981-19-4178-8_18

314              18 Gemstones, Jade, and Rare Earth Optical Glasses and Ceramics

(3) Synthetic and artificial gemstones: cubic zirconia, silicon carbide, yttrium aluminum garnet.

(4) Imitation gemstones: cubic zirconia, silicon carbide, and natural zircon are used to mimic diamonds, although their crystal structures and physical properties are completely different. Most rubies, sapphires, and agate sold internationally have been artificially improved.

Most natural minerals are crystalline, having been slowly formed from amorphous matter.

Jade is generally a collection of single or multiple mineral species with beautiful color, fine structure, and transparency. Nephrite is a collection of interwoven felty form or fibrous microcrystalline tremolite. Xiuyan jade is a single mineral aggregate composed of serpentine. Jadeite is a single-mineral aggregate with interwoven fibers (Zhao 1999; Chen 2013; AJSGEM; Yu et al. 2019).

2. Geological genesis of common gemstones and jade

(1) Magma genesis

Most terrestrial generation of gemstones is related to magmatic rocks, and the magmatic process from mantle to earth's crust involves magmatic, pegmatite, and hydrothermal stages. During the different stages of magma evolution, minerals, especially gemstones, crystallized separately under specific conditions.

In the magmatic stage, various magmas formed magmatic rock masses through differentiation and crystallization. Some gem/jade crystals and deposits were formed in this process. For example, diamond, olivine, magnesia-bauxite-garnet, and other gemstones can be produced in basic and ultrabasic magma; quartz and potassium feldspar can be produced in intermediate acidity magma; and various gemstones, such as sapphire, zircon, garnet, jasper, moonstone, agate, and obsidian, can be produced in extrusion magma.

According to the species of natural gemstones, most were formed during the pegmatite stage. In the hydrothermal stage, gemstone deposits were alternatively produced from residual hydrothermal solution and granite, such as emerald, ruby, sapphire, jade, nephrite, amethyst, agate, topaz, opal, and Xiuyan jade.

(2) Metamorphism

Metamorphism is the process of forming new rocks by physical and chemical changes in the original rock under the conditions of a new specific environment. Precious stones formed in this way include ruby, sapphire, jadeite, nephrite, and jasper.

(3) Sedimentary genesis

Sedimentary rock is formed by external geological action and diagenesis under normal temperature and pressure. High quality gems are common in sedimentary rocks, such as diamond, ruby, sapphire, green chalcedony, glass, crystal, opal,

## 18.2 Origin and Classification of Precious Gemstones/Jade

nephrite, and jadeite. These gemstones appear in sedimentary rocks because of weathering, denudation, and transport, which enrich their contents in sediments compared to the original rocks.

3. Types of compounds commonly found in gemstones

(1) Natural elements
(2) Oxides: corundum (recrystallized $Al_2O_3$), rutile ($TiO_2$), opal ($SiO_2 \cdot nH_2O$), crystal ($SiO_2$), etc.
(3) Silicates
   Island-like olivine [$(Mg, Fe)_2SiO_2$].
   Chain-like: jade $NaAl[Si_2O_6]$, $Ca_2Mg_5[Si_4O_{11}]_2(OH)_2$.
   Layered: Xiuyu $Mg_6[Si_4O_{10}](OH)_8$.
   Frame: such as moonstone $(K, Na)[AlSi_3O_9]$.
(4) Phosphates: turquoise $CuAl_6(PO_4)_4(OH)_8 \cdot 5H_2O$.
(5) Carbonates
   Malachite $Cu_2[CO_3](OH)_2$.
   Pearl, coral, white marble (all $CaCO_3$).
(6) Sulfates
   Barite, $Ba(SO_4)_2$, etc.

4. Hardness and toughness of gemstones

The hardness of a gem is a measure of its resistance to external forces, that is, its performance when carved, pressed, or ground. Generally, a gemstone has high hardness in the vertical cleavage direction. The hardness of diamond in the vertical octahedral crystal plane direction is greater than that in the other directions. The heterogeneity of hardness is often exploited in gemstone cutting and grinding.

The resistance of a gemstone to breakage by external forces is referred to as its toughness, and the toughness and hardness of gemstones are not proportional to one another. When name of gemstone at hardness, diamond is 10, ruby is 9, berg crystal is 7, jadeite is 6.5–7, and jade is 6.5–7; when name of gemstone at tenacity, diamond is 7.5, ruby is 8, berg crystal is 7.5, jadeite is 8, and jade is 3.5.

5. Mechanistic origin of gemstone coloring

(1) Elemental coloring by transition metals

In the case of ruby, the pure mineral, $Al_2O_3$, is colorless. When it contains a trace of $Cr_2O_3$, it becomes red, because the $Cr^{3+}$ of $Cr_2O_3$ replaces part of the $Al^{3+}$. The d electrons of the transition metal absorb purple light, and the residual wavelength combination forms the red color of the ruby.

Emerald, $Be_3Al_2Si_6O_{13}$, has more $Be^{2+}$ and $Si^{4+}$ ions than corundum, thus weakening the intensity of the surrounding coordination field. Its energy level is lower than that of corundum, and the transition from the energy level A to C can be excited by an absorption of only 2.05 eV, corresponding to the red band, that is, red light is absorbed, and the residual color is beautiful emerald.

With unpaired single electrons in d orbitals, most ions have colors, including some rare earth ions.

(2) Color centers

In some natural and artificial gemstones, color centers may be created by radiation. Some colors are stable, whereas others are unstable (Jadeite-Atelier.com).

- An electronic color center is a center formed when an electron resides in a vacancy of a crystal defect.
- Hole color centers are electron vacancies arising from an absence of cations.

6. Other gems

- Zircon ($ZrSiO_4$): A gemstone with a high dispersion value, second only to diamond in terms of refractive index.
- Spinel: Spinel with Mg, Fe is often used as a gemstone material; red spinel like a ruby, for example, $MgAl_2O_3$ containing Cr, Fe, Zn, Mn, and other elements.
- Crystal ($SiO_2$): This kind of jade contains Ti, Fe, Al, and so on, can form color centers, make the crystal look different, for example, jasper, agate, and chalcedony.
- Jade: A multi-mineral gemstone, mainly composed of jadeite minerals, mostly pyroxene. Its chemical composition is $XY[Si_2O_6]$, X = Ca, Mg, Fe, Mn, Na, Li; Y = Mg, Fe, Mn, Al, Cr. There are natural, synthetic, and imitation versions.
- Nephrite: A multi-mineral aggregate composed of major components such as $(Ca_2Mg_5[Si_4O_{11}]_2(OH)_2F)_2$.
- Pearl: The chemical composition of pearl includes organic and inorganic components, generally at levels of 91–96% inorganic, 2.5–7% organic, and 0.5–2% water. The main body of the organic component is keratin (or solid protein). The inorganic component is primarily $CaCO_3$ with a small amount of $MgCO_3$, and more than ten trace elements, such as Na, K, Cr, Cu, Zn, Ba, Pb, and Fe.

## 18.3 Rare Earth Optical Glasses

1. Optical glass

Common optical glass mainly refers to colorless optical glass used in various optical instruments (such as optical lenses), and colored optical glass used in filters in the traditional sense.

Among optical glasses, those with a PbO content of less than 3% are called coronal glasses, whereas those with a PbO content of more than 3% are called flint glasses. The chemical composition of commonly used optical glass is summarized (Li 2012). Glass crystals include light crown (LK), crown (K), Barium crown (BaK), heavy crown (SK), crown flint (KF), light flint (LF), flint (F), Barium flint (BaF), heavy

## 18.3 Rare Earth Optical Glasses

Barium flint (BaSF), and heavy flint (SF); the compounds include $SiO_2$, $B_2O_3$, RO, PbO, $R_2O$, and RF, and the unit is %. R is positive divalent or positive univalent metal ions.

Colored optical glass is obtained by introducing colorants into the composition. According to their spectral characteristics, colored optical glasses can be divided into three categories: colloidal colored glass (selenium cadmium glass), ion-colored selective absorption glass, and ion colored neutral dark glass (Massera et al. 2019).

2. Silicate structure and composition

Almost all commonly used optical glasses contain predominantly $SiO_2$, namely silicate, and silicate structures obey several rules (Chen 2013):

- The basic units forming silicate are $SiO_4$ tetrahedra, which are connected through the common angle.
- The Si-O-Si bond angle at oxygen is around 145°.
- According to the ratio of silicon to oxygen, silicon-oxygen tetrahedron takes the highest spatial dimension in a stable silicate lattice to combine each other.
- The spatial dimension is an eigenvalue of the combination mode. When tetrahedra are connected into a chain, the number of dimensions is one; a layered structure has two dimensions; a stereo lattice has three dimensions; and a single silicon-oxygen tetrahedron is regarded as 0-dimensional.
- When silicon-oxygen tetrahedron is connected to each other, the more compact structure is preferred, with the energy state of each tetrahedron as similar as possible.
- One silicate structure contains only silicon-oxygen tetrahedron, at most only one oxygen bridge atom difference for each tetrahedron.

Silicate classification and examples are summarized, for instance, the type is single tetrahedron, shape is single pair, dimension is 0, when silicone containing anion is $[SiO_4]^{4-}$, O:Si is 4.0, e.g. magnesite olivine, molecular formula is $Mg_2[SiO_4]$; when silicone containing anion is $[Si_2O_7]^{6-}$, O:Si is 3.5, e.g. Calcium silicate, molecular formula is $Ca_3[Si_2O_7]$. Other types include ring (e.g. Three single rings, six single rings, six double rings), chain (e.g. single-chain, double-strand), layer (e.g. monolayer), and space lattice.

Some diagrams further illustrate the silicate structure features, for example, (1) combination form of silicon-oxygen tetrahedron, the tetrahedral mutually combinations are shown in more detail, and according to cycle length, naming them as Section I (a) to Section VI (f); (2) a two-section monolayer structure, as found in clay minerals; (3) two-section monolayer structure combined with octahedral layer; (4) structures such as layered kaolinite, scaly quartz, and calcite, (a) is kaolinite-like structure combined with octahedral layer, (b) means mica-like structure; and (5) spatial stereo structures of several tetrahedral, there are structure of 2, 3, and 4 sections (The chain that makes up the spatial network is marked with thick black strokes, the chart is omitted here).

318                    18   Gemstones, Jade, and Rare Earth Optical Glasses and Ceramics

3.   Several important crystal forms of $SiO_2$

$SiO_2$ is the simplest silicon-oxygen compound. It adopts a variety of crystal forms, with α and β denoting high-temperature and low-temperature quartz, respectively. The properties of $SiO_2$ are summarized. The items of properties of various $SiO_2$ crystal forms include crystal form, crystal system, lattice constant /Å, temperature /°C, Si–O spacing /Å, bond angle /(°), density (20 °C) /(g/cm$^3$), refractive rate ($n_D$), linear coefficient of expansion ($\alpha_0$) / ($100 \times 10^6$/°).

The crystal forms include low temperature quartz, high temperature quartz, low temperature cristobalite, high temperature cristobalite, low temperature scaly quartz, high temperature scaly quartz, keatite, coesite, super quartz, sulfur quartz, $SiO_2$, fibrous, and quartz glass. The crystal systems include triangle, hexagonal, square, cube, monoclinic, oblique, and vitrescence.

4.   Glass

Glass is characterized by a disordered state of atoms, or a state of close-range order and remote disorder. Many compounds have low viscosity at temperatures above their melting point, releasing heat of solidification at the melting point, and cooling the liquid state to form an ordered crystalline solid. Glass or glassy matter retains high viscosity at the melting point. As the temperature is decreased, the atoms maintain an irregular state upon solidification. The glass state is a nonequilibrium state. A schematic diagram of volume-temperature relationship between glass and crystal illustrates the formation of crystals and glass from liquids (Shen 1974; Pisarska et al. 2012; Karmakar 2017).

Transformation from a glassy state to a crystalline state is called anti-glassing, and indicates that glass is a metastable substance that tends to change to a stable crystalline state under suitable conditions. The glass state is an amorphous state obtained by rapid cooling of the melt, that is, freezing of the melt with a chaotic structure at high temperature. In the process of cooling, the arrangement of atoms varies according to the cooling rate, hence, the structure and properties of glass are determined not only by its chemical composition, but also by the heat-treatment process.

Therefore, attention must be paid to quenching and annealing conditions in the production of glass. As described in Sect. 18.2, gemstones and jade are crystals that have evolved over a long geological period.

A glass composition must have a high viscosity (about $10^{13}$ P) near the melting point. In terms of structure, there are many kinds of liquids that can satisfy this requirement. Among inorganic glasses, more than 90% are oxide glasses. Therein, some oxides make up the glass skeleton network, while other fusible oxides enter voids in the network. The former oxides include $SiO_2$, $B_2O_3$, and $P_2O_5$, with atoms of high valence and electronegativity, and small ionic radius.

The fusible oxide is an alkali metal or alkaline earth metal oxide. When such an oxide is added to the glass, the original network is disconnected into the pores

# 18.3 Rare Earth Optical Glasses

of the structure. As this process develops, the fluidity of the glass increases and its melting temperature decreases. Therefore, for oxygen polyhedron (or ion forming the network) must have more than three bonding oxygen atoms to bond with it, which is a necessary condition for forming a three-dimensional glass network and for easily forming glass.

Borate glass is composed of a network of triangular $BO_3$ units, to which alkalimetal oxides, such as $Na_2O$, are added to form borate with $BO_4$ tetrahedra. The product has the lowest coefficient of thermal expansion of any glass, and this property is exploited in several applications.

The aforementioned characteristics of glassesendow them with the following advantages:

- The composition ratio can be changed and adjusted arbitrarily over a large range, thus changing the electrical, optical, thermal, and other properties of the glass.
- Glass is a homogeneous substance that maintains a liquid state and transparency. Its transmittance and refractive index can be adjusted by changing its composition to facilitate the fabrication of optical materials for various purposes.
- Products of any shape, such as sheet, pipe, film, or fiber, can be made by various processing methods, including blowing, rolling, drawing, and molding.
- Glasses have good chemical stability, and are widely used as acid–alkali corrosionresistant coatings on metals and acidresistant container protective layers.
- Good density, allowing for use as sealing containers and welding materials.
- Good insulation.
- Good heat resistance, allowingfor use at temperatures above 1000 °C.

Inorganic glassescan be classified according to chemical composition as follows:

(1) Pure oxide glasses, based on $SiO_2$, $B_2O_3$, $P_2O_5$, $GeO_2$, etc.
(2) Composite oxide glasses, such as those based on silicate, borate, phosphate, germanate, tellurite, borosilicate, aluminate, aluminum phosphate, aluminum titanate, or aluminum niobate.
(3) Singlecomponent chalcogen system glasses, such as $As_2S_3$, $As_2Se_3$, $As_2Te_3$, As–Te, or Ge–Se.
(4) Multiple component sulfur system glasses, such as As–Si–Ge–S, As–Te–Ge–S, or $Li_2S$–$CdS$–$GeS_2$.
(5) Sulfur oxide glasses, such as $As_2S_3$–$Sb_2O_3$ or $As_2S_3$–$CuO$.
(6) Metallic glasses, such as Pd–Si, Te–Ge, Cu–Zr, Pd–Si–Cu, Fe–P–C, or Fe–Ni–P–C.

5. Compositions and structures of rare earth glasses

Colored optical glass is made by introducing a colorant into the composition. The resultant absorption depends on the nature and quantity of the colorant in the glass.

$RE_2O_3$ (RE = rare earth) are network modifiers with high $RE^{3+}$ field strength and strong agglomeration. RE binary glasses form over a small range and their system variety is limited. RE ternary glass systems are mainly based on borate, silicate, phosphate, germanate, tellurite, and halide.

The dissolution capacity of $RE_2O_3$ in borate systems is greater than that in silicate systems, and the range of glass formation is wider. RE borate systems are the main component systems of RE optical glasses.

Borate-based glass systems are characterized by low viscosity and ease of crystallization. The introduction of high-valence $RE^{3+}$ further reduces the high-temperature viscosity of the glass, and induces strong ionic agglomeration and ease of crystallization. The crystalline phase consists mainly of RE borate compounds, such as $La(BO_2)_3$, $LaBO_3$, $YBO_3$, or $GdBO_3$. In systems containing a large number of crystals, both $La(BO_2)_3$ and $LaBO_3$ may precipitate as crystalline phases. To reduce $La_2O_3$ crystallization, a certain amount of $Y_2O_3 \cdot Gd_2O_3$ can be used instead of $La_2O_3$ to complicate the composition and improve the high-temperature viscosity.

## (1)  Rare earth colored glasses

Rare earths are added to glass as colorants to change the transmittance or adjust the refractive index and dispersion. REs impart color by virtue of the specific arrangement of their electron layers. Their f-f excitations are less affected by the external field, such that their absorption peaks in the visible range are sharp and almost unaffected by external factors. RE compounds, except for those of La, Y, and Lu, have obvious absorption bands in the ultraviolet, visible, and infrared regions (380–780 nm). $RE^{3+}$ cations have complex absorption spectra that make their colors be changeful under different lights. $RE^{3+}$ shows fine color reproducibility in stained glass. Colored glass incorporating REs has pure tone, good transmittance, and strong luster.

### a.  Cerium glasses

The coloring of cerium glass is imparted by Ce and Ti, Mn, Cu, or V. Ce and Ti mixed oxides form cerium titanate and make glass yellow; Ce, Mn, and Ti mixed oxides make glass orange; a small amount of CuO in potassium glass and titanium cerate makes glass sapphire; golden and green glass contains 2–3% $CeO_2$ along with 3–4% $TiO_2$, 0.3% $V_2O_3$, and 0–0.45% CuO. Anti-radiation colored glass is made by adding an appropriate amount of $CeO_2$ to ordinary glass to remove its color centers after long irradiation. Anti-ultraviolet glass can be made by exploiting the characteristics of $Ce^{3+}$ absorption near 310 nm.

Photosensitive glass contains $Ag^+$ and $CeO_2$. After repeated irradiation with purple light and heating, transparent images with different color depths can be formed. The complete spectrum of red, yellow, green, blue, and purple can be covered.

### b.  Neodymium glasses

Neodymium ions are characterized by pure tone and strong coloring power, and the color of neodymium glass is stable at 4–64 °C. A bright red neodymium glass made from $Nd_2O_3$ is used in navigational instruments. Using the absorption characteristics of $Nd^{3+}$ near 585 nm, the color of the glass can be changed under different light sources, eliciting dual colors of blue and purplish-red.

For example, by adding 0.2% $Nd_2O_3$ to a standard glass consisting of 70.8% $SiO_2$, 15.5% $Na_2O$, and 13.7% CaO (mole fraction), and the resulting neodymium glass

18.3 Rare Earth Optical Glasses

has the highest transmittance and a fluorescence absorption maximum at 555 nm. It almost completely absorbs sodium yellow light for signal sign recognition, and can be used in glasses for fluorescent screen operators, safety lights for darkroom operations, and X-ray equipment. The addition of $Nd_2O_3$, Se, and $MnO_2$ can impart the glass with light and dark changes in clove-color. $MnO_2$ imparts a yellow color, whereas Se imparts a rose color. Therefore, red-purple-rose-and lilac-rose-colored glasses can be prepared with formulae of 1–3% $Nd_2O_3$, 0–0.1% $MnO_2$, and 0–0.5% $SeO_2$.

c. Praseodymium glass

The $Pr^{3+}$ absorption peak near 470 nm can impart glass with a green color. High purity $Pr_6O_{11}$ glass appears green in sunlight, but is almost colorless in candlelight. Through this property, it can be used to emulate gems or other ornaments.

Reasonably collocating $Pr^{3+}$ and $Nd^{3+}$ provides a series of blue praseodymium-neodymium glasses for colored lenses. A mixture of $Pr_6O_{11}$ and $Nd_2O_3$ in a glass remains stable, and the melting temperature, medium, and time have no effect on the color. Erbium-neodymium glass is pure purple. Glass containing $Eu_2O_3$ is orange-red, whereas that containing $Er_2O_3$ is pink. Mixtures of REs with Cr, Cu, Ti, or Ni cations are also commonly used as colorants in the manufacture of various colored glasses.

(2) Rare earth optical glasses

La has characteristics of a large number of dissolved in glass, especially borate glass (sometimes more than 60%). By adding 10–50% $La_2O_3$, $Y_2O_3$, $Gd_2O_3$, etc., to borosilicate or borate systems, excellent lanthanide optical glasses are obtained.

La can markedly improve the chemical stability of glass, prevent surface deterioration caused by water or acid, increases the durability of borate glass, and increase glass hardness and softening temperature. The low thermal expansion coefficient of lanthanum glass makes it an indispensable material for high-grade cameras with large apertures and fields of view, as well as the optical lenses of periscopes.

The main problem associated with lanthanum glass is unwanted coloring. It is difficult to prepare pure colorless optical glass owing to impurities in the raw material. Hence, impurities such as Fe, Ce, Pr, and Nd should be rigorously removed when the raw material is treated and purified. $La_2O_3$, $Y_2O_3$, and $Gd_2O_3$ have been used to make optical fibers and improve fiber performance. Ho is used in filter glass. Er is used in eyeglasses and crystal glass decolorization and coloring.

(3) Rare earth photochromic glasses

Photochromicity refers to changes in the visible region of the absorption spectrum when a material is touched or light impinges on it. A material that displays such versible or irreversible chromogenic, achromatic phenomenon is referred to as photochromic glasses.

Photochromic glass can change color repeatedly over a long time without fatigue aging. Complex products with high mechanical strength, good chemical stability, simple preparation, and stable shape can be obtained. They can be roughly divided into three categories: high-soda silicate glasses incorporating $Ce^{3+}$ or $Eu^{3+}$, glasses containing AgX or TiX, and silicate glasses with structural defects constituting color centers. At present, AgX alkaline aluminate, borate, and phosphate glasses are most widely used.

Photochromic glass can be made by adding $CeO_2$, $Sb_2O_3$, and $SnO_2$ to $RE_2O_3$-$B_2O_3$-$SiO_2$ systems. When the $Sb_2O_3/CeO_2$ ratio is about 1.5 and the $SnO_2$ loading is about 0.3% (mole fraction), the photochromic properties are good. High-sodium silicate RE-AgX photochromic glass incorporating $Ce^{3+}$ and $Eu^{3+}$ can be used to make automatic dimming window glass for sunglasses, cars, locomotives, aircraft, ships, and large buildings. Optical fiber panels made of photochromic glass have been used in computer technology, display technology, and holographic recording media.

AgX-sensitized photochromic glass has many advantages, the most prominent of which is that it is not subject to fatigue during use.

(4) Rare earth luminous glasses

Luminous glass incorporates material that can transition from a low to a high energy level by external excitation. A small amount of RE is added to the matrix glass, often forming luminescent centers in the form of $RE^{3+}$, and different RE activators emit light of different colors. In order to improve the luminous brightness, red or infrared light emitted by a glass with high luminous efficiency should be converted into high-brightness light, such as by double-doping with $Yb^{3+} + Er^{3+}$, $Ce^{3+} + Tb^{3+}$, or $Ce^{3+} + Tm^{3+}$.

Rare earth luminescent glass is manufactured in the following types: fluorescent, thermoluminescence dose, neutron dose, reference (standard), scintillation, and tracer glass.

(5) Rare earth optical function glasses

a. Nonlinear optical functional glasses

Optical nonlinearity of a glass is primarily related to the structure and composition of the matrix itself, that is, it is an intrinsic effect. It is a direct result of the interaction between propagation and electron charge distribution in the structural units. Optical nonlinearity of glass doped with $RE^{n+}$, other metal ions, and organic dyes mainly depends on the nature, concentration, and properties of the dopants. Here, the glass serves only as a matrix, and the nonlinearity is an extrinsic effect.

In all glasses, there is a certain level of third-order nonlinear optical effects, and a glass with high density, high linear refractive index, and low dispersion coefficient should have a high nonlinear refractive index. This is achieved by adding an adjusting substance with high refractive index or an easily polarizable material, such as PbO, $Bi_2O_3$, $Nb_2O_3$, $ZnO$, $TeO_2$, or $RE_2O_3$(RE $=$ La, Pr, Nd, or Sm).

Among such materials, the tellurium-niobium-zinc system has excellent stability, infrared permeability, and linear optical properties. Its third-order nonlinear optical

## 18.3 Rare Earth Optical Glasses

properties are far superior to those of traditional glass systems, such as silicate and borate, making it a new type of nonlinear optical glass material. Indeed, tellurium-niobium-zinc glass is one of the important materials of all-optical switching devices. After doping with $RE^{3+}$, for example, $Ce^{3+}$, $Ho^{3+}$, or $Y^{3+}$, excellent optical nonlinearity is attained.

$RE^{3+}$ doping can improve the third-order nonlinear optical properties of tellurium-niobium-zinc glass, and the contributions decrease in the order Ce > Ho > Y > Nd > Eu > Tb > Er > Pr > La.

### b. Preparation method for rare earth nonlinear optical functional glasses

At present, the preparation methods for nonlinear optical glasses are principally high temperature melting, sol-gel, and ion implantation methods, RF magnetron sputtering, and chemical vapor deposition (CVD).

The high temperature melting method is one of the most common methods for preparing glasses. The sol-gel method does not require high temperature, thereby avoiding the volatilization and oxidation of compounds, but it is limited by a narrow range of raw materials. Ion implantation is a better preparation method in terms of controlling the parameters of ion species, dosage, energy, and so on. Sputtering method is that target atom is sputtered out, attached to the upper surface of the matrix.

### c. Rare earth infrared and visible light up-conversion glasses

RE infrared-visible light up-conversion materials are mainly solid compounds doped with REs. By exploiting the metastable energy level characteristics, low energy long wave radiation can be absorbed. High energy shortwave radiation is emitted after multiple photon addition, converting invisible infrared light into visible light.

By using the up-conversion properties of $RE^{3+}$ in fluoride, many up-converting materials working at room temperature, suitable for use in lasers, can be obtained. $RE^{3+}$ energy levels have longer lifetimes in fluoride, forming more metastable levels that facilitate abundant laser transitions. As a starter $RE^{3+}$ can be easier to doping into fluoride glass matrix. Fluoride glass has a lower phonon energy than quartz glass (about $500 \text{ cm}^{-1}$). Because the matrix has high phonon energy in quartz glass, the probability of $RE^{3+}$ undergoing radiationless transitions is increased and the lifetimes of energy levels are decreased. If radiative transitions occur, typically not less than $4000 \text{ cm}^{-1}$, while this spacing is reduced to $2500$–$3000 \text{ cm}^{-1}$ in fluoride glass.

### d. Rare earth glass fiber optics

At present, optical communication is mainly reliant on quartz glass fiber. RE ($Er^{3+}$, $Tm^{3+}$, $Pr^{3+}$, $Dy^{3+}$, $Nd^{3+}$, or $Ho^{3+}$) compounds are incorporated into the cores of quartz fibers to provide novel media for laser oscillation or laser amplification. After that, these rare earths were incorporated into the optical fiber to become a new material that can make laser excitation and amplification (Qiu 2004).

Rare earth elements, such as Er, Tm, Pr, Dy, Nd, and Ho, possess the specific energy level structure. For example, a diagram displays $Er^{3+}$ energy level, in which energy $/ \times 10^3$ cm is listed from 0 to 20, the exciting light is divided into 1.46–1.48, 0.98, 0.8, 0.6, and 0.5 μm.

Light propagates along the waveguide path in an optical fiber. The physical parameters and the guided wave, polarization, and dispersion characteristics are the same as those of ordinary single-mode fibers. After adding REs, the optical fiber performance is modified; the ion absorption peak is increased, thus causing enhanced light absorption and scattering. The RE fiber loss characteristics are actually determined by the energy-level distribution of the $RE^{n+}$ dopant. Losses in the absorption band of $RE^{n+}$ in the fiber are large, whereas those in the stimulated radiation band are small; doping with $RE^{n+}$ can reduce the specific threshold power.

The manufacturing process of RE quartz fibers permits the incorporation of Nd, Ho, Eu, Er, Yb, Tb, and Dy, and the resultant fibers have been woven into single films, multi-films, and so on. Erbium-doped fiber can be used as an optical amplifier in optical communication technology, especially in long distance optical communication, and can satisfy all of the necessary conditions for a direct optical amplifier.

Optical fiber sensor sutilize the photon as the information carrier, and have advantages of integrated small volume, remote telemetry, and a fiber network to realize multi-parameter measurements, information processing, and centralized monitoring. They have very good application prospects in industrial waste gas monitoring, infrared imaging, program control, and missile fiber guidance.

## 18.4 Rare Earth Polishing Materials

Polishing materials for glass consist of crystalline powders. The main base material has been progressively improved by switching from $Fe_2O_3$ to $ZrO_2$ to $CeO_2$. In the polishing of optical glass, $CeO_2$ has advantages of good finish, high speed, and long life. It is widely used to finish flat glass, optical instrument glass, picture tubes, and panel glass, as well as some gemstone materials. RE polishing agents are based on $RE_2O_3$ polishing powder, and certain physical properties are imparted by adding $RE_2(CO_3)_3$, $RECO_3F$, $RE_2(C_2O_4)_3$, $RE_2(SO_4)_3$, and $RE(OH)_3$, which are termed intermediates, followed by high-temperature roasting. Glass polishing experiments have proven that $Ce_2O_3$ oxidized to $CeO_2$ has similar hardness to glass. The calcination temperature has a great influence on the polishing ability of the polishing material. Hence, the hardness of $CeO_2$ can be fine-tuned over a range of calcination temperatures (800–1000 °C) according to the polishing process.

Typically, $CeO_2$ crystals suitable for polishing materials are spheroids with a diameter of 45 nm equi-axed crystal system. There is a schematic diagram of equi-axed spheroids, which the tetrahedron is not suitable as a polishing powder, while the hexagonal is good polished powder.

$La_2O_3$ and $Nd_2O_3$ mixed REs co-existing with $CeO_2$ belong to the hexagonal crystal system, but have no polishing ability. At present, polishing mechanism of RE polishing agents for glass remains somewhat unclear. It is generally believed that the efficiency with which RE polishing powder acts on the surface of glass is due to both mechanical grinding and chemical dissolution. That is to say, it is a joint action of physical and chemical grinding. Physical grinding refers to the mechanical grinding process of the glass surface by the RE polishing agent; chemical grinding mainly refers to the micro-grinding of the convex part of the glass surface by the RE polishing agent.

At the same time, the RE polishing slurry converts the glass surface into a hydrated softening layer and imparts it with some degree of plasticity. The hydration softening layer first fills the low lying glass surface to form a smooth surface, which is further finished by mechanical grinding with the RE polishing powder. At present, chemical grinding theory dominates, but a polishing mechanism that combines physical and chemical grinding should be more reasonable.

## 18.5 Rare Earth Ceramic Color Glazes

1. Glaze

The idiozome of calcined ceramic products is formed by the crystalline phase in the glass phase, and the surface is rough. In order to make the surface of the billet glossy and beautiful, an important and common method is to glaze (Schultz 1975; Zhou 2015).

Glazing is the formation of a glass layer on ceramic products after firing, and its composition should be close to that of the glass in the billet. With increasing $SiO_2$ content, both the glaze temperature and the viscosity are increased. Low temperature firing glaze contains mostly $B_2O_3$. In order to ensure firm combination of the billet and glaze during and after calcination, and good mechanical and chemical properties of the glaze, its chemical composition should be carefully tailored.

2. Rare earth ceramic colored glazes

REs have an unfilled 4f electron layer. When excited by light of different wavelengths, the 4f electron layer exhibits selective absorption or reflection; after absorbing light of one wavelength, it may emit light of another wavelength. By virtue of this characteristic, REs can be used as colorants, color aids, discoloration agents, or gloss agents in ceramic color glazes to prepare various ceramic pigments with stable color, pure tone, or photochromic color (Huang 1996; Massera et al. 2019). RE coupling produces many energy level sublayers, and 4f-4f electronic transitions result in the selective absorption of visible light and coloration. High $RE^{n+}$ electricity has expensive and large radii, are easily polarized, and have high refractive indices. By using the high refractive indices of $RE^{n+}$ in ceramic pigments, the color of everyday or craft porcelain can be rendered bright and elegant. RE ceramic color glazes are applied to the surfaces of semi-finished ceramic products, and produce a bright color after high-temperature firing.

$La^{3+}$ is colorless, and has the largest radius among the rare earths. Its polarization coefficient is the highest, and so it can increase the refractive index of a glaze. Adding a small amount of $La_2O_3$ to milky white glaze can enhance its gloss, rendering it opalescent. As such, it is an excellent gloss agent for ceramic glazes. When mixed with other ceramic pigments, it can give rise to gem and moisture effects.

$CeO_2$ is a good opacifier in enamel. It can be used to prepare opacified glazes with high whiteness and strong covering, which can cover variegated porcelain and improve its whiteness. It can also reduce the appearance of cracks in glaze.

The addition of 1–2% $Sm^{3+}$ to ceramic black pigment as a color aid can make the black glaze color pure and bright. Used in the reduction atmosphere, to make up for the defects of Fe, Cr, Co, Al and other synthetic black pigments showing insufficient color.

3. Rare earth high temperature colored ceramic glazes

The range of high temperature colored ceramic glazes is limited by the high roasting temperatures of up to 1300 °C. Therefore, such coloring agents are composed of a colored metal oxide, sometimes in combination with silicate, which can withstand high temperatures and is not easily eroded by from the glaze.

(1) High temperature praseodymium yellow glaze

Praseodymium yellow ceramic pigment is a widely used high-temperature ceramic glaze with good color. It is a zirconium silicate-based pigment, and can be mixed with ceramic glaze to make color glaze, or directly mixed with base glaze to make high temperature yellow glaze.

The praseodymium yellow ceramic pigment is pale yellow. It is characterized by a bright and uniform color, stability, good glossiness of glaze, heat resistance, corrosion resistance, moderate high-temperature flow performance, wide applicable temperature range, and low sensitivity to the atmosphere of a furnace. The color is less affected by glaze composition; and does not react to other colors. It can be mixed with various other pigments. For example, yellow-vanadium-zirconium is rendered light-green by $Pr_6O_{11}$, andadjusting its content modifies the tone depth, making the product suitable for art porcelain, antique porcelain, building porcelain, and everyday ceramics.

The preparation method of yellow ceramic pigment involves mixing $SiO_2$ and $ZrO_2$ in a 1:1 mass ratio, and then adding 3–6% $Pr_6O_{11}$ and 10–20% of one or more mineralizer, such as NaF, NaCl, or $Na_2MoO_4$. The firing temperature is about 1100 °C. With Ca, Mg, and Zn white glaze as the basis glaze, the addition of about 0.08% $Pr_6O_{11}$ imparts a green color grass cyan, bright, pure, uniform color glaze, glistening, and good glossiness.

(2) High temperature lanthanum/cerium gold glaze

Golden light glaze is applied to porcelain surfaces as an imitation gold decoration. Its tone and luster are similar to those of copper gold glaze. The glaze is bright, rich, and elegant, providing a unique artistic effect, which is widely used in architectural or

18.6 Rare Earth Heating Materials 327

garden art decoration; using Li-Pb-Mn gold glazes, devitrification through the glaze during firing, forming glaze surface with metallic luster.

Experiments have shown that the glaze layer precipitated on a porcelain surface is as smooth as a mirror when the new metal glaze prepared from $CeO_2$ and $La_2O_3$ is incorporated. This not only improves the stability and glossiness of the golden glaze, but also imparts it with good chemical erosion resistance and a gold imitation effect. The chemical stability, glaze hardness, and other technical parameters are superior to those of titanium gold film.

(3) High temperature neodymium discoloration glaze

Discoloration glaze, also known as heterotopic variegated glaze, is a ceramic art glaze produced by $RE_2O_3$ for a special decorative effect. The photosensitive characteristics of discolored glaze are reversible. Under excitation by different light sources, and with variations in the intensity of the light, the glaze surface presents different colors.

Discoloration glazes are obtained by adding a colorant to $ZnSiO_3$ transparent crystalline glaze. The main component of such discoloration glazes is $Nd_2O_3$, but its intrinsic coloring ability is poor, and one or several other $RE_2O_3$ is typically added to promote its coloring and discoloration sensitivity. The added $RE_2O_3$ are $Pr_2O_3$, $CeO_2$, $Sm_2O_3$, $La_2O_3$, $Y_2O_3$, and $Yb_2O_3$, used in conjunction with $Li_2CO_3$, $CaCO_3$, borax, or fused quartz.

$Nd_2O_3$ is an important and unique colorant for enamel. The coloring effect of neodymium is almost unaffected by sintering temperature and atmosphere, and so the color reproducibility is good. The specific electronic structure of $Nd^{3+}$ endows it with complex spectral characteristics, with a series of stable sharp absorption peaks in the infrared, visible, and ultraviolet regions. The f electrons of $Nd^{3+}$ can exist in long-lived metastable excited states, making it an excellent laser material. It gives a good color rendering effect and decolorizes crystalline glaze. According to the Lambert–Beer law, the combination of multiple ions should produce an addition effect in the absorption spectrum and form new spectral lines. According to this principle, the discoloration effect of Nd is enhanced and enriched by modulating the light absorption through combination with Ce, Pr, Sm, La, Y, or Yb.

## 18.6 Rare Earth Heating Materials

1. $LaCrO_3$ heating material

For high temperature physical chemistry studies in metallurgy and materials science, especially those conducted at above 1600 °C, commercially available stoves often fail to meet the particular experimental requirements. Therefore, furnaces have to be designed for specific purposes or requested in customized forms from manufacturers. A chapter on "High-temperature acquisition" in Wang CZ's "Research Methods in Metallurgical Physical Chemistry" introduced all kinds of furnaces used in experimental research according to temperature, but the heating principle of the $LaCrO_3$

heating body was not extensively explained. A more detailed discussion is given here.

$LaCrO_3$ has a melting point of 2490 °C, a radiation coefficient of about 0.9, and can directly conduct electricity at room temperature. An $LaCrO_3$ heating body is a novel type of heating element developed in the 1970s that can be used in an oxidizing atmosphere. $LaCrO_3$ is a perovskite $ABO_3$-type composite oxide. The A-site element in $ABO_3$ is easily replaced by other elements, that is, dopants. When the chemical valence of the dopant is different from that of the A atom, it will lead to the formation of $V_{\ddot{O}}$ in the crystal or change the valence of the B atom, thereby rendering the material semi-conductive, superconducting, catalytic, etc.

Some of the La in $LaCrO_3$ can be replaced by doping with alkaline earth metals, such as Ca, Sr, and Ba. To maintain electrical neutrality in the crystal, Cr changes from the original $+3$ to $+4$ valence. As a result, electron vacancies appear in the Cr positions, making the material an n-type semiconductor. Its electrical conductivity is affected by the content of the doped alkaline earth metal, the oxygen partial pressure of the working environment atmosphere, and the ambient temperature.

To improve the electrical properties of $LaCrO_3$, appropriate amounts of alkaline earth metal oxides, such as those of Ca or Sr, are added during the synthesis process to partially replace $La^{3+}$ to form $La_{1-x}M_xCrO_3 (M = Ca^{2+}, Sr^{2+})$.

Electron-transfer between $Cr^{3+}$ and $Cr^{4+}$ significantly increases the electrical conductivity of $La_{1-x}Ca_xCr_{1-x}^{3+}Cr_x^{4+}O_3$, resulting in n-type semiconductor characteristics. This reaction must be carried out in an oxidizing atmosphere. In a reducing atmosphere, the RE reacts at high temperatures, significantly reducing the conductivity of the material. The reaction is:

$$La_{1-x}Ca_xCr_{1-x}^{3+}Cr_x^{4+}O_3 \rightarrow La_{1-x}Ca_xCr_{1-x+2y}^{3+}Cr_{x-2y}^{4+}O_{3-y} + YO^{2-}$$

where

$$0 \leq x \leq 0.2, 0 \leq y \leq x/2$$

As $x$ is increased, the conductivity gradually increases. The room-temperature resistance and high-temperature volume instability of $LaCrO_3$ can be significantly improved by adding an appropriate amount of $Y_2O_3$.

## 2. Preparation of $LaCrO_3$ heating material

$LaCrO_3$ is synthesized and sintered in two steps. $LaCrO_3$ synthesis processes include solid-phase, chemical co-precipitation, sol-gel, and hydrothermal methods. The solid phase method is the most basic approach for preparing $LaCrO_3$ heating material. The raw materials of $La_2O_3$, $Cr_2O_3$, and $CaCO_3$ are first dehydrated and pretreated. They are mixed according to the required $La_{1-x}Ca_xCrO_3$ stoichiometry, then pressed and formed. The synthesis is conducted at above 1300 °C.

The chemical reaction is as follows:

$$La_2O_3 + 2CaCO_3 + Cr_2O_3 \rightarrow 2La_{1-x}Ca_xCrO_3 + O_2 \uparrow + 2CO_2 \uparrow$$

## 18.6 Rare Earth Heating Materials

The solid phase method is simple and suitable for industrial production, but it is not easy to obtain uniform ultrafine powders. When sintering at high temperature, Cr components readily volatilize, such that the raw materials ratio and sintering atmosphere are not easy to control.

The raw materials used in the chemical co-precipitation method are similar to those for the solid-phase method. The principle is that $La^{3+}$ and $Cr^{3+}$ readily precipitate from $NH_3 \cdot H_2O$. In the presence of $CO_3^{2-}$, $La^{3+}$, $Cr^{3+}$, and $Ca^{2+}$ can produce insoluble carbonate precipitates. The chemical co-precipitation method is simple and easy, making it amenable to industrial-scale production.

The raw materials for the sol-gel method are high purity $La(NO_3)_3$, $Ca(NO_3)_2$, and $Cr(NO_3)_3$. After drying, the gel powder is heated to a certain temperature, whereupon it undergoes a spontaneous self-sustaining exothermic reaction until completion. The product is calcined at 850 °C to obtain an $La_{1-x}Ca_xCrO_3$ powder.

The sol-gel method is a dust-free process when conducted by uniformly mixing each component on a colloidal or molecular scale, and proceeds at a low reaction temperature. $LaCrO_3$ powder is also prepared by a hydrothermal method by the decomposition of $La\{Cr[CH_2(COO)_2]_3\} \cdot 6H_2O$:

$$La\{Cr[CH_2(COO)_2]_3\} \cdot 6H_2O \rightarrow La\{Cr[CH_2(COO)_2]_3\}$$
$$\rightarrow LaCrO_x(CO_3)_y \rightarrow LaCrO_4 \rightarrow LaCrO_3$$

The formation temperature of $LaCrO_3$ powder is about 800 °C. The obtained powder has high crystallinity, a low degree of agglomeration, and high sintering activity.

3. Properties of $LaCrO_3$ heating material

$LaCrO_3$ is unaffected by air or an oxygen-rich atmosphere, either at room temperature or at high temperature, demonstrating its chemical stability. Aqueous HCl, $H_2SO_4$, $HNO_3$, NaOH, KOH, and $Na_2CO_3$ solutions do not erode $LaCrO_3$ at room temperature, nor do boiling HCl, $H_2SO_4$, or $HNO_3$. Molten NaOH, KOH, or $Na_2CO_3$, however, have an eroding effect.

The physical properties of $LaCrO_3$ heating material are reflected in several aspects, such as color, melting point $/°C$, relative density, porosity $/\%$, flexural strength $/MPa$, radiation rate, heat conductivity (room temperature – 1000 °C) $/[W/m \cdot K)]$, thermal expansion coefficient (room temperature – 1000 °C), and resistivity (1500–1800 °C) $/\Omega \cdot m$.

4. $LaCrO_3$ heater furnace

Generally, the $LaCrO_3$ heating body is rod-shaped, made to fit large and small tube furnaces, but is also suitable for box furnaces. Both ends of rod $LaCrO_3$ heater are in close combination with the heat part in the middle. The electrode is partially coated with silver paste, which forms an electrical contact, as shown in Rod KERAMAX heater diagram.

Taking the KERAMAX (commercial name of lanthanum chromate heating agent) furnace of Nikkato Co. (Japan) as an example, its dimensions and operational parameters are shown in a table of the specifications and properties of KERAMAX lanthanum chromit heaters. The model is B14-230, B14-380, B14-450, B16-550, B18-650B, B18-650, and B22-900. The specifications (mm) include diameter ($D$), overall length ($L$), heating (a), terminal (b), and electrode (c).

The heating body of the jacketed corundum tube is shown in a diagram of $LaCrO_3$ heater of alundum tube.

During operation of the heating element, a small amount of $Cr_2O_3$ volatilization is incurred at high temperatures. In order to prevent volatilized $Cr_2O_3$ from spreading to the product, the heating element and product can be separated by high purity porous corundum. In a tube furnace, the tube itself acts as an isolator. In a box furnace, the product can be sealed in a short corundum tube or placed in a closed vessel to prevent contamination.

# References

Chen JJ (2013) Rock mineral of typical jadeite, geochemical characteristics and gemology significance. China University of Geosciences, Beijing

Gems AJS, Jade, Jadeite and Nephrite. https://www.ajsgem.com/articles/jade-jadeite-and-nephrite.html

Gonçalves MC, Santos LF, Almeida RM (2002) Rare-earth-doped transparent glass ceramics. Comptes Rendus Chimie 5(12):845–854

Huang LZ (1996) Rare earth colored ceramic glaze. Chin Rare Earths 5:69–71

Jadeite Atelier, What is jade? the difference between jadeite and nephrite. https://www.jadeite-atelier.com

Karmakar B (2017) Functional glasses and glass-ceramics: processing, properties and applications, 1st edn. Elsevier

Li WM (2012) Current situation and development of rare earth optical glass. Rare Earth Inf 5:21–24

Liu GH (2007) Rare earth materials. Chemical Industry Press, Beijing, pp 246–264

Massera J et al (2019) Fabrication and characterization of new phosphate glasses and glass-ceramics suitable for drawing optical and biophotonic libers. In: 2019 Conference on lasers and electro-optics Europe & European quantum electronics conference (CLEO/Europe-EQEC). IEEE 23–27 June 2019

Pisarska J et al (2012) Heavy metal glasses and transparent glass-ceramics: preparation, local structure and optical properties. Optica Applicata XLII(2)

Qiu GM (2004) Rare earth optical glass fiber. In: Proceedings of the 10th China rare earth entrepreneurs association conference, pp 64–77

Schultz H (1975) The physics and chemistry of ceramics. Translated by Huang ZB. Revision 5. China Construction Industry Press, Beijing, pp 3–77

Shen HN (1974) How to look at the silicate phase diagram. China Construction Industry Press, Beijing, pp 1–153

Sun LC et al (2002) Research status and application fields of Lanthanum Chromate ($LaCrO_3$) materials. Chinese Rare Earths 6:55–58

Wang CZ (2013) Metallurgical physical and chemical research methods. Version 4. Metallurgical Industry Press, Beijing, pp 1–11

Yan HZ (2012) Development status and suggestions of rare earth new material industry. Northern Economy 15:25–27

# References

Yu J, Zhi Y (2002) Key applications of rare earth materials. Nonferrous Metall Equipment 6:22–25

Yu HY, Jia ZY, Lei W (2019) Geochemical characteristics and influencing factors of rare earth elements in Chinese nephrite. Modern Mining 3:6

Zhang SQ (1985) Introduction to new inorganic materials. Shanghai Science and Technology Press, Shanghai, pp 65–82

Zhang Y, Qiu Y (2007) Soft chemistry synthesis and characterization of rare earth spinel $ZnY_2O_3$, Bulletin of Mineralogy, Petrology and Geochemistry, Chinese Society for Mineralogy Petrology and Geochemistry. In: Proceedings of the 11th annual academic conference, p 254

Zhao JL (1999) Coloring function of rare-earth elements in gemstone glass. J Qiqihar Univ 1:21–24

Zhou ZH (2015) Application of rare earth minerals in ceramic glaze and pigment. Shandong Ceram 3:27–29

# Chapter 19
# Scandium and Its Materials Applications

## 19.1 Brief Description

Scandium (Sc), yttrium (Y), and the 15 lanthanides belong to group 3 in the periodic table of elements; the main chemical properties are similar to those of lanthanum. These 17 elements are called rare earth elements. However, there are no 4f electrons in the Sc atomic structure, and with the small number of electron shells in the atom, the ionic radius is much smaller than for the other rare earth elements. As such, the properties are quite different from those of other rare earth elements. Additionally, the lanthanide minerals that are rarely associated with scandium, and if any exists, it is dispersed and difficult to recycle, therefore, it is difficult to find scandium products in the general rare earth production process (Liu 2007, 2009).

Referrig to Los Alamos National Laboratory's decription to scandium, it is as follows:

- Scandium is a silver-white hard metal which develops a slightly yellowish or pinkish cast upon exposure to air.
- Atomic Number: 21
- Atomic Radius: 211 pm (Van der Waals)
- Atomic Symbol: Sc
- Melting Point: 1541 °C
- Atomic Weight: 44.96
- Boiling Point: 2836 °C
- Electron Configuration: [Ar] $4s^2 3d^1$
- Oxidation States: 3, 2, 1 (an amphoteric oxide)

© Science Press 2023
C. Wang, *Theory and Application of Rare Earth Materials*,
https://doi.org/10.1007/978-981-19-4178-8_19

## 19.2 Scandium Resources

Scandium is widely distributed in nature, existing in the moon, meteorites, atmosphere, crust, animals, and plants. Its average abundance in the crust is $3.6 \times 10^{-3}\%$, which is more abundant than gold, silver, aluminum, antimony, molybdenum, mercury, and bismuth. Because its existence is extremely scattered, it gives the impression of being "rare." It follows calcium in the periodic table, so it is a typical lithophile element. Scandium can be found in almost all of the by-products of granitic pegmatite types, but for scandium oxide, there are very few minerals with a concentration of more than 0.05%. At present, there are only a few independent minerals, such as scandium–yttrium ore, iron–silicon–scandium ore, and water–phosphorus–scandium ore, and resources are scarce. In addition, it has high chemical reactivity, so it is difficult to produce high-purity metals. Although known as early as 1879 when Swedish chemists discovered scandium in euxenite and gadolinite, 95% pure scandium metal was prepared with electrolyzed mixed molten salts of scandium, potassium, and lithium chlorides by Fischer only in 1937. Spedding prepared 99.9% pure scandium metal in 1973. In recent years, scandium research, applications, and resource development have been undertaken more widely.

China is a country with abundant scandium resources, and its related mineral reserves are very large, such as bauxite and phosphate rock deposits, South China porphyry and quartz vein-type tungsten ore, South China rare earth ore, Inner Mongolia Baiyun-Obo rare earth–iron ore, and Sichuan Panzhihua vanadium titanomagnetite. The main distribution is as follows.

(1) Bauxite and phosphate rock deposits

The $Sc_2O_3$ content of bauxite in North China (mainly Shandong, Henan, and Shanxi) and the western margin of the Yangtze platform (mainly Yunnan, Guizhou, and Sichuan) is 40–150 $\mu g/g$; and $Sc_2O_3$ content of Kaiyang phosphorite in Guizhou and Zhijin Xinhua phosphate rock is 10 and 10–25 $\mu g/g$, respectively.

(2) Tungsten ore

There is high scandium content in South China porphyry-type and quartz vein tungsten ore: $Sc_2O_3$ in wolframite is 78–377 $\mu g/g$, with some up to 1000 $\mu g/g$.

(3) Rare earth minerals

Large-scale scandium-rich deposits have been found in the massive ion-adsorbed rare earth deposits in South China: $Sc_2O_3$ at 20–50 $\mu g/g$ as associated scandium ore deposits, with more than 50 $\mu g/g$ as independent ones, and an average of 50 $\mu g/g$ in the Bayan-Obo REE deposit.

(4) Vanadium titanomagnetite

$Sc_2O_3$ content of ultramafic rock and magnesite rock in Panzhihua vanadium titanomagnetite is 13–40 $\mu g/g$, scandium mainly occurs in ordinary titanium pyroxene, ilmenite, and titanomagnetite.

19.3 Scandium Extraction and Purification 335

(5) Other minerals

Guangxi poor manganese ore contains a considerable amount of scandium, at about 181 $\mu$g/g, in the form of ion adsorption.

Bauxite and phosphate rock deposits are dominant in China's scandium resources, followed by tungsten ore, vanadium titanomagnetite, rare earth ore, and rare earth–iron ore. Scandium reserves of bauxite and phosphate rock deposits are about 290,000 tonnes, accounting for 51% of the total reserves of all scandium-ore types, and 1–4 times the world average content. China has the world's largest tungsten resources, and scandium accounts for 0.02% of smelted tungsten slag. In addition, uranium and titanium tailings are also important sources for industrial scandium production.

A summary of world scandium resources is as follows:

The world's scandium reserves are about 2000 kt, of which 90–95% are found in bauxite, phosphate rock, and magnetite, and a few in the uranium–thorium, tungsten–tin, tantalum–niobium, and rare earth ores. The scandium resources of industrial significance are mainly by-products of the above-mentioned ores and ilmenite. These scandium minerals are mainly distributed in Russia, China, Tadzhikistan, the United States, Madagascar, Norway, Mozambique, Canada, and Australia.

The major scandium distribution abroad is as follows:

Russia: The Russian Platform and the Kola Peninsula are the largest scandium distribution areas. The apatite of the Kola Peninsula contains scandium 16 $\mu$g/g, the total reserves are 16,000 tonnes; $Sc_2O_3$ average content is 650 $\mu$g/g, and up to 1400 $\mu$g/g in the weathered crust leaching phosphate rock deposit at Tomtor, Russia.

U.S.: The uranium-bearing sandstone deposits in the Colorado Plateau of the United States contain $Sc_2O_3$ 100 $\mu$g/g; $Sc_2O_3$ 300–1500 $\mu$g/g in phosphorite-bearing deposits of Fairfield; hydro phosphorite and lucinite have been discovered and exploited as scandium deposits.

Madagascar, Norway, and Mozambique: scandium is concentrated in granite, being rich in thortveitite.

Australia: $Sc_2O_3$ content in the hydrothermal uranium titanomagnetite deposit of Radium Hill is 3000 $\mu$g/g, while the world's largest and highest grade scandium ore is Australia's Syerston ore, which has reported the scandium content as 11,819 tonnes.

## 19.3 Scandium Extraction and Purification

Independent scandium deposits are extremely rare and cannot be used as industrial raw materials for its extraction. As a companion element, however, it is often dispersed in minerals such as rare earths, vanadium, titanium, zirconium, tungsten, tin, uranium, and coal, and can be recovered as a by-product from the industrial production of waste residue, sludge, and waste liquid. Therefore, industrially, scandium is reclaimed from the integrated treatment of nonferrous and rare-metal minerals, and its feasibility depends on several aspects of the raw materials, such as scandium content, production scale of main metals, and enrichment degree in waste materials (Lu et al. 1995, Liu 2009).

For example:

(1)  Bauxite

Bauxite is an important scandium extraction resource, accounting for 75–85% of the total scandium amount obtained from other metal ores. More than 98% of scandium is enriched in bauxite red mud.

(2)  Titanium-bearing minerals

Melted ilmenite is almost all in the slag, scandium can be recovered from the slag (mainly from the chlorinated soot); also from sulfuric acid waste liquor of titanium dioxide production.

(3)  Wolframite and cassiterite

China is rich in tungsten (from wolframite) and tin (from cassiterite) ores, so it is also an important way to recover scandium in China.

(4)  Ion-adsorption rare earth ore

Recycling scandium from the leaching solution, the process is relatively simple.

(5)  Zirconium-containing

After acid leaching of zirconium concentrate, about 60% of scandium is in the mother liquor, and this scandium can be extracted.

(6)  Containing uranium

When uranium is extracted, scandium can also be extracted because it is easy to separate from uranium.

Four stages are typically involved when extracting scandium:

- From the low content of raw materials, the initial scandium enrichment;
- Separation of impurities for industrial crude scandium oxidation;
- Refining and purification to produce high-purity scandium oxide (purity over 99%);
- The scandium metal is prepared from high-purity scandium oxide.

Because the main impurities in the scandium extraction materials are Ti, W, Zr, Th, U, rare earths, Fe, Ca, and Si, the separation and extraction of coexisting elements are mainly by extraction, ion exchange, and precipitation. By these methods, 99.9999% ultra-high-purity $Sc_2O_3$ can be achieved. Scandium metal ingot and its distillation purification products have been produced using calcium reduction, electrolysis, and vacuum distillation methods (Li and Liu 2000).

## 19.4 Methods of Oxidized Hydrate Extraction and Separation

1. Extraction of scandium raw materials

As mentioned above, the raw materials for industrial extraction of scandium are generally by-products for the extraction of other valuable metals, so the treatment of scandium-containing raw materials can save the step of ore decomposition in the usual metallurgical process, only starting from leaching. For example, the optimum conditions for extraction of scandium from titanium chloride dust are as follows: concentration of leaching solution HCl = 1.0 mol/L; leaching solid–liquid ratio 1:1.2; temperature 70–80 °C; time 30–60 min. Under these conditions, it had a high leaching extent (84–89%), but also a high extraction extent (87.5%).

When leaching with industrial hydrochloric acid, the leaching extents of Sc and Fe are 90% and 95%, respectively, and scandium in the leach liquor is 200–400 mg/L, HCl1.5–2.5 mol/L, the consumption of hydrochloric acid per ton of tungsten residue is 3–4 tonnes.

When the aluminum ore is leached, 98–100% of scandium is left in the red mud, and when the red mud is reduced to smelting iron, scandium has completely entered into the slag (scandium content is 0.012%).

When leaching residue with sodium hydroxide, 95–98% of scandium is left in the solid phase, and the scandium content of the white mud is 1.65 times larger than that in the red mud. The white mud is dissolved in acid and then scandium is recovered by solvent extraction.

2. Separation and purification

Because of the complex composition of the scandium leaching solution, it is necessary to separate the impurities effectively for scandium extraction. The most widely used methods are solvent extraction; other commonly used methods are ion exchange and precipitation. However, because of the complexity of the composition of the scandium leaching solution, it is difficult to separate scandium from other coexisting elements using only one separation method, these methods should be used alternately according to the specific conditions to ensure the economical and effective separation to obtain the desired purity of products when establishing a scandium separation and purification process.

(1) Solvent extraction

The solvent extraction method is widely used in various stages of scandium extraction (including enrichment, purification, and pretreatment for analytical determination) for its advantages of simple operation, low cost, good separation effect, low consumption of raw materials, and large handling amount. The scandium extraction mechanism is similar to that of other rare earths, but the scandium ionic radius is the smallest of the rare earths and the electronic structure is the simplest, which make the structure of the extracted species different. Extractants used in separating scandium are

338                              

mainly phosphorus-containing, acid-containing phosphorus, carboxylic acid, amine, chelate extractant, etc.

a. Neutral phosphorus extractant

Using neutral phosphorus-based extractant extracts metal ions; the extracted substance binds to the extractant to form a neutral complex and enters the organic phase.

Tributyl phosphate (TBP) can extract scandium from hydrochloric acid or nitric acid solutions and separate it from other rare earth elements, aluminum, zirconium, and thorium. 100% TBP extracts scandium from 6 mol/L HCl solution. The extraction extent can reach more than 99%. TBP for scandium extraction is carried out according to a solvation mechanism, which can be expressed as follows:

$$Sc^{3+} + 3A^- + (2-3)TBP \rightleftharpoons ScA_3 \cdot (2-3)TBP$$

where $A^-$ represents $NO_3^-$, $Cl^-$, $CNS^-$, or $[Sc(SO_4)_2]^-$; TBP extraction must be carried out under the condition of high anion concentrations. Usually, anion concentrations in the system can be maintained by a large number of acids or salts. When the anion concentration increases, the reaction proceeds in the forward direction, which is the extraction process; conversely, it is the baci-extraction process. So, simply using the difference in the solution acidity can achieve the purpose of extraction or reverse extraction.

According to data published by the Japan Atomic Energy Commission, the partition coefficients of some elements in the extraction with 100% TBP in 9 mol/L [$Cl^-$], are shown in a table of partition coefficient of elements under certain conditions, for example, when element is Sc, Ti, $Mn^{2+}$, $Fe^{3+}$, Ca, Al, Mg, and Mo, the partition coefficient is 100, 0.32, 0.32, 6000, 0.03, 0.001, 0.003, and 100; when element is Cu, U, Th, Lu, La, Ce, Eu, and Y, the partition coefficient is 0.63, 10, 0.32, 0.05, 0.001, 0.001, 0.05, and 0.1.

For example, the chlorinated soot leaching solution of titanium contains many chloride ions, and it also contains metal elements that differ greatly from the partition coefficient of scandium when extracted with TBP. TBP is an ideal extractant for extracting scandium from titanium chloride solution.

TBP can also be used for refining scandium. The impurity zirconium was separated from 100% TBP in a perchloric acid medium. Listolayov et al. used TBP to extract scandium in $HNO_3$ medium, and achieved the separation of scandium from trace impurities (ytterbium, zirconium, etc.). Zhang ZH et al. reported that the TBP extraction method extracts scandium from Panzhihua titanium tailings. Zhao YL et al. studied the mechanism of $P_{350}$ n-heptane solution extraction of scandium from hydrochloric acid solution.

The extraction reaction is:

$$Sc^{3+} + 3Cl^- + 3P_{350(0)} = [ScCl_3 \cdot 3P_{350}]_{(0)}$$

19.4 Methods of Oxidized Hydrate Extraction and Separation

$$k_{ex} = [ScCl_3 \cdot 3P_{350}]_{(0)} \, / \, [Sc^{3+}][Cl^-]^3 \gamma_{\pm}^3 [P_{350}]_{(0)}^3$$

where $\gamma \pm$ is the mean ionic activity coefficient of $Sc^{3+}$ and chloride ions, $k_{ex}$ is independent of $P_{350}$ concentration but related to HCl concentration.

Huang GW and Zeng XR reported the extraction of scandium from high-purity Scandia by $P_{350}$ separation from the leaching solution of scandium metal waste residue, and discussed in detail the influencing factors for the separation of scandium from 31 metal impurity elements.

b. Acid phosphorus extractant

The extractant with acid phosphate-containing ore is usually a cation exchange reaction of scandium replacement of extractant $H^+$. At high acidity, the phosphate group on the extractant molecule shows solvation.

$P_{204}$ is a commonly used extractant for scandium separation from tungstate residue. The extraction extent is related to the phase ratio, also to the scandium concentration of the feed liquid.

c. Carboxylic acid extractant

Carboxylic acid is a weak acid extractant that can effectively extract metal ions from slightly acidic or alkaline solutions. The extraction mechanism is similar to that of acidic phosphorus-containing extractants and is also carried out by cation exchange.

Aryl carboxylic acid is an effective scandium extractant.

In the n-heptane system, naphthenic acid and scandium form $Sc(OH)A_2 \cdot 2HA$, namely,

$$Sc(OH)^{2+} + 4HA = Sc(OH)A_2 \cdot 2HA + 2H^+$$

In another example, Zebreva et al. extracted scandium using naphthenic acid–$P_{204}$ kerosene solution to achieve scandium separation from other impurity metal ions.

d. Amine extractant

Amine extractant has a complex extraction mechanism. It is generally considered that the organic amine acts as a Lewis alkali that is "protonated" in an acidic medium to form cations, and the metal complex anions in the aqueous phase attract each other by electrostatic action to form ionic complexes.

An amine extractant is the best extractant with a primary amine and quaternary ammonium salt.

Gorski studied the extraction and separation of rare earth elements by several types of quaternary ammonium salts in the presence of various salting-out agents. With the increase in the atomic number of the elements, the separation coefficient of rare earth elements decreased, and scandium was located between samarium and yttrium.

### e. Chelating extractant

The chelating extractant is usually an organic weak acid with multiple functional groups, often containing acidic functional groups ($-OH$, $= NOH$, $SH$, etc.) and coordination functional groups ($=C = O$, $\equiv N-$, $= N-$, etc.). During the extraction process, the metal ions replace the acid group of the chelating agent, and at the same time, hydrophobic metal chelates $MA_n$ with a ring structure form with the coordination group to form a coordination bond, which is insoluble in water and easily soluble in organic solvents.

For example, $\beta$-diketones are widely used as chelating agents for scandium extraction. Most commonly used is thiophene formyl trifluoropropanone.

### (2) Liquid membrane extraction

Liquid membrane extraction is essentially a combination of liquid–liquid extraction and reverse extraction, especially suitable for separating and extracting substances in dilute solution. Wang YC, Wang XK, Li YD, Xu TW, Yang X, etc. carried out a large number of related studies on scandium extraction.

### (3) Ion exchange process

The ion exchange method is an important means to purify and refine scandium. In most of the separation methods, the separation factors of scandium and other rare earth elements are less than 20, while those of the ion exchange method are as high as hundreds or even thousands, so they can be used for final scandium purification. In dilute hydrochloric acid, sulfuric acid, or nitric acid solution, scandium can be firmly adsorbed on acidic cation exchangers. The loose structure of oxyphosphate or carboxylic acid cation exchange resin is most suitable for scandium separation.

The U.S. GTE (General Telephone and Electronics Corporation) loaded the scandium feed liquid on a hydrogen chelating resin (Amberlite IRC-718), washed the impurities with an inorganic acid, adsorbed scandium with strong cation exchange resin (Amberlite IRC-118), with deionized water, $NH_4NO_3$ solution to drip wash scandium, obtaining scandia by oxalate refining.

Moreover, TBP leaching resin can effectively separate trace amounts of scandium from a large number of impurities in the perchloric acid medium; extraction resin can separate trace amounts of zirconium, PMBP (1-Phenyl-3-Methyl-4-Benzoyl-5-Pyrazolone) quenched resin can be loaded on foam plastics to extract and separate scandium from other rare earths. This kind of new separation method will occupy a more prominent position in scandium recovery and extraction in future.

### (4) Precipitation method

The precipitation and coprecipitation methods were the main methods used in the past to separate and enrich scandium. They are still commonly used, and the process is simple. Many anions can precipitate scandium, the most widely used being oxalate anion. The dosage of oxalic acid is very important when precipitating scandium. The molecular formula of scandium oxalate is $Sc_2(C_2O_4)_3 \cdot nH_2O$ ($n = 3$ or $6$), the solubility in water is 0.06 g/L, and solubility increases with the increase of HCl, $NH_4^+$, Cl, and excess $C_2O_4^{2-}$ concentration in solution. The optimal conditions for precipitation of scandium oxalate are pH $\approx$ 3, temperature 90 °C, oxalic acid addition amount is $H_2C_2O_4$:Sc (mass ratio) = 16:1. Concentration of oxalic acid solution is 1 mol/L, scandium precipitation extent is about 97.6%, purity 99%. The mixture often contains Fe, Ti, Mn, Mg, and other impurities, with purity of about 95% of the raw materials containing scandium. After two hydrochloric acid dissolution and oxalic acid precipitation treatments, the final product's purity can reach 99.5%.

In addition, capillary isokinetic electrophoresis, chemical gaseous phase transport, and electrodialysis can be used for the enrichment and determination of trace scandium. However, because of the complex composition of the scandium leaching solution, it is difficult to separate scandium completely from its coexisting elements using only one of the above methods. Therefore, it is necessary to use these methods alternately to ensure effective separation, and obtain the desired products.

## 19.5 Preparation of Scandium Metal

1. Electrolytic process

Fischer et al. (1937) first prepared scandium by electrolysis of $SrCl_3$ with LiCl–KCl in the melt.

During electrolysis, scandium was separated on the zinc cathode and then distilled to remove the zinc to produce 99% scandium metal.

2. Metallothermic reduction

The metal was thermally reduced to anhydrous $ScCl_3$ or $ScF_3$ to prepare the crude scandium and then purified by vacuum distillation. The purity of the prepared scandium can only reach 99.98% at present. At every step of preparing high-purity scandium metal, it is easy to introduce impurity, so it is necessary strictly to operate and use the raw material of high-purity scandium oxide, which demands the impurity content should be less than $3 \times 10^{-3}\%$, except for F, Si, Ca, and Ta (Liu et al. 2016). The process of calcium thermal reduction to prepare scandium metal is shown in a diagram.

The preparation process of scandium metal can be divided into three stages: the preparation of scandium fluoride, calcium thermal reduction of scandium, and the vacuum distillation purification of scandium. This is summarized below.

(1) Preparation of scandium fluoride

$Sc_2O_3$ was placed in a platinum boat and heated at 600–750 °C for 16 h in an anhydrous HF–Ar mixed atmosphere. The higher the temperature, the lower the $ScF_3$ oxygen content, but the evaporation loss is greaterat higher temperatures. This process usually reduces the oxygen content in $ScF_3$ to $(1–2) \times 10^{-6}$ (note: 1 mg/L $= 1 \times 10^{-6}$). The scandium fluoride is obtained by mixing ammonium acid fluoride with $Sc_2O_3$ at 300–400 °C as follows:

$$Sc_2O_3 + 6NH_4F \cdot HF = 2ScF_3 + 6NH_4F + 3H_2O$$

342                                            19 Scandium and Its Materials Applications

The high-purity scandium oxide was prepared proportionally as $Sc_2O_3:NH_4HF_2$ = 1:(2–2.5) (mole ratio) with ammonium acid fluoride, each 2–3 kg batch of $Sc_2O_3$ was reacted in a vacuum fluorination furnace. The heating-up in the whole process of fluorination and deammoniumation is by stages, the obtained scandium fluoride as good quality, high yield, and good fluorination effect. Scandium fluoride is for the next process.

(2) Scandium metal preparation by thermal reduction

Scandium metal was prepared by reducing anhydrous $ScF_3$ with reductants such as potassium, calcium, and magnesium, and the crude scandium metal was prepared by calcium thermal reduction of $ScF_3$ in a vacuum frequency induction furnace.

$$2ScF_3 + 3Ca = 2Sc + 3CaF_2$$

The reduction crucible uses 1–1.5 mm tantalum, and argon arc welding supports the upper-thin and lower-thick (1.5–2 mm) tantalum crucible. This shape crucible plays a major role in preventing scandium metal loss caused by the strong erosion of the crucible during reduction. The reaction began at 700–800 °C, and the reaction was vigorous at 850 °C. After the reaction, it was cooled to room temperature under the protection of argon gas, and then the slag was separated using amechanical method.

In theory, the density of scandium metal is similar to that of the reducing product calcium fluoride, and it is difficult to separate the calcium fluoride slag from the scandium metal after reduction. However, calcium fluoride contains calcium with less residual density, so the actual density of slag becomes smaller. Furthermore, actual scandium density increases due to the density of scandium metal being increased by dissolving 8–10% tantalum with higher density. Therefore, the density difference between scandium metal and calcium fluoride increases, which is beneficial to scandium metal sinking to the bottom of the crucible, and the slag floating above the metal, so that the scandium metal and slag are well separated.

(3) Scandium distillation and purification

A reduction-obtained crude scandium contains 8–10% tantalum and impurities, due to the high vapor pressure at its melting point temperature, the scandium metal with high purity can be prepared by distillation purification. Steaming scandium is carried out in a vacuum medium frequency induction furnace or vacuum carbon tube furnace. The distillation crucible and scandium collection crucible are made of tantalum sheet metal. The vacuum degree reaches more than $133.3 \times 10^{-4}$ Pa, the distillation temperature is 1700–1800 °C, and the distillation time depends on the amount of metal being distilled.

When the scandium metal is distilled, the affecting factors are as follows: as the scandium metal evaporates, the tantalum concentration increases; and the evaporation rate of scandium metal decreases as tantalum concentration increases at the same distillation temperature.

For the distillation of high-purity scandium metal, it is important to determine the temperature gradient of the condensing area of the scandium metal collection

19.5 Preparation of Scandium Metal 343

crucible, which is beneficial to improve the purification efficiency and the purity of the metal. With such scandium metal, its rare earth impurities are very low (13 $\mu g/g$), non-rare earth impurities are also low (157 $\mu g/g$ or so).

The difference between the vapor pressure of various impurity elements of refined scandium and the vapor pressure of element scandium is the basis for analyzing the possibility and reaching the limit of element separation by vacuum distillation. The vapor pressure data of each element given in the literature can be determined. Near scandium's melting point (1550 °C), the elements with higher vapor pressure than scandium are Cd, As, Zn, Mg, Ca, Bi, Pb, and Mn in order; Ti, Mo, Zr, Nb, Ta, and W are those with lower vapor pressure. Only Al, Cu, Co, Cr, Fe, Ni, and Si are very close to the vapor pressure of scandium. Only by improving and innovating the process of vacuum distillation to purify scandium metal can the preparation level and product quality be improved.

3. Scandium sublimation and purification

Scandium can be purified by steps using the chloride or acetylacetone, and the separation effect is good. The boiling point of $ScCl_3$ is 967 °C, while the boiling points of Fe, W, Ti, Ni, Zr, Hf, Nb, and Ta chlorides are less than 350 °C; the rare earth element chloride is sublimed above 1200 °C, so the sublimation temperature can be controlled to separate them. $ThCl_4$ and $MnCl_2$ sublimation temperatures are close to that of $ScCl_3$ and difficult to separate. Scandium acetylacetonate has a low melting point [(187.25 ± 0.25) °C], sublimation point 157 °C, decomposing at 360 °C. It is soluble in ethanol, ether, chloroform, and benzene, while acetylacetonates of the rare earth metals zirconium and hafnium are less soluble in organic solvents and are difficult to volatilize or decompose with heat, so they can be separated by these properties, and prepared to obtain spectroscopically pure scandium.

4. Scandium single-crystal preparation

The preparation of scandium as a single crystal by pulling the crystal is carried out in a sealed crucible under an argon atmosphere. Arc-melted scandium metal is pulled into single crystals and annealed in a high vacuum and 1200 °C for 5 days. After annealing at $10^{-7}$ Pa, 1250 °C, and 30 h, there is sufficient grain growth without a long annealing time. The single crystal is cut using a spark in the specified direction and shape, but not by a saw, to avoid the twinning phenomenon and damage to the single crystal.

5. Scandium film preparation

The thickness of scandium films used for a nuclear reaction is 0.0003–0.3 mm. By evaporating scandium onto aluminum, and dissolving aluminum in NaOH solution, the scandium can be evaporated onto various substrates. Foil thicker than 0.0012 mm is evaporated onto carbon and the carbon substrate is not removed in nuclear studies. When the scandium film is 0.0012 mm thick it is supported by the film itself, and then the scandium film is separated from the Ni substrate by an aqueous solution with salt. With vacuum sublimation scandium, 0.0003 mm scandium film can be prepared on a sapphire substrate; the substrate surface is cleaned by chemical polishing. It is baked for 1 h at 1000 °C, then baked for 1 h at 600 °C in $10^{-5}$ Pa vacuum, and then the scandium is evaporated onto the substrate.

### 6. Scandium alloy preparation

Scandium alloy has important applications. Many countries such as Russia, the United States, Japan, Germany, and China have successively carried out research, development, and applications of scandium alloy, and the most studied is scandium–aluminum alloy. The alloyed aluminum with low levels of scandium has high strength, toughness, weldability, and corrosion resistance. It is a new generation of aerospace structural materials.

Scandium's melting point is 1541 °C, and the chemical is reactive. When preparing scandium–aluminum alloy, it must be added in the form of Al–Sc, Mg–Sc, or Al–Mg–Sc intermediate alloy. Intermediate alloy containing scandium is mainly prepared with three methods: mixing, hot metal reduction, and molten salt electrolysis. The first two methods are the main ones.

### (1) Scandium alloy preparation by admixture

Direct mixing is a traditional method for scandium interalloy preparation. It takes a certain proportion of high-purity scandium metal wrapped with aluminum foil, incorporates it into molten aluminum under argon protection, and heat preservation with enough time, after fully stirring the cast into an iron or cold copper mold. Smelting can be done in a high-purity graphite or corundum crucible, heated by resistance furnace or medium frequency induction furnace.

In the process of preparation, the scandium metal is mixed evenly with dispersant, aluminum powder, flux in advance, then pressed into a mass, then added to the molten metal, the dispersant is decomposed at high temperature, and the mass is automatically crushed so that a uniform alloy can be prepared.

### (2) Scandium alloy preparation by metallothermic reduction

#### a. Vacuum aluminum thermal reduction of scandium fluoride

The method is based on scandium fluoride as raw material, active aluminum powder as a reducing agent, under vacuum reduction:

$$ScF_3 + Al \rightarrow Sc + AlF_3$$

One preparation method is to mix 99.8% $ScF_3$ with aluminum powder in a mechanical mixer for 30 min, after compaction at 400–500 MPa into a corundum or high-purity graphite crucible, then in a quartz reactor, vacuum to $1.33 \times 10^{-2}$ Pa, 900–920 °C, thermal reduction for 30–60 min, $ScF_3$ conversion is 87–92%.

#### b. Thermal direct reduction method of scandia–alumina

## 19.6 Scandium Properties

The process of thermal direct reduction of scandia–alumina is shown in a diagram, which this method uses powdery scandia as raw material, mixed with active aluminum powder to form small balls, and then immersing them in molten aluminum as reducing agent, aluminum powder as the dispersant, and reducing scandia to scandium metal at high temperature to form an intermediate alloy in the aluminum liquid. Under the above conditions, $Sc_2O_3$ reacts with Al to form a series of scandium–aluminum intermetallic compounds such as $Sc_2Al$, $ScAl$, $ScAl_2$, and finally $ScAl_3$.

$$Sc_2O_3 + 8Al = 2ScAl_3 + Al_2O_3$$

This reflects the process flow of preparing Al-Sc alloy by oxidation of scandium-aluminum heat direct reduction. As the scandium–aluminum intermetallic compound is formed, $ScAl_3$ gradually dissolves in the aluminum melt to form an intermediate alloy. The technical route of this method is simple, but it is necessary to control the partial pressure of gas-phase oxygen in some steps to avoid oxidation of the scandium–aluminum alloy.

c.   Thermal direct reduction method of scandium–aluminum–magnesium chloride

This method uses very pure $Sc_2O_3$ as raw material, dissolved in hydrochloric acid, and transformed into $ScCl_3$ solution, which is transformed into $ScCl_3$ molten salt by evaporation, vacuum dehydration, and high-temperature heating. Then, at 900 °C, the molten salt is placed in the melt of molten aluminum–magnesium alloy. The $ScCl_3$ is reduced to metal in the crucible by magnesium, and scandium metal is trapped by aluminum to form Al–Mg–Sc alloy.

# 19.6   Scandium Properties

The chemical properties of scandium are similar to those of yttrium and the lanthanides, very reactive, and easily react with oxygen, carbon dioxide, water, and so on in the air. Oxidized scandium films at room temperature can prevent further scandium oxidation, which is still stable in air at 200 °C, but there is strong oxidation above 250 °C. Scandium reacts readily with halogens at room temperature, and at a higher temperature with nitrogen, phosphorus, arsenic, and other gases. Powdered scandium metal starts to react with nitrogen above 600 °C. The reaction with carbon, silicon, boron, and hydrogen needs to be carried out at a high temperature.

Scandium ions in aqueous solution are positive trivalent, not in the form of simple $Sc^{3+}$, rather forming stable complexions. $Sc^{3+}$ can also form complexes with $SO_4^{2-}$, $CO_3^{2-}$, $HCO_3^-$, $F^-$, amines, and other ligands.

Scandium can react with all inorganic acids, but slowly with chromic acid (the surface forms insoluble chromates). Scandium forms alloys or intermetallic compounds with other metals, such as scandium and rhenium forming the high melting point compound $Sc_5Re_{24}$, whose melting point is as high as 2575 °C, second only to ScN (melting point 2600 °C). Scandium can form a solid solution with magnesium, yttrium, zirconium, lanthanum, and gadolinium. It can form compounds with the IIIB (Mn and congeners) family and elements on its right (except noble gases). Scandium chalcogenides $Sc_2X_3$ (X is S, Se, and Te), as well as $ScCuS_2$ and $ScCrS_3$, are semiconductor materials: mixed silicides of it and other metals, such as $ScNi_2Si_2$, are important superconducting materials.

## 19.7 Application of Scandium and Its Materials

Although the amount of scandium and its compounds is not large, the application fields are very wide, and almost all materials fields are involved (Liu 2007).

Sumitomo Metal Mining Co. (2016) have summarized the potential usages of scandium, for example, Fuel cells: used in electrolytes/anodes, high efficiency (high ion conductivity), long life (low operation temperature); Aluminum/scandium alloys (scandium: 0.1–0.5%): aircraft and aerospace, automobile, luxury sporting goods; Emerging applications: LED, laser, electronics device, additive for structural metal, novel refining process.

1. Scandium application in metal materials

Scandium is an excellent modified addition element for many nonferrous metals. As long as a few parts of scandium are added to aluminum, the $Al_3Sc$ phase can be formed, which will deteriorate the aluminum alloy and make the structure and properties of the alloy change noticeably.

The addition of 0.2–0.4% scandium can improve the high-temperature strength, structural stability, welding performance, and corrosion resistance of the alloy, and can avoid the brittle phenomenon easily produced during long-term high-temperature work. Based on the existing aluminum alloy, adding small amounts of scandium develops a new generation of aluminum alloy materials, such as ultra-high-strength tough alloy, new high-strength corrosion-resistant weldable aluminum alloy, new high-temperature aluminum alloy, high-strength radiation-resistant aluminum alloy, and so on.

These alloys are widely used in aerospace, aviation, ships, nuclear reactors, light vehicles, and high-speed trains. For example, in Russia, scandium–aluminum alloy has been developed in the 1420 alloy, widely used as the structure of MiG-29, Tu-204, and Yak-36 vertical takeoff and landing aircraft. 1421 alloy is an aluminum–lithium–magnesium–zirconium alloy containing scandium, which is used in the longitudinal girder of transport aircraft fuselage to form an extruded profile.

In the U.S., scandium–aluminum alloy has been used to manufacture welding wires and sports equipment, such as baseball and softball sticks, baseball bats, and bicycle frames. Scandium has a melting point higher than that of aluminum, and the density is similar, so it can be used instead of aluminum in some structural materials for producing rockets and spacecraft. Materials of low density, high strength, and

19.7 Application of Scandium and Its Materials 347

corrosion resistant below 920 °C are required for structural materials in spacecraft, scandium–titanium alloy and scandium–magnesium alloy are ideal materials.

Scandium can also be used as an additive for special thermosensitive alloys; scandium addition to steel can improve the performance of iron and steel so that the graphite spheroidization of cast iron is more effective than with other rare earth elements. Adding to titanium-based alloys or nickel, chromium and tungsten-based high-temperature-resistant alloys can significantly improve the antioxidant properties of the material. Adding scandium can remove scars and scales when refining aluminum, magnesium (Liu 2009), copper, and other metals. Using high-temperature electrolysis makes scandium incorporate base metal, which can form a beneficial coating structure on the base metal.

2. Application of Sc in special ceramic materials

Carbonized scandium (ScC) can significantly increase the hardness of transition metal carbides, such as adding 20% (mole fraction) ScC to TiC can increase TiC hardness from 3060 to 5680 $kg/mm^2$, a hardness that is second only to that of diamond.

ScN, with a melting point of 2650 °C, can be used to make semiconductor single-crystal crucibles for drawing GaAs and GaP. $Sc_2O_3$ is better than some of the four refractory materials commonly used, namely, BeO, MgO, $Al_2O_3$, and $ZrO_2$. This is because of its thermal impact properties, for example, which can be used for flame spraying glass components. Adding $Sc_2O_3$ to $ZrO_2$ and $Y_2O_3$ produces a conductive refractory. It is used in solid oxide fuel cells as a promising new medium-temperature solid-electrolyte material.

Iwahara et al. (1981) reported that perovskite $SrCeO_3$-based sintered oxides $SrCe_{0.95}Yb_{0.05}O_{3-\alpha}$ exhibit proton ($H^+$) conductivity in a hydrogen or aqueous vapor atmosphere above 600 °C. They studied this kind of proton conductor oxide again, and the results show that $CaZrO_3$-base solid solution with In, Sc, or Ga partially substituting Zr provides good proton conductivity and high chemical stability and strength (Iwahara et al. 1983).

$CaZrO_3$ electrolyte material was prepared using the direct synthesis method. $CaCO_3$, $ZrO_2$, and $In_2O_3$ or $Sc_2O_3$ were mixed with grinding and pressing in appropriate proportions, and the synthetic material was heated at 1100–1400 °C for 10 h in an air atmosphere (Fukatsu et al. 1987). After grinding, the sheet or tube was made as needed and burned at 1400–1650 °C in air. The results of conductivity measurement by impedance spectrum show that the doped scandium–indium material has higher conductivity than doped Ga. Because the price of $In_2O_3$ is lower than that of $Sc_2O_3$, Iwahara has launched a $CaZr_{0.9}In_{0.1}O_{3-\alpha}$ proton conductor electrolyte tube, and first applied it to a fast online liquid aluminum hydrogen sensing measurement (Yajima and Iwahara 1992).

Wang CZ's team found that the material underwent loss and segregation during the sintering process of the proton conductor tube (also including synthesis). Scanning electron microscopy and electron probe analysis proved that the solid-electrolyte tube was sometimes not measured as expected and the In amount was insufficient.

Fedorov et al. (2000) reported that $In_2O_3$ begins to dissociate to $In_2O + O_2$, $In_2O$ sublimesat temperatures higher than 1200 °C, and the relation between dissociation and temperature (expressed as $p_{O2}$) is:

$$\lg p_{O_2} = 8.49 - 6314/T(1323-1573 \text{ K}) (1050-1300 \text{ °C})$$

$In_2O_3$ can be reduced to In metal by $H_2$, CO, or C at 760–800 °C. Therefore, when making $CaZr_{0.9}In_{0.1}O_{3-\alpha}$ proton conductor tubes, protective measures should be taken in the process of synthesis and sintering, which not only ensure $CO_2$ produced by $CaCO_3$ decomposition of raw materials can fully escape, but also avoid the loss of $In_2O_3$ by dissociation, but it is often not easy to do, and there is still In segregation. Huang (2013) carried out the preparation and studied the properties of proton conductors doped with In and scandium, measured the effect of different doping amounts on the conductivity of the samples, and discussed the variation of electron vacancy conductivity under different oxygen partial pressures and different doping amounts. The results show that the conductivity of $CaZr_{1-x}Sc_xO_{3-\alpha}(0.05 \leq x \leq 0.15)$ is largest at $x = 0.10$, which is higher than that of In-doped $CaZrO_3$ (Huang 2013).

$Sc_2O_3$ is stable and not volatile, so it should be a more suitable proton conductor electrolyte tube material. Negative thermal expansion materials made of composite oxides of scandium plus rare earth elements are widely used. They can not only be combined with conventional materials to form high-temperature ceramic devices with a thermal expansion coefficient close to zero but are also used in aerospace, engine components, integrated circuit boards, optical devices, etc. The refractive index of silicate and borate glass with scandium is clearly higher than that without scandium, and is suitable for making optical glass.

3. Application of scandium in petrochemical catalytic materials and electronic communication materials

The petrochemical industry is one of the major scandium sectors in industry. Pt–Al catalyst containing $Sc_2O_3$ is used for heavy oil hydrogenation and purification in refined petroleum, for dehydrogenation and dehydrating agents of ethanol or isopropanol, as well as for CO and $N_2O$ oxidation and ammonia synthesis, and for efficient catalysts for ethylene production and chlorine production with hydrochloric acid. In isopropylbenzene cracking, the activity of ScY zeolite catalyst is 1000 times larger than that of aluminum silicate.

Scandium is suitable for making a variety of semiconductor materials. The sulfide of $Sc(Sc_2S_3)$ can be used as a thermistor and thermoelectric generator, $ScB_6$ can be used as the cathode material for an electronic tube, and $Sc_2O_3$ single crystal can be used in instrument fabrication. By replacing iron oxide in the middle of a ferrite with $Sc_2O_3$, the coercivity can be improved, and the performance of computer memory elements can be improved, which can be used to convert computer storage cores quickly. Adding a small amount of scandium to yttrium iron garnet can improve its magnetic properties. Using scandium instead of iron, its magnetic moment and magnetic conductivity are enhanced and its Curie temperature is reduced, which is

## 19.7 Application of Scandium and Its Materials

beneficial to its application in microwave technology. Scandium can also be used to make superconducting materials.

4. Application of scandium in electrooptic sources and laser materials

Some countries such as America and Japan use a high-brightness electric light source—the scandium–sodium lamp. This electrooptic source excites $ScI_3$ and $NaI$ under high voltage discharge conditions. A certain wavelength of light is emitted when scandium and sodium atoms jump back to a low-energy level from the high-energy excited state. Scandium spectral lines are in the range 361.3–424.7 nm, which is the near UV and blue light. Sodium has a spectral line of 589.6 nm, yellow light. Scandium and sodium match close to sunlight. Back to the ground state, scandium and sodium atoms can form iodides with iodization. This cycle can maintain a high atomic concentration in the lamp tube and prolong the service life. The scandium–sodium lamp is also a green energy-saving lighting source, which has the advantages of high luminous efficiency, good color rendering, and long life, strong fog-breaking ability, small volume, and convenient use. Scandium–sodium lamps can save 80% of electricity compared with ordinary incandescent lamps with the same illumination, 50% more than mercury lamps, and the service life is 5000–25,000 h, while incandescent lamps last 1000 h. The scandium–sodium lamp is widely used in high-brightness lighting in film and television cameras, stadiums, airports, docks, stations, squares, and other places.

Applications of scandium in the field of laser technology have increased rapidly in recent years. 30% of the total consumption of Japanese scandium is used in the laser field. Since 1983, gadolinium scandium gallium garnet (Cr, Nd; $Gd_3Sc_2Ga_3O_{12}$) laser crystals have been manufactured, the excitation efficiency of which is 3.5 times that of yttrium aluminum garnet. America, Japan, and Russia have successively made large-scale laser crystals (called GSGG crystals) that are suitable for the manufacture of kilowatt-class high-average-output slab lasers, which can be used for submarine underwater lasers and so on.

5. Application of scandium in nuclear materials

Scandium can be used in controllable thermonuclear reactions. The effective cross section of scandium for capturing neutrons is 25 barn (nuclear reaction cross section unit target), and the structural material suitable for the reactor nuclear fuel shell and other parts. $Sc_2O_3$ can be used as a special refractory material for a reactor without oxygen. Hydrogenated scandium can be used as a target in the ion accelerator. Scandium metal can be used as a neutron filter, which allows electrons with an energy of 2 keV to pass through and prevents neutrons with other energies. Adding a small amount of $Sc_2O_3$ to fuel a $UO_2$ high-temperature reactor can avoid the transformation of $UO_2$ into $U_3O_8$, to avoid the volume increase and the appearance of cracks.

Among the ceramic insulating materials commonly used in nuclear reactors, $Sc_2O_3$ is the best positioning ceramic. Locating ceramics require small radiation expansion, low thermal conductivity, small expansion coefficient, resistance to wear, etc. Scandium radioisotopes are generated by irradiation, [46]Sc can be used as a γ-ray source and tracer atom for all aspects of research, medical treatment, and production. Scandium deuteride ($ScD_3$) and tritide ($ScT_3$) are used for uranium mineral detectors.

For example, according to Los Alamos National Laboratory's studies, on uses, about 20 kg of scandium (as $Sc_2O_3$) are used yearly in the U.S. to produce high-intensity lights. The radioactive isotope [46]Sc is used as a tracing agent in refinery crackers for crude oil, etc. Scandium iodide added to mercury vapor lamps produces a highly efficient light source resembling sunlight, which is important for indoor or night-time color TV. On handling, little is yet known about the toxicity of scandium; therefore it should be handled with care.

# References

Fischer W et al (1937) On metal scandium, ztachr, anorg, allgem. Chem 231:54–62

Fukatsu N, Yamashita K, Ohashi T et al (1987) Electromotive force of the hydrogen concentration cell based on $SrCe_{0.95}Yb_{0.05}O_{3-\delta}$ solid electrolyte. J Jpn Inst Met 51(9):848–857

Huang WL (2013) Preparation and performance characterization of calcium zirconate based solid electrolyte materials. Master thesis, Northeast University, Shenyang

Iwahara H, Esaka T, Uchida H et al (1981) Proton conduction in sintered oxides and its application to steam electrolysis for hydrogen-production. Solid State Ionics 3–4:359

Iwahara H, Uchida H, Tanaka S (1983) High temperature type proton conductor based on $SrCeO_3$ and its application to solid electrolyte fuel cells. Solid State Ionics 9–10:1021–1025

Li GD, Liu YL (2000) Study on technics and optimozation of purifying scandium by vacuum distillation. J Chin Earths Soc 18(2):183–186

Liu GH (2007) Rare earth materials. Chemical Industry Press, Beijing, pp 100–418

Liu XZ (2009) Rare earth fine chemical chemistry. Chemical Industry Press, Beijing, pp 7–13

Liu Y, Chen HQ et al (2016) Preparation of high-purity metal scandium. Hunan Nonferrous Met 5:46–49

Los Alamos National Laboratory, Scandium, Periodic Table of Elements: LANL. https://periodic.lanl.gov/21.shtml

Lu QT, Ni FS et al (1995) Scandium application and progress in extraction metallurgy. Shanghai Nonferrous Met 3:160–166

Scandium. https://en.wikipedia.org/wiki/Scandium

Sumitomo Metal Mining Co., Ltd. (2016) Establishment of Scandium Recovery Business, April 28. https://www.smm.co.jp/en/news/release/uploaded_files/20160428en.pdf

The world's largest and highest grade rare earth ore. https://www.sohu.com/a/164872941_788233. https://www.cnmn.com.cn/ShowNews1.aspx?id=341736&page=2

Yajima T, Iwahara H (1992) Studies on proton behavior in doped perovskite-type oxides: (II) dependence of equilibrium hydrogen concentration and mobility on dopant content in Yb-doped $SrCeO_3$. Solid State Ionics 53–56:983

# Chapter 20
# Rare Earth Research, Production, Policy, and Future Development

## 20.1 Current Situation

1. Basic situation

Since scientists discovered mixed-rare earth "yttrium soil" in 1787, 234 years have passed. The research, production, development, and application of rare earths are becoming ever more extensive. The value of rare earths as strategic resources and as part of environmental resources, and how to develop, utilize, and protect them, are not only technical issues, but also economic and security issues.

From the many application fields of rare earths, it is clear why the world pays so much attention to them, giving rise to many issues and debates. For example, many vehicles control air pollution through the use of rare earth catalysts in their exhaust systems; rare earth metals are used in important alloys; glass, granite, and marble, and precious stones are often polished with cerium oxide powder. Additionally, new energy powered vehicles and generators use magnets made of rare earth elements; phosphorescent bodies for digital displays, monitors, and televisions are produced by rare earth oxides; and most computers, mobile phones, and motor batteries incorporate rare earth metals.

According to the "Mineral Commodity Summaries 2021" of the United States Geological Survey (USGS), data showed that "the estimated distribution of rare earths by end use was as follows: catalysts, 75%; ceramics and glass, 6%; polishing, 5%; metallurgical applications and alloys, 4%; and others, 10%". In 2013, these figures were chemical catalysts, 65%; metallurgy and alloys, 19%; glass and polishing agents, 6%; permanent magnets/magnets, 9%; and others, 1%.

Association of China Rare Earth Industry (ACREI) information shows that in 2019, the world's total rare earth production was about 210,000 tonnes, and China's rare earth mining was about 132,000 tonnes, that is, around 63% of the global total. The second largest producer is the United States, which produced about 26,000 tonnes of rare earths in 2019, accounting for more than 12% of the global total. China, the

---

© Science Press 2023
C. Wang, *Theory and Application of Rare Earth Materials*,
https://doi.org/10.1007/978-981-19-4178-8_20

United States, Myanmar, and Australia account for 95% of global rare earth production. Thus, the high volume shows that the international market is still dependent on China's rare earth products.

China remained the world's largest exporter of rare earths in 2019, exporting more than 46,000 tonnes, which was about twice the total annual production of rare earths in the United States. Japan, the United States, the Netherlands, Germany, and Italy are some of China's largest trading partners in rare earth exports. Taking the United States and Japan as examples, the United States has begun to increase its own mining and production, while Japan has adopted measures such as diversification of rare earth import channels.

In 2019, China imported about 41,000 tonnes of rare earth compounds, which was slightly lower than its exports, but is still the world's largest importer of rare earths, accounting for 56.5% of global rare earth consumption. Myanmar and the United States are the main channels through which China imports rare earths. Indeed, the United States is both the main channel for China's exports of rare earths, and the main channel for its imports.

According to the USGS "Investigation of U.S. foreign reliance on critical minerals", the following statements apply to lanthanum:

- Its forms are mischmetal; lanthanum oxide, carbonate, and other compounds; and the pure metal.
- Foreign reliance considerations (based on 2018–19): lanthanum is mined and enriched domestically, but, as with other REs, lanthanum compounds are produced largely outside of the domestic supply chain. Catalyst compounds containing lanthanum are produced domestically.
- Trends (looking back 5–15 years): China's market share of mined rare earth compounds has decreased in the past 5 years, but is still greater than 50%. U.S. lanthanum oxide production has occurred sporadically within the past decade.
- Technical options: exploration, research, and other actions are currently being taken to address the risks for RE.

2. American case

According to USGS (U.S. Geological Survey, 2019) website reported, "Rare Earth Element Mineral Deposits in the United States" (Gosen et al. 2019), "Because of their unique special chemical properties, many of the metals in the group of rare earth elements have essential applications in twenty-first century technologies. Examples of products that use RE are cell phones, computers, fluorescent and light-emitting-diode lights, flat-screen television and computer monitors, and in high-strength magnets used by clean energy technologies such as the generators of wind turbines and batteries of hybrid and electric vehicles. RE is used in many defense applications, such as in components of jet engines, missile guidance systems, antimissile defense systems, satellites, and communication systems".

Since 2009, the U.S. Geological Survey (USGS) has conducted numerous studies focused on the distribution, geology, and potential resources for RE-bearing mineral deposits in the United States. Between 2011 and 2017, China produced approximately

## 20.1 Current Situation

84% of the world's RE, and during this time the United States only produced RE between 2012 and 2015. The U.S. production came entirely from the Mountain Pass mine in California, providing only about 4% of global RE supply. Because REs are essential for technological applications and are primarily supplied by one nation, there has been increased concern in identifying new sources of RE, including economic RE deposits.

In November 2020, the Department of Defense announced contracts and agreements with several rare earth element producers aimed at strengthening the domestic rare earths supply chain. Three of the awards were made under the authorities of Title III of the Defense Production Act (DPA). MP Materials, which owns the largest rare earth element mining operation outside of China, was awarded a DPA Title III technology investment agreement to establish domestic processing capabilities for light rare earth elements. MP Materials will refine its current mixed rare earth concentrate production, which represents approximately 12% of global rare earth oxide supply, into separated rare earth products at its site in Mountain Pass, California.

3. Exploration and recovery

Because of the significant value of rare earths, many countries have accelerated the exploration and recovery of rare earth minerals.

(1) Japan's prospecting and recycling/Germany's recycling

A research group composed of collaborators from Yamaguchi University and Tokyo University in Japan announced in April 2015 that two new minerals with lanthanum as the main rare earth component had been discovered and recognized by the International Mineralogy Association. These were named Ferriakasakaite-(La) and Ferriandorosite-(La). Although this discovery did not meet the criteria for industrial production, it has been shown that there is a possibility that manganese deposits throughout Japan contain such lanthanides.

In terms of recycling, Japan's Ministry of Economy, Trade and Industry proposed a plan to make Japan a world rare earth recycling power. Germany is also encouraging its companies to recycle metals, including rare earths, from electronics, in order to find ways of reducing reliance on rare earths from China and other countries. "If companies recycle electronics such as mobile phones and hard drives, they can significantly reduce rare earth imports", said environmental officials. The goal of resource efficiency in Germany was to increase its efficiency to twice the 1994 level by 2020. Europe, the United States, and Japan have begun their efforts to diversify rare earth procurement. Some companies have begun recycling metals such as rare earths or have switched to alternative products. The head of the United Nations Special Committee on International Resources stated that 99% of consumer products containing rare earths are discarded after use.

(2) Development of rare earth industry in Canada

By Canada.ca (the official website of the Government of Canada), Canada has some of the largest known reserves and resources (measured and indicated) of rare earths in the world, estimated at over 15 million tonnes of rare earth oxides. Many of Canada's most advanced rare earth elements exploration projects contain high concentrations of the globally valued heavy rare earth elements used in high-technology and clean-energy applications.

The Calgary Herald reported on January 7, 2014 that Canadian mining companies, the federal government, research institutions, and other partners have jointly established the Canadian Rare Earth Elements Network (CREEN). More than 200 rare earth projects have been explored, more than half of the world's rare earth projects. The Natural Resources Committee of the House of Commons of Canada initiated this work. Canadian rare earth mines contain a higher proportion of heavy rare earths, favoring the development of rare earth resources.

According to a paper by TMX Group Limited in April 2021, rare earths companies listed on the TSX and TSXV (Toronto Stock Exchange (TSX) and TSX Venture Exchange (TSXV)) offer exposure to non-Chinese rare earth resources, such as Search Minerals, Avalon Advanced Materials, Defense Metals, Medallion Resources, and Geomega Resources. All companies mentioned had market caps of at least C\$10 million.

(3) India

The Indian rare earths company Indian Rare Earths Ltd. (IREL) was established on 18 August 1950, and operated in partnership with some private companies for basic commercial rare earth processing. In 1963, it was taken over by the Indian central government, as part of the Department of Atomic Energy (DAE), and later IREL acquired a large number of private companies for mining.

4. China's mine areas and enterprises

Rare earths in China are mainly distributed in Inner Mongolia, Shandong, Jiangxi, Guangxi, Guangdong, Hunan, Fujian, and Yunnan, among other places. The applications of rare earth metals are mainly in magnetic materials, hydrogen-storage materials, luminescent materials, and so on.

Examples of China's rare earth enterprises:

- Inner Mongolia Baotou Steel Rare-Earth (Group)Hi-Tech Co., Ltd.;
- China Minmetals Rare Earth Co., Ltd.;
- Jiangxi Tungsten Industry Group Co., Ltd.;
- China Nonferrous Metal Industry's Foreign Engineering and Construction Co., Ltd.;
- Rising Nonferrous Metals Group Co., Ltd.;
- Xiamen Tungsten Co., Ltd.;
- Hunan Rare-Earth Research Institute;
- Leshan Shenghe Rare Earth Technology Co., Ltd.;
- China Rare Earth Holdings Ltd.

Examples of China's new materials enterprises for rare earths:

- Beijing Zhong Ke San Huan High-Tech Co., Ltd.;

## 20.1 Current Situation

- Ningbo Yunsheng Co., Ltd.;
- Taiyuan Twin Tower Aluminum Oxide Co., Ltd.;
- Jiangmen Kanhoo Industry Co., Ltd.;
- Shannxi IRICO Fluorescent Materials Co., Ltd.

5. Russia

The Rare Earth Magnets in Russia: Production, Market and Forecast, 2015, analyzed the future market and forecasts for rare earth magnets (by 2020) in Russia, and provided information on Russian rare earth magnet manufacturers and major importers and buyers.

Examples of Russia's producers of rare earth magnets:

- LLC: Waltari Magnet; Technomage; Wend; Magnisep;
- JSC: Progmat; ELMAT-PM;
- NPO: Magneton; ERG.

Other companies:

- LLC: Elemash-Magnet; Tula factory permanent magnets; Himstalkomplekt;
- JSC: Machine-Building Plant; Siberian Chemical Plant.

6. Toxic contamination of rare earths

The types and long-term nature of rare earth pollution are very complex. The intrinsic toxicity of rare earth elements is not high, and although europium and yttrium may have carcinogenic effects, most of the other rare earth elements are considered to be of low toxicity. Nevertheless, their mining and refining processes create pollution. Rare earth minerals are usually accompanied by radioactive materials, such as thorium and uranium. For example, monazite has long been the most important rare earth mining mineral, but the same mines are also used to extract radioactive elements and even to make plutonium. Rare earth extraction also requires the large-scale use of strong acids and bases, and so on. Ion adsorption rare earth ores allow for easy extraction, which has led to a large number of small enterprises and individual mining concerns. However, the extraction process produces large volumes of waste water, which poses a serious threat to the local soil and water sources.

7. Reserves

Taking Japan as an example, it set up the "Special Metal Reserve Association" in 1976, which adopts the form of government-funded, private management to ensure the accumulation of private reserves of mineral resources. This amounts to 60 days of domestic consumption of rare-metal reserves, including 42 days of national reserves and 18 days of private reserves. Japan has acquired a large amount of rare earth resources for its national strategic reserves, making it the world's largest reserve of rare earth resources.

356         20   Rare Earth Research, Production, Policy, and Future Development

8. Other examples

Australia/Norway/Canada: Vital Metals signed a five-year, 1000 tonnes per annum supply contract with REEtec for rare earth oxides, except for that of cerium (2 February 2021), which is supplied by Nechalacho rare earth mines in Canada. The two companies have the option of increasing this supply to up to 5000 tonnes per annum in 10 years.

According to information from 1 February 2021, Myanmar is China's largest supplier of rare earths, and a total of 12,986.216 tonnes of rare earth carbonate was imported from Myanmar in 2019, falling to 6225.179 tonnes in 2020, and equating to $18,337,530 and $21,809,636, respectively. The mines in Myanmar are mainly directly managed by local forces and have little to do with government forces. The central government has weak authority over remote areas in northern Myanmar; changes in this unstable region can lead to reductions in the number of mine employees, reduce investment in enterprises, and so on.

Indonesia's Ministry of Energy and Mineral Resources Geological Department survey results show that 28 of the country's mine sites found rare earths according to information released on 22 January 2021. It has been confirmed that there are also rare earth-related materials associated with coal mining activities.

On January 25, 2021, British Pensana Rare Earths submitted a permit application for the construction of rare earth oxide separation facilities in Yorkshire, U.K. After completion, it will become the world's largest rare earth oxide production facility.

## 20.2 Research Areas

In recent years, researchers have continued to carry out in-depth research on rare earth resources from the perspectives of function, material, extraction, recovery, and development.

Of 31 rare metals, the term "rare earth" applies to 17 elements, for example, neodymium and dysprosium used as rare earth magnet materials for next-generation automobiles and the abrasive used for hard disk drive (HDD) glass substrates, and cerium and lanthanum used for automobile exhaust catalysts. Their applications include TVs, digital cameras, mobile phones, computers, new generation cars, electronic components such as semiconductors and integrated circuits, small and lightweight products, energy saving and environmental measures such as liquid–crystal displays (e.g., cerium), small batteries, and exhaust purification (e.g., cerium).

Some research examples are as follows:

According to China cre.net information, British and Australian researchers have released a study that could help find new undiscovered rare earth deposits (neodymium-dysprosium). Experiments have shown that sodium and potassium cations are key components leading to the dissolution of rare earth elements, rather than chloride or fluoride anions as previously thought.

Magnetostrictive materials have the intelligent characteristic of sensing a magnetic field and generating drive, making them indispensable in many high-technology fields, such as high-power transducers, micro-displacement control systems, and high-precision machining equipment. Existing high-performance magnetostrictive materials are highly dependent on expensive, scarce, heavy rare earth strategic elements. There is an urgent need to develop highly sensitive magnetostrictive materials reliant on other rare earths. Researchers at Xi'an Jiaotong University have devised a new approach for developing highly sensitive rare earth magnetostrictive materials.

Researchers at the Lanzhou Institute of Chemical Physics, Chinese Academy of Sciences, have prepared nitrogen-doped nanoporous graphene membranes with high separation selectivity for rare earth ions.

Researchers at the Fujian Institute of Physical Structure, Chinese Academy of Sciences, have demonstrated immediate detection of salivary tumor markers based on rare earth nano-fluorescent probes to increase the labeling ratio of analyses.

Researchers at the Mapuche Optical Institute (MPL) have demonstrated precise location of single rare earth ions, and accurately measured the quantum mechanical states of such cations, making a great contribution to the research and development of quantum computers.

American researchers have used different rare earth elements to make a set of probes with different sensitivities to various cancers, allowing them to obtain accurate images of the distribution and development of diseased cells.

Japanese researchers have developed a new technology for extracting the rare earth neodymium from waste magnetite at ambient temperature and atmospheric pressure, allowing almost 100% recoveries. At the same time, this system permits the recovery of other rare earths besides neodymium.

Other examples of studies include research on China's rare earth export policy; the research and application of China's rare earth resource yield prediction method (Wang 2015); green preparation of rare earth functional material precursors based on controlled crystallization of rare earth carbonates; the Japanese development of fluorescent materials with reduced rare earth content; the recovery of rare earths from monazite ore by-product (Ding 2015); study on the optimal allocation of rare earth resources; and analysis and research on rare earth industry development.

## 20.3  Production and Market

The rare earth industry includes prospecting and mining, R&D, production and application, and marketing, pricing, and trading. These are intertwined and closely related. Deutsche Bank "Rare earth industry—attracted by magnets: the value in rare earths" in 2015 predicted that oversupply of rare earths would continue. Magnets play a strong driving role, but investors should focus on downstream producers. Such production areas have high growth demand, such as electric vehicles.

According to cre.net information in December 2020, Russia plans to significantly increase the exploitation of rare and rare earth metals in the next few years, increasing their production to about 20,000 tonnes by 2024 and to more than 70,000 tonnes by 2030. This would make Russia the world's largest exporter of rare and rare earth metals.

Vital Metals Ltd. (Australia) prospects for and develops rare earths, technological metals, and gold projects in Canada, Africa, and Germany. It holds 100% interest in the Nechalacho rare earths project located at Nechalacho in the Northwest Territories of Canada. It plans to commence production in 2021, and aims to produce a minimum of 5000 tonnes of rare earth oxides (REO) by 2025. It aims to become the lowest cost producer of mixed-rare earth oxides outside of China by developing one of the highest grade rare earth deposits in the world and the only rare earth project capable of beneficiation solely by ore sorting.

An ananlysis from Reports and Data shows that the rare earth metal compounds market is projected to grow at a rate of 16.8% in terms of value, from 2019 to reach USD 9.19 billion by 2027; the valuation was USD 3.14 Billion in 2019, and it is forecasted to reach USD 9.19 Billion by 2027. The global market is currently observing a significant pace owing to the massive rise in the consumption by different industry verticals and propelling number of new use cases, which predominantly is replacing many conventional applications.

The abbreviation "CRE index" denotes the composite index of rare earth listed companies. It is a rare earth stock index compiled by the Chinese rare earth website (www.cre.net), for which a total of 26 listed companies closely related to the rare earth industry were selected. As of December 1, 2016, they were composed of upstream rare earth companies and middle- and lower-reach rare earth companies.

There are differences in the scarcities of rare earth elements. The relative scarcity of heavy rare earths makes them more sensitive to rare earth supply than relatively sufficient elements, such as praseodymium and neodymium.

According to Business Wire's analysis, the global rare earth metals market is projected to grow from USD 5.3 billion in 2021 to USD 9.6 billion by 2026, at a CAGR (compound annual growth rate) of 12.33% during the forecast period. In terms of value, the Asia rare earth metals market is projected to grow at the highest CAGR during the forecast period.

Problems faced by the rare earth industry in 2021 include adjustment after price increases, supply security, and international utilization of rare earth resources. Promote the sustained and stable recovery of demand, including new energy vehicles, production restrictions, market forces, and other measures.

## 20.4 Policy and Management

The development and utilization of rare earth resources need to be managed by laws, regulations, and policies. In the following, we take several countries as examples.

1. China

According to the 2020 market analysis of the China rare earth network, the development of rare earth industry often brings unforeseen environmental impacts and a waste of resources. On February 18, 2020 and July 1, 2020, the Ministry of Industry and Information Technology and the Ministry of Natural Resources promulgated the "total control index of rare earth mining and smelting separation". The target increase was mainly allocated to China Southern Rare Earth Group Co., Ltd., China Rare Earth Co., Ltd., and China Northern Rare Earth (Group) High-Tech Co., Ltd.

On April 30, 2020, the Ministry of Industry and Information Technology issued a circular on the implementation of the policy of preferential enterprises in the rare earth industry, and entrusted the China Nonferrous Metals Industry Association and the China Rare Earth Industry Association to carry out a survey and to study the relevant policies to assist enterprises in reporting.

On September 1, 2020, the Resources Tax Law of the People's Republic of China was formally implemented. According to the new law, the tax rate of light rare earth mineral processing is 7–12%, and that for medium and heavy minerals is 20%, lower than the original tax of 27%.

Member units of 12 interministerial coordination mechanisms for rare metals, including the Ministry of Industry and Information Technology, the Development and Reform Commission, and the Ministry of Natural Resources, jointly issued the Notice on Continuous Strengthening of Order Rectification in Rare Earth Industry [2018] No. 265.

On January 15, 2021, the Ministry of Industry and Information Technology publicly advised on the Regulations on the Administration of Rare Earths. For example, illegal acts include mining and separation without quota, illegal smelting separation, illicit sales, violation of product traceability, unauthorized use of reserves, and so on.

In February 2021, the Ministry of Industry and Information Technology and the Ministry of Natural Resources issued a notice concerning the total control index for the first batch of rare earth mining and smelting separation companies in [2021] No. 16. For example, China's Rare Earth Co., ltd. has 1500 tones and 8730 tones of heavy- and light-rare earth quota, respectively; China Southern Rare Earth Group Co., ltd. has 5100 tones and 19,650 tones of heavy- and light-rare earth quota, respectively. It was stated that the quotas should be centralized and allocated to enterprises with advanced technical equipment and a high level of safety and environmental protection (the first batch was allocated to six rare earth groups), whereas for some non-compliant enterprises, the quota should not be allocated, such as enterprises that fail to meet the Emission Standards of Rare Earth Industrial Pollutants and environmental requirements such as radioactivity protection.

## 2. U.S.

CNBC reported in February 2021 that U.S. President Biden had instructed his government to review key USA supply chains, including semiconductors, high-capacity batteries, medical supplies, and rare earth metals, to assess the domestic manufacturing gap in the USA, and the dependence of key supplies, such as semiconductors and rare earths, on overseas sources.

The Rare Earth Industry and Technology Alliance (REITA) were established in October 2012. Its members include: Arnold Magnetic Technologies, Boulder Wind Power (BWP), and the Colorado School of Mines, General Electric, and Global Tungsten & Powders (GTP). This alliance includes not only rare earth producers, but also users and research institutions; the aim is to provide relevant training and information services to members, the media, and policy makers, as well as rare earth producers and users.

The DPA Title III awards follow a series of rare earth element actions that the Department of Defense has taken in recent years to ensure supply and to strengthen defense supply chains. Specific actions include stockpiling, implementing Defense Federal Acquisition Regulation Supplement (DFARS) rules to transition defense supply chains to non-Chinese sources of rare earth element magnets, launching engineering studies with the Industrial Base Analysis and Sustainment program focused on re-establishing domestic heavy rare earth element processing, partnering with industry to re-establish domestic neodymium-iron-boron magnet production, and leveraging the Small Business Innovative Research and Rapid Innovation Funds to accelerate development of new rare earth element processing technologies.

In regards to strategies for reducing net import reliance, the USGS report states that some actions may have some unforeseen or unintended consequences. This relates to thrifting, that is, the use of less of a mineral commodity. Although thrifting may decrease demand for a mineral commodity, it may also make end-of-life recycling less economical because less of the material is available to recycle. Investigations into each of these strategies may reveal their potential effectiveness and highlight unintended consequences. Various strategies may be implemented to reduce the net import reliance of the United States or any other country, many of which may require, or result in, improvement of the comparative advantage of domestic mineral supply chains.

These strategies include increasing domestic primary and secondary (that is, recycling) production; diversifying and reinforcing global supply chains to eliminate single points of failure and bottlenecks; securing supplies with reliable partners through collaborative agreements and stronger trade ties; developing alternative materials that have lower supply risk and supply chains that are largely independent of those of the mineral commodity in question; maintaining strategic inventories; utilizing less of the mineral commodity through material thrifting; and employing enhanced manufacturing techniques, such as additive manufacturing and near-net shape-forging processes that can reduce the amount of waste generated and thus the amount of mineral commodity needed.

## 20.4 Policy and Management

In 2019, these and other initiatives were identified as "calls to action" as part of the federal strategy entitled "A Federal Strategy to Ensure Secure and Reliable Supplies of Critical Minerals" (U.S. Department of Commerce) that was due to be implemented in 2020 by various U.S. Federal Government agencies and coordinated through the U.S. National Science and Technology Council's Critical Minerals Subcommittee.

The most effective strategies will vary according to the specific mineral commodity. When the target net import reliance is achieved, industry and government strategic inventories may be maintained at a commensurate level to provide the necessary buffer in the event of supply disruption.

3. Japan

The Ministry of Economy, Trade and Industry issued a supplementary provision in 2011 to reduce the use of rare earths and metals and to support alternative parts schemes, noting that companies using rare metals support high-value-added industries and are a source of industrial strength. As the procurement environment deteriorated, Japan became concerned that companies' overseas transfers would lead to technology outflows and a loss of domestic markets and employment opportunities. It was deemed necessary to minimize the use of rare earth magnets containing dysprosium and to establish technologies that can be installed applied to the final product.

In November 2015, six countries, including Japan, formed a professional committee to determine the international standard specifications for rare earths. Japan's New Metal Association, made up of economic and industrial provinces and companies involved in rare earth business, promotes related work. It decided to participate in the setting of standards in order to prevent its rare earth products from failing to meet new ISO specifications, and requested that the rare earth method developed by Japan should be included in the standard.

In order to make Japan the foremost recycling country, it has implemented measures such as technology development related to recycling and promoting capital investment. In order to ensure stable business continuity in Japan for advanced technology industries, it will support the introduction of facilities needed to increase industry's resistance to supply risks such as rare earths.

Relevant policy and support institutions include JOGMEC (Japan Oil, Gas, and Metals National Corporation), JBIC (Japan Bank for International Cooperation), NEXI (Nippon Export and Investment Insurance), and JICA (Japan International Cooperation Agency).

JOGMEC provides financing and debt guarantee, technology development and technical support, information collection and provision, geological structure survey, and so on.

## 4. Other examples

Interest rate subsidies

According to a Russian report on February 19, 2021, the Ministry of Industry and Trade of Russia proposed a bill to extend the full capacity of the rare earth project from 2025 to 2030, as a condition for companies engaged in the development of rare earth deposits to receive interest subsidies on investment loans from the state budget. Interest rate subsidies for rare earth development will be a significant support policy, given the expected rise in major policy rates in the future.

Refining support

According to an Australian report on 25 February 2021, the mineral sands giant, Iluka Resources, has stated that the idea of refining monazite concentrate produced by the company into rare earth products at Eneabba mines is likely to be supported by the Australian and overseas governments.

## 20.5 Further Discussion

### 1. Brief

The world has recognized the value and importance of the 17 rare earth elements, and a lot of work has been done in the fields of research and development, technological application, and the trading market. The challenges are becoming increasingly fierce. In an era when Japan, the United States, and other rare earth users have begun to solve production needs through technology and policy, the challenges faced by other countries are obvious.

Rare earths have excellent physical properties, such as optical, electrical, and magnetic properties, and in combination with other materials can give rise to new composites. For example, the tactical performance of steel, aluminum alloy, magnesium alloy, and titanium alloy used in the manufacture of tanks, aircraft, and missiles is greatly improved. Rare earths also find use in high-tech lubricants for electronics, lasers, the nuclear industry, and super conductivity. They are widely used in military technology, which often leads to subsequent advances in related fields.

From a mining perspective, the USGS (2019) has stated that the economic development of rare earth mineral deposits is affected by many factors beyond mining, such as commodity prices and mineral processing costs. Most of the rare earths are hosted by minerals that have complex chemical formulae, which presents more challenges for processing and extracting them. Continued advancements and refinements in mineral processing techniques may allow rare earth deposits with complex mineralogy to be economically developed in the future.

## 20.5 Further Discussion

2. Analysis and prospect by the RAND Corporation (RAND)

The importance of rare earths is very clear in the United States; the official website of the United States Geological Department has a detailed introduction to rare earths. The government, military, geologists, and so on, clearly understand the industrial contribution of rare earths (including extensive research and development and application in military technology), and the role of rare earths in other fields (such as agricultural and medical research).

David L. An of RAND presented a long report on the significance of rare earth policies and national defense in the United States and China at the Pardee RAND Graduate School in May 2015, linking rare earths, national security, and the interaction between the United States and China through analysis of the policy design of the rare earth dysprosium. According to his analysis, dysprosium is an essential material in American military technology, and the degree of government policy support for the military's expected budget increase seems to depend on an understanding of the rare earth industry. RAND recommends that the United States should strive to prevent sudden disruptions in the supply of vital resources.

The report recommends that policymakers first consider the political, economic, and geographical generality of rare earth elements, as well as the specificity of dysprosium. They should then systematically evaluate the effectiveness and cost of the new policy, and provide adequate budgetary support for policies to mitigate supply disruptions. The same should be true of policies concerning other rare earths and materials. It is hoped that the relevant methods and research findings will become a common template for American thinking and policy advice.

The analytical and exploratory perspectives of the report mainly include: political and economic aspects of rare earth elements; the application and geographical nature of rare earth elements; projections of supply and demand for rare earths on the global market; policy assessment and optimization; policy options; policy formulation and strategy; military applications of rare earth permanent magnets; promotion of rare earth projects outside of China; level of technical and commercial preparedness of the Ministry of Energy; level of R&D of rare earths; judicial activities; national strategies and important mineral production laws; reserves of experts and expertise; key research areas and research publications of the Institute of Key Materials; the Trust Fund for the Defense Microelectronics Activities project; WTO survey process; and other potential timing planning options.

3. Developments in the field of new materials

In today's world, the science and application of new materials is most representative of the development situation of cross-departmental, interdisciplinary, cross-industry collaboration. A recent report looks forward to the development trend of the new materials industry, including rare earth materials from this perspective.

New materials are products with strong application potential, and the integration of cross-departmental, interdisciplinary, and cross-industry collaboration is becoming ever more important in their development. In the future, the combination of new materials in upstream and downstream industries will be closer. From the point of view of new material enterprises, integration and development with downstream industries should reduce the risk of R&D and production, and from the perspective of downstream industries, cooperation with upstream industries may improve market response time and adaptability.

Taking rare earth functional materials as an example, it is believed that the formulation of a national incandescent lamp development strategy road map, national energy saving and environmental protection, and an enhanced consumer electronic support policy will greatly promote the development of rare earth luminescent materials and rare earth permanent magnet materials.

4. The application value of rare earths in the defense field will continue

In recent years, the extensive application of rare earths in military technology has made their national defense role ever more important. For example, rare earths find use in night-vision goggles for military purposes, precision-guided weapons, communications equipment, GPS equipment, batteries, and other defense electronics. These provide great advantages for military activities. Rare earth metals are key elements in making hard alloys for armored vehicles and rockets. In defense applications, some substances can replace rare earths, but these alternatives are usually less effective and reduce the military advantage. Therefore, it can be predicted that the value advantage of rare earth application in the defense field will continue compared with alternative materials.

A summary of military uses of several rare earth elements was presented by An (2015). Taking Nd, Pr, Sm, Dy, and Tb as examples, the technology is on permanet magnet, functions include guidance and control, electric motors and, actuators, stealth/noise, and cancellation, the application examples are smart bombs, joint direct attack munition (JDAM), joint air to ground missile (JAGM), cruise missiles, unmanned aerial vehicles (UAVs), AIM-9x, AIM-120 AMRAAM, helicopter acoustic signature reduction (NdFeB plus Terfenol-D); when function is electric drive motors, the application examples are zumwalt DDG 1000, joint strike fighter (JSF), hub mounted electric traction drive, integrated starter generator, combat hybrid power system (CHPS).

An also pointed out other military uses of Y, Eu, Tb, Nd, La, Lu, Eu, Ce, and various: Nd-doped yttrium aluminium garnet (YAG) laser for targeting and underwater mine detection (e.g. magic lantern), laser targeting (air- and ground-based), counter-improved explosive device (IED) (e.g. laser avenger), SaberShot photonic disrupter; sonar transducers, radar, enhanced radiation detection, multi-purpose integrated chemical agent alarm (MICAD), microwave amplification for satellite communication, high-capacity fiber optics; driver's vision enhancer (DVE), avionics displays; and jamming devices, electromagnetic railgun, ni metal hydride battery, area denial system (e.g. long range acoustic device or LRAD).

However, because of the role of rare earths in military technology, their development and utilization also gives rise to deeper issues. On the one hand, evermore countries and military forces are beginning to enter and participate in the field of rare earth competition and research and development; on the other hand, the countries that acquire this ability are more inclined to use this tool to initiate/resolve disputes.

## 20.5 Further Discussion

There is a view that rare earths make weapons more cold-blooded. Rare earths are strategic metals related to world peace and national security. About 4 kg of Sm-Co magnets and Nd-Fe-B magnets are used in the precision guidance system of the Patriot missile for electron-beam focusing. Rare earth alloys are also used in key parts, such as body-controlled wing surfaces. An M1 tank equipped with a neodymium-doped yttrium aluminum garnet laser rangefinder has a viewing range of almost 4000 m on a clear day, as compared to only 2000 m for a T-72 tank. The F-22 Fighter has a supersonic cruise function because both the powerful engine and the light and strong fuselage rely on rare earth technology to create the special material; the fuselage is made of rare earth-strengthened magnesium titanium alloy. F-119 engine blades and the combustion chamber are made of flame-retardant titanium alloy incorporating rare earths. Rare earths also play an important role in fiber laser weapons.

5. Resource requirements for low-carbon societies

Taking Japan as an example, the meeting entitled "Mineral Resources Policy for Carbon Inclusion and Society 2050" on 15 February 2021 was presented in the meeting paper of the Subcommittee on Mining of the Subcommittee on Resources and Fuel of the Committee on Resources and Energy Survey.

- Dependence on rare metals and many other mineral resources from specific countries is a problem from a security perspective. In addition, future technological innovation, circular economy, and responsible procurement need to be considered. It is important to visualize the use of natural resources, how they recycle, and the flow of materials, including the movement of used products.
- Achieving carbon–neutral societies, while requiring new resources, does not require significant amounts thereof. The development of new mines looks set to become uneconomical because of over-supply. For example, of the 17 rare earths, only the heavy rare earths have been extensively applied. In addition, if the risk is high and consumption decreases, the parts manufacturing industry that uses these resources will decline, and the related technology development will stagnate, resulting in the concentration of technology development in places such as China, where these resources can be used.
- Long-term contracts are good from a risk-hedging perspective, but as new risks, consideration of how to sell the remaining minerals is also important. If acquired as a reserve, new financial risk reduction measures also need to be followed up. At the same time, equipment and human resources are also important.
- The history of rare earths is not long, starting around 1970. Among them, the "star" element ceaselessly changes with demand. If the industry is steady, it can deal with the demand. If attention is paid to historical changes and technology progress, changes in future demands can be accommodated. Support of domestic smelters is essential to preserve the Japanese industrial base. Only Japan and China have such an industrial base, with that of Japan being broader.
- Although the private sector should take the lead in participation, there are a variety of international rules, including ISO, IEC, OECD, and Copper Mark,

which require the full attention of governments. Inter-mining measures could be supported by governments to encourage private enterprises to take comprehensive measures.

- With regard to the development of materials technologies, material substitution is important, but it should not worsen the balance of future demand and should be coordinated with a circular economy and measures to combat climate change.

In the future, different countries will have clearer rare earth strategies. For example, the USA will place emphasis on new material R&D and exploration/mining for rare earths, while Japan looks set to focus on rare metal reserves and recycling. However, a common goal will be to develop a countermeasure plan for rare earths and other rare metals. The development of technologies that do not use rare earths will help to strengthen supply chains because of the high risk of disruption to rare earth supplies. For heavy rare earths, efforts will be made to achieve the same level of performance from smaller amounts. Meanwhile, more applications of the light rare earths will be sought.

## 6. Problems

Any resource and policy measures taken will bring about some new problems to be solved, such as obtaining heavy rare earths, but also finding a balanced way to exploit, produce, and apply light rare earths. For the issue of reducing dependence on rare earths from other countries, there are also issues of separation and balance of advantages and disadvantages.

This is exemplified by the strategic considerations analyzed in the USGS report "Investigation of U.S. foreign reliance on critical minerals":

The effectiveness of each individual strategy in reducing net import reliance will vary by mineral commodity. Moreover, each of these strategies has strengths but also significant limitations. For example, the ability of viable substitutes to reduce net import reliance (and the overall supply risk) may be limited by supply constraints of the substitute material. This may occur because mineral commodities that are good substitutes are those with similar chemical and physical properties and are thus typically found in the same ore deposits. The co-produced REs can substitute one another in certain applications, as can tellurium and selenium, cobalt and nickel, and tantalum and niobium.

The substitution of co-produced mineral commodities may create (or exacerbate existing) supply and demand mismatches if the demand for one of the co-produced mineral commodities leads to over-production of the other as a result of their relative proportions in the source ore deposits. For example, meeting the future demand for dysprosium may result in the co-production of cerium and other rare earths in excess of their projected demand.

Other net import reliance reducing strategies have their own unique challenges (for example, low collection rates impede post-consumer recycling, and lack of geological knowledge hinders mineral asset development). If a specific limitation goal is established, then scenarios can be developed to determine how much each of these strategies can contribute (given their limitations) toward that goal under various assumptions of future domestic demand.

# References

An DL (2015) Critical rare earths, national security, and U.S.–China interactions: a portfolio approach to dysprosium policy design. Rand Pardee Rand Graduate School, Santa Monica

American media says arms dealers do not want to provoke China, the best weapons rely on China rare earth. http://mil.huanqiu.com/observation/2014-06/5031203.html. 23 June 2014

Based on December 1, 2016, they were composed of upstream rare earth index and middle and lower reaches rare earth index. 2021-03-01. http://cre.net/show.php?contentid=149066

Canada will develop the rare earth industry. http://www.cre.net/show.php?contentid=111798. 09 Jan 2014

Cardenes I (2018) Rare earth metals: challenge for a low carbon future. Oxford Policy Management Limited, December

Chemical Industry "low carbon society implementation plan" (2020 target). https://www.meti.go.jp/shingikai/sankoshin/sangyo_gijutsu/chikyu_kankyo/kagaku_wg/pdf/2020_01_04_02.pdf

China Rare Earth Industry Report, 2014—2018, Research in China. http://www.gii.co.jp/report/rinc302440-china-rare-earth-industry-report.html. 09 June 2015

China rare earth new policy issued——illustrated china rare earth resources tax reform. http://www.cre.net/show.php?contentid=119515. 21 May 2015

Ding JX (2015) Medallion company metallurgical test rare earth recovery 91%. Rare Earth Inf 8:22

Experts calling for a reassessment of the impact of the tea rare earth limit on exports. http://www.chanyeguihua.com/646.html. 17 June 2015

Exposing the military use of rare earths: China rare earths are cheaper than pork. http://bbs.tiexue.net/post2_4454713_1.html. 30 Aug 2010

Fiber-Laser weapons are favored with rare earth elements will become the protagonists in the future war. http://www.fyme.cn/a/news/fenxishi/2015/0602/29284.html. 02 June 2015

Gosen BSV, Verplanck PL, Emsbo P (2019) Rare earth element mineral deposits in the United States, U.S. Geological Survey, Mineral Resources Program, 11 Apr 2019. https://pubs.er.usgs.gov/publication/cir1454

Indonesia: 28 rare earth potential sites identified, Mineral Resources Information, JOGMEC, 2 Feb 2021. http://mric.jogmec.go.jp/news_flash/?me=%E3%83%AC%E3%82%A2%E3%82%A2%E3%83%BC%E3%82%B9/%E5%B8%8C%E5%9C%9F%E9%A1%9E

Investigation of U.S. Foreign Reliance on Critical Minerals—U.S. Geological Survey Technical Input, Document in Response to Executive Order No. 13953 Signed 30 Sept 2020. https://pubs.usgs.gov/of/2020/1127/ofr20201127.pdf

Japan has developed new technologies for efficient recovery of rare earth neodymium from spent magnets. http://www.cre.net/show.php?contentid=113525. 07 May 2014

Japan hopes to become a world leader in rare earth recycling. http://news.5rchina.com/news/50607.shtml. 25 May 2015

Japan plans to increase its "rare earth self-sufficiency rate" to 50% by 2030. http://www.cre.net/show.php?contentid=119563. 26 May 2015

Japan releases of new fluorescent materials that can reduce rare earth amount. http://www.fpdisplay.com/news/2013-05/info-159518-920.htm. 24 May 2013

Li YG, Directed energy weapons in major countries of the world. http://military.china.com.cn/2015-03/16/content_35062248.htm. 16 Mar 2015

Myanmar: implications for imports in rare earth minerals, Mineral Resources Information, JOGMEC, 3 February 2021. http://mric.jogmec.go.jp/news_flash/20210203/152935/

Pensana Rare Earth, obtains the first stage of support from the British government's Automotive Transformation Fund, Mineral Resources Information, JOGMEC, 15 Apr 2021. http://mric.jogmec.go.jp/news_flash/20210415/155007/

Pistilli M, Top Canadian rare earths stocks, TMX Group Limited, 1 Apr 2021. https://investingnews.com/daily/resource-investing/critical-metals-investing/rare-earth-investing/top-canadian-rare-earths-stocks/

Prospects for the development of the new materials industry. http://www.chanyeguihua.com/1514.Html. 02 Jan 2016

Rare earth elements facts, The official website of the Government of Canada. https://www.nrcan.gc.ca/our-natural-resources/minerals-mining/minerals-metals-facts/rare-earth-elements-facts/20522

Rare-earth elements in the CIS: Production, market and forecast (8th edn). Market Research Report. http://www.gii.co.jp/report/info169812-cis-rare-earth-element.html. 08 July 2014

Rare earth magnets in Russia: production, market and forecast, Market research report, INFOMINE Research Group, 2015-05-20

Rare metal reserve strategies. https://www.meti.go.jp/main/yosan/yosan_fy2020/hosei/pdf/hosei_yosan_pr.pdf

Reports and Data, Market Summary, Rare Earth Metal Market. https://reportsanddata.com/report-detail/rare-earth-metal-market

ResearchAndMarkets.com (2021) Rare-Earth Metals (Lanthanum, Cerium, Neodymium, Praseodymium, Samarium, Europium, Others) Market—Global Forecast to 22 Apr 2026

Resource companies provide rare earth concentrate samples to customers. http://www.cre.net/show.php?contentid=121882. 29 Oct 2015

Russia: Ministry of Industry and Trade proposes a five-year extension of the full capacity of rare earth projects. http://mric.jogmec.go.jp/

Summary of proceedings, mineral sub-committee of the resources• fuel sub-committee on mining (7th) https://www.meti.go.jp/shingikai/enecho/shigen_nenryo/kogyo/pdf/007_gijiyoshi.pdf

The Pentagon: Market forces change supply and demand patterns for rare earths. http://www.cre.net/show.php?contentid=111494.html. 20 Dec 2013

The United States established a rare earth technology alliance under ACC management. http://xianhuo.hexun.com/2012-10-25/147198451.html. 25 Oct 2012

The University of Japan has developed new technologies to recycle rare earths using salmon DNA. http://www.cre.net/show.php?contentid=109804.html. 23 Aug 2013

The US military will use laser and microwave weapons more widely. https://www.laserfair.com/yingyong/201507/30/61492.html

United States Geological Survey, USGS, Mineral Commodity Summary (2013). https://pubs.usgs.gov

United States Geological Survey, USGS, Mineral Commodity Summary (2021). https://pubs.usgs.gov/periodicals/mcs2021/mcs2021.pdf

USGS Reports use REE or REEs to express rare earth elements, here simplified as RE. Title III of the Defense Production Act (DPA): DOD announces rare earth element awards to strengthen domestic industrial base, 17 Nov 2020

U.S. Resources Corp. patented rare-earth elements. http://cre.net/show.php?contentid=148802

Wang XB (2015) Research and application of production forecasting method of rare earth resources in China. Dissertation of China University of Geosciences, Beijing